Humic Substances: Structures, Properties and Uses

Humic Substances
Structures, Properties and Uses

Edited by

Geoffrey Davies
Northeastern University, Boston, USA

Elham A. Ghabbour
Northeastern University, Boston, USA

With the assistance of

Khaled A. Khairy
Northeastern University, Boston, USA

ROYAL SOCIETY OF CHEMISTRY

The proceedings of the second Humic Substances Seminar held on 27 March 1998 at the Northeastern University, Boston, Massachusetts.

The front cover illustration is taken from the contribution by L.G. Akim, G.W. Bailey and S.M. Shevchenko, p. 133.

Special Publication No. 228

ISBN 0-85404-704-2

A catalogue record for this book is available from the British Library

Published by The Royal Society of Chemistry,
Thomas Graham House, Science Park, Milton Road,
Cambridge CB4 0WF, UK

For further information see our web site at www.rsc.org

Printed and bound by MPG Books Ltd, Bodmin, Cornwall, UK.

Humic substances may not be beautiful, but they do beautiful things.

Fritz Frimmel (with permission)

Preface

Humic substances (HSs) are nature's least understood materials. These brown or black biopolymers exist in animals, plants, sediments, soils and water. HSs contain more carbon than all living things. They seem to be purpose built for many life-sustaining functions. As solids, they retain water. In addition, they are effective acid–base buffers, they bind metals, molecules, ions and other biopolymers, they are redox-active, and they stick firmly to clays and minerals. The roles of HSs as catalysts and regulators are largely unexplored.

1 DEFINITIONS

Humic substances are operationally classified by their aqueous solubility. This classification is arbitrary because HS solubilities depend on their molecular weight, their state of aggregation and their metal and mineral contents.

Fulvic acids (FAs, observed $<M_W>$ limit *ca.* 5 kDa) are the smallest members of the HS family. They are soluble at all pHs, mobile and part of so-called dissolved organic matter (DOM). Some scientists regard FAs as oxidized fragments of larger HSs, while others see them as HS precursors. In any event, FAs generally exist at low concentrations in natural waters and they are surface- and redox-active. These properties make FAs hard to isolate, purify and understand. FA light absorption and photochemistry control natural UV penetration of surface waters and their other properties are undoubtedly vital.

Humic acids (HAs), the adults of the HS family, are highly functionalized carbon-rich biopolymers that stabilize soils as soil organic matter (SOM). They are anchored by metal binding and by attachment to clays and minerals, which decreases their solubility at a given pH. HA gels retain water very strongly. Gel drying by different methods gives solids with quite different morphologies and properties.

Humins (HUs), the most 'coal-like' HS family members, are insoluble at all pHs. HUs are weaker water retainers, sorbents and metal binders than FAs and HAs, and they are more aromatic in character. HUs are further along in the natural progression from live animals and plants towards 'dead' coals and carbon.

2 HS MOLECULAR STRUCTURES

HS molecular building blocks and higher structures have been sought after and debated from almost the day they were first isolated. Early proposed HA building blocks were aromatic and 'coal-like'. Present knowledge indicates a hydrophobic framework of aromatic rings linked by more flexible carbon chains, with alcohol, amide, amine, carboxylic, carbonyl, phenol and quinone functional groups.

No other biopolymers with this range of properties are so widely distributed in nature.

But HS molecular structures still are unknown. For more than two centuries, these fascinating materials have been taken to be intractable mixtures that can be studied and wondered at but never fully understood. How could they ever be understood, coming as they do from so many forms of animal and plant life?

The same could be said half a century ago when the structures of nucleic acids and proteins also were unknown. The classes of biopolymers we now call nucleic acids and proteins could in reality have been intractable mixtures. But some inspired scientists felt otherwise. Progress depended on purifying samples to the point where their properties no longer varied. Then it could be and shown that purified materials from different sources had recognizably similar analytical characteristics and other properties that set them apart from other biopolymers.

Nucleic acids, proteins and other important biopolymers are no mystery now because our predecessors with faith, intuition and determination applied the best tools at their disposal to find their molecular structures.

3 THE NATURE OF THE PROBLEM

The problem with fully understanding HSs has at least three parts.

The first is that HSs are mostly found in sediments and soils along with fauna, flora, decaying animal and plant remains, metals, clays and minerals. Sediment or soil samples from different locations could never be the same because they have different constituent proportions resulting from different animal, mineral and plant inputs and variable reaction conditions. For this reason agronomists, environmentalists, geologists and soil scientists regard the Earth holistically. Over the centuries they have classified sediment and soil samples for historical and predictive purposes. This approach is practical and good but it does not address the question of HS molecular structures.

We know that HSs dominate sediment and soil properties even at low proportions. Also, humic-like substances have been found in animals, plants and sewage, and studies of composting and humification indicate a point at which the properties of the products change very slowly, if at all. Such HS products are called 'refractory' or 'recalcitrant', and they seem appropriate for isolation, purification and close examination. Could it be that, given moisture for reactant mobility and sufficient time for reaction completion, the products of sediment and soil processes have similar or related molecular structures? Animals and plants have different biochemistries, but could common sediment and soil biochemistries lead to the same class of biopolymers?

The second part of the problem comes from the properties of HSs themselves. As noted, they bind metals, molecules, ions and other biopolymers, are redox-active, and they stick to clays and minerals. The result is that they are difficult to purify to the point where their properties no longer change. As in the purification of nucleic acids and proteins, we certainly have to remove metals, molecules, ions, other biopolymers, clays and minerals from an isolated HS sample. Only then can we understand the fundamental biopolymers that HSs contain.

A third part of the problem is that HSs have a strong tendency to aggregate through physical and chemical crosslinking. HSs exist in high oxidation states with a range of functional groups. This makes it difficult to identify HS building blocks and structures from gel permeation chromatography, light scattering, viscosity and other measurements that give useful information on other polymers.

As regards knowing when HS sample purification is complete, we should highlight their characteristic association with polysaccharides and proteins. The constituents of glycoproteins are proteins and polysaccharides that can be separated, characterized and reconnected. Could the fundamental biopolymers of HSs be a different class of macromolecules that associates with microbially vulnerable polysaccharides and proteins and somehow protects them over long periods of time? More importantly, can a new, fundamental biopolymer class be separated from the polysaccharides and proteins, identified and reassembled like a glycoprotein?

We have to know how to conduct HS isolation and when HS purification is complete. Otherwise, we will lose vital information on the biopolymer class that associates with polysaccharides and proteins in HSs. Perhaps the fundamental biopolymers are immutably bound to polysaccharides and cannot exist as separate entities? Other factors to be borne in mind are that HSs are fragmented by long exposure to strong acids and bases, and that they can be oxidized and reduced.

4 THE CHALLENGE

Given their wide distribution and multitude of functions, the problem of knowing HS molecular structures is one of the great challenges of nature. As with any material characterization, the basic problem is *analytical*. Having an idea that the molecules we collectively call HSs are composites of polysaccharides, proteins and an as yet unknown class of biopolymers provides a focus of effort at proof or disproof and the promise of progress. The literature of humic substance research is substantial but there are too few points of connection to give deep understanding. Given the wide and powerful range of current analytical tools and evidence that purified HAs from different sources and locations are not drastically different.

The study of humic substances is a worldwide activity with many implications for our survival, the environment and human health. Much has happened in humic substance research recently, particularly with regard to trying to deduce HS structures with the help of sophisticated analysis and molecular modelling. Humic acids may be oxidized glycoproteins. Studies of highly purified HAs indicate that they have reproducible and accountable properties and appear to have common metal binding sites. Humic substances are natural fertilizers and an industry based on HSs continues to grow.

5 THIS BOOK

This book is a collection of papers from Humic Substances Seminar II, which was held at Northeastern University, Boston, Massachusetts, USA on March 27, 1998. We were especially pleased to have the company of Drs. James Alberts (President), Michael Hayes (Immediate Past President) and C. Edward Clapp (Treasurer) of the International Humic Substances Society, together with other eminent authors from ten countries. Also included are papers invited by the Editors from other experts in the field. The emphasis is on very recent work that bears directly on the relationships between the origins, structures, properties and applications of humic substances. The high rate of recent progress in understanding HSs is due to application of techniques that are well known in biology, chemistry, engineering, geology and physics but have yet to fulfil their full potential in HS research. We feel that the mysteries of HS structures and properties will be solved by approaches and techniques that work well in other disciplines. Determination of the first 'refractory' HS molecular structure will attract wide interest and major recognition.

We want to encourage experts from other disciplines with a different view of nature to work on solving this problem. Thus, the objectives of publication of this book by The Royal Society of Chemistry are to highlight the best work in the search for HS structures and to encourage new effort with new ideas and approaches to this field. The stage seems set for substantial progress.

ACKNOWLEDGEMENTS

We sincerely thank the authors for their co-operation and fine contributions. It is a pleasure to acknowledge financial support and encouragement from Arctech, Inc. and our other sponsors and to thank Northeastern University for providing its fine facilities for the Humic Substances Seminar series. Michael Feeney of the Northeastern University Media Center managed the Seminar II presentations. Michelle Bowman, Beth Rushton and the staff of the Barnett Institute handled mounds of correspondence and countless details. Sam Abronowitz and Brian Gadoury cheerfully converted e-mailed manuscripts into readable text. We also acknowledge a debt of gratitude to Janet Dean and Janet Freshwater of The Royal Society of Chemistry for helping us to bring this book to timely fruition.

<table>
<tr><td>Boston, Massachusetts</td><td>Geoffrey Davies</td></tr>
<tr><td>August, 1998</td><td>Elham A. Ghabbour</td></tr>
<tr><td></td><td>Editors</td></tr>
</table>

Contents

Humic Substances: Progress Towards More Realistic Concepts of Structures   1
    M.H.B. Hayes

Use of ^{13}C NMR and FTIR for Elucidation of Degradation Pathways During Natural   29
Litter Decomposition and Composting. II. Changes in Leaf Composition After
Senescence
    Robert L. Wershaw, Kay R. Kennedy and James E. Henrich

Use of ^{13}C NMR and FTIR for Eucidation of Degradation Pathways During Natural   47
Litter Decomposition and Composting. III. Characterization of Leachate from
Different Types of Leaves
    Robert L. Wershaw, Jerry A. Leenheer and Kay R. Kennedy

Use of ^{13}C NMR and FTIR for Elucidation of Degradation Pathways During Natural   61
Litter Decomposition and Composting. IV. Characterization of Humic and Fulvic
Acids Extracted from Senescent Leaves
    Robert L. Wershaw and Kay R. Kennedy

Characterization and Properties of Humic Substances Originating from an Activated   69
Sludge Wastewater Treatment Plant
    Benny Chefetz, Jorge Tarchitzky, Naama Benny, Patrick G. Hatcher,
    Jacqueline Bortiatynski and Yona Chen

Structure and Elemental Composition of Humic Acids: Comparison of Solid-State   79
^{13}C NMR Calculations and Chemical Analyses
    J. Mao, W. Hu, K. Schmidt-Rohr, G. Davies, E.A. Ghabbour and B. Xing

Comparison of Desorption Mass Spectrometry Techniques for the Characterization   91
of Fulvic Acid
    Teresa L. Brown, Frank J. Novotny and James A. Rice

The Relative Importance of Molecular Size and Charge Differences in Capillary   109
Electrophoresis of Humic Substances of Different Origin
    Maria De Nobili, G. Bragato and A. Mori

Fluorescence Decay of Humic Substances. A Comparative Study   113
    Fritz H. Frimmel and Michael U. Kumke

Effect of Lime Additions to Lake Water on Natural Organic Matter (NOM) in   123
Lake Terjevann, SE Norway: FTIR and Fluorescence Spectral Changes
    James J. Alberts, Dag O. Andersen and Monika Takács

A Computational Chemistry Approach to Study the Interactions of Humic   133
Substances with Mineral Surfaces
    Leonid G. Akim, George W. Bailey and Sergey M. Shevchenko

Determination of Trace Metals Bound to Soil Humic Acid Species by Size   147
Exclusion Chromatography and Inductively Coupled Plasma-mass Spectrometry
    Peter Ruiz-Haas, Dula Amarasiriwardena and Baoshan Xing

Formation and Voltammetric Characterization of Iron–Humate Complexes of   165
Different Molecular Weight
    L. Leita, M. De Nobili, L. Catalano and A. Mori

Nonlinearity and Competitive Sorption of Hydrophobic Organic Compounds in   173
Humic Substances
    Baoshan Xing

Adsorption of a Plant- and a Soil-derived Humic Acid on the Common Clay   185
Kaolinite
    E.A. Ghabbour, G. Davies, K. O'Donaughy, T.L. Smith and M.E. Goodwillie

Effect of Dissolved Organic Matter on the Movement of Pesticides in Soil   195
Columns
    K.M. Spark and R.S. Swift

Generation of Free Radicals by Humic Acid: Implications for Biological Activity 203
    Mark D. Paciolla, Santha Kolla, Lawrence T. Sein, Jr., James M. Varnum,
    Damien L. Malfara, Geoffrey Davies, Elham A. Ghabbour and Susan A. Jansen

Humic Acid as a Substrate for Alkylation   215
    Santha Kolla, Mark D. Paciolla, Lawrence T. Sein, Jr., John Moyer,
    Daman Walia, Harley Heaton and Susan A. Jansen

Humic Substances for Enhancing Turfgrass Growth   227
    C.E. Clapp, R. Liu, V.W. Cline, Y. Chen and M.H.B. Hayes

Greenhouse Gas Dilemma and Humic Acid Solution   235
    D.S. Walia, A.K. Fataftah and K.C. Srivastava

Subject Index   243

HUMIC SUBSTANCES: PROGRESS TOWARDS MORE REALISTIC CONCEPTS OF STRUCTURES

M. H. B. Hayes

Department of Industrial Chemistry, The University of Limerick, Ireland

1 INTRODUCTION

1.1 Humic Substances Origins and Definitions

Humic substances (HSs) are found in all soils and waters that contain organic matter (OM).[1] HSs are the result of biological and chemical processes, and the amounts generated in soil are several times greater than those in waters. The origins of HSs in deep sea environments are almost entirely *autochthonous* (*i.e.* produced within the aquatic system),[2] as are those in soils, but the HSs in streams and in fast flowing waterways are largely *allochthonous* (*i.e.* produced outside the system, usually in the soils of the watersheds).[3] HSs in lakes, reservoirs, estuaries and coastal waterways are both autochthonous and allochthonous.[1,2]

Labile plant materials decompose rapidly on entering aerobic soil environments with adequate water supplies, but more resistant components transform slowly in the same environment. Because of the compositional diversities and the differences in the transformation modes of the components, it is impossible to define the gross mixtures that compose soil organic matter (SOM), or the dissolved organic matter (DOM), or the particulate organic matter (POM) of waters.

Hayes and Swift[4] have distinguished between recognizable plant/animal debris and the highly transformed materials that contain no recognizable plant, animal, or microbial structures. All of the recognizable plant debris and identifiable classes of organic macromolecules such as carbohydrates, peptides and nucleic acids are considered to be non-humic substances. The highly transformed, amorphous, dark coloured materials are classified as HSs.

The classical definitions of HSs are operational and based on aqueous solubility. Aiken *et al.*[1] state that *humic substances* are *"a general category of naturally occurring heterogeneous organic substances that can generally be characterized as being yellow to black in color, of high molecular weight, and refractory"*. This definition is still broadly relevant but, as discussed below, the classical interpretations suggesting that HSs have high molecular weight (HMW) values may not be as acceptable now. Also, HSs are refractory only when contained in protected environments.

Recognizable plant remains constitute a small percentage of the SOM of mineral soils and float as the 'light fraction' when soils are dispersed in liquids of known densities. In mineral soils the transformed materials, or *humus*, can be more than 90% of the SOM, and HSs can amount to 70–80% of the total. Carbohydrates, peptides, and lipophilic materials from plant and microbial remains and processes are also present.

HSs have a degree of resistance to further microbial degradation, which is attributable to self associations of the molecules, to protection through associations with the soil mineral colloids, and to entrapment in soil aggregates. Availability to transforming microorganisms and their degradative enzymes is impeded in these situations. Thus some labile molecules can survive for a time, but all are eventually biodegraded. Without biodegradation the surface of the earth would be deeply covered in HSs, and the waters would be dark coloured 'soups'. HSs are by far the most abundant of the organic molecules of nature. Their carbon contents can be expected to be two to three times greater than in all living matter.

The compositions and amounts of HSs in terrestrial waters are related to the HSs of the watershed soils.[3] HSs in estuarine environments are related to the terrestrial aquatic HSs, but those in the oceans have different origins and compositions.[2] The generalized terms *humic acids* (HAs), *fulvic acids* (FAs), and *humins* cover the major fractions still used to describe HS components, but the boundary between these fractions is not clear. HAs and FAs can be considered as a continuum, and the cut off point (soil derived HAs are insoluble at pH 1) is arbitrary.

HAs are defined by Aiken *et al.*[1] as *"the fraction of HSs that is not soluble in water under acid conditions but becomes soluble at greater pH"*. Water scientists take precipitation at pH 2 as their standard, whereas pH 1 is that used by soil scientists.

The term 'fraction of humic substances' conveys the fact that the HSs do not include biomolecules such as peptides, sugars, nucleic acid residues and fats. However, such molecules can be sorbed to or co-precipitated (at pH 1 or 2) with the HAs. Hence the humic fraction isolated may not consist wholly of HSs.

Häusler and Hayes[5] have shown that dissolving HAs in dimethyl sulfoxide (DMSO) containing 1% HCl, and passing the solution on to an XAD-8 [poly(methyl methacrylate)] resin column, significantly decreases the amino acid and neutral sugar NS contents of the HAs. Ping *et al.*[6] used XAD-8 resins to remove non-humic molecules associated (but not by covalent links) with HAs. The process involved dissolving the fraction precipitated at pH 1 in 0.1M NaOH, diluting the solution to < 100 ppm, and loading it onto an XAD-8 resin. The polar non-humic components passed through the resin, and the HAs that sorbed were recovered by back elution in base, H^+-exchanged by passing through a cation exchange resin (in the H^+ form), and recovered by freeze drying. Hayes *et al.*[7] found it necessary to dilute the solute to < 20 ppm to maintain solubility.

Fulvic Acids (FAs) are defined[1] as *"fraction of humic substances that is soluble under all pH conditions"*. Soil scientists take FAs to be the fraction that stays in solution when basic soil extracts are adjusted to pH 1. Supernatants from the acidified base will, of course, contain non-humic materials and are best defined as the *"fulvic acid fraction"*. In the IHSS (International Humic Substances Society) procedure for isolating FAs,[8] the acidic FA fraction is passed on to XAD-8 resin and then treated as described for the HAs. *Humin* is defined[1] as *"that fraction of humic substances that is not soluble in water at any pH value"*. On that basis, humin can include any humic-type material that is dissolved in non-aqueous solvents after the soil has been exhaustively extracted with basic aqueous solvents. Humins are often considered to consist mainly of humic materials strongly associated with the soil inorganic colloids. In the sequential procedure of Clapp and Hayes,[9] soil was exhaustively extracted at one pH value before passing to an aqueous solvent at a higher pH. After application of the final aqueous solution (0.1 M NaOH at pH 12.6), a DMSO/HCl medium was used. Although the DMSO-soluble substances would normally be classified as humin, the materials isolated were similar to HAs and FAs. The materials did not dissolve in the aqueous media because the polar 'faces' of the molecules were bound to the inorganic colloids, and the hydrophobic moieties faced outwards. Humic associations and protective coatings of non-polar moieties, such as fatty acids and long chain hydrocarbons, could greatly decrease solubility even in strong aqueous base.

Advances in humic sciences warrant changes in the classical definitions of HS fractions. Recent work has used XAD-8 and XAD-4 (styrene–divinylbenzene) resins to isolate HAs, FAs and the so-called XAD-4 acids from waters, and Malcolm and MacCarthy[10] have used XAD-8 and XAD-4 resins in tandem to isolate HSs from water. Water at pH 2 (at which the acidic functionalities are not dissociated) is loaded on to the resin. Hydrophobic interactions occur between the solute materials and the resins under these conditions. The less polar fractions (which include the HAs and the FAs) are sorbed to the XAD-8, and the more hydrophilic materials are sorbed by XAD-4. Fractionation based on charge density differences is achieved by back eluting with aqueous solvents of different pH values. The material not eluted in aqueous base can be removed in ethanol and in acetonitrile, and this organic soluble fraction is called the *"Hydrophobic Neutrals"*.

The above studies[9,10] show that HAs can be fractionated using the XAD-8 resin approach. However, the author is not convinced that the XAD-4 acids represent a legitimate fraction of HSs. These do contain some brown materials that can be classified as HSs, but they also have high neutral sugar (NS) and amino acid contents suggesting major carbohydrate- and peptide-related components.

1.2 Roles of HSs in Soil and Water Environments

Hayes and Swift[4,11] and Stevenson[12] list properties of HSs that are important for soil conservation and for crop growth. Even though the SOM composition of mineral soils is usually in the range of 1–5%, it would not be possible to achieve the agricultural productivity that is needed to sustain the world population in its absence. SOM is vital to the formation and stabilization of soil aggregates, and Swift[13] has outlined how poly-saccharides and HSs have different roles in the formation and in the stabilization, respectively, of the aggregates.

HSs are important as cation exchangers, in the release of plant nutrients (especially N, P, and S) when mineralized, and in the binding of anthropogenic organic chemicals (AOCs). The biological activity of most aromatic AOCs is decreased or lost on contact with HSs, and HSs in waters are not hazardous to health. Some research indicates that HSs have healing effects.[14] However, dissolved HSs are chlorinated during water treatment, and the products can have deleterious health effects.[3,15]

1.3 The Importance of an Awareness of Humic Structures

HSs are heterogeneous mixtures. The evidence indicates that the HSs from the same soil types and formed under the same environmental conditions are broadly similar. However, there are distinct compositional differences between HSs from different sources, soil types, and climates. Nevertheless, some of the general reactivities, regardless of source, are broadly similar.

It is important to know the components of HSs responsible for some of the major processes in soils and in waters, and the proportions of these components in a humic source. To understand process mechanisms it is important to have an awareness of structure. For meaningful structural studies it is desirable to deal with pure substances. However, HSs are gross mixtures, and some degree of homogeneity with respect to size and to charge density is as much as can be hoped for. Even that goal has not yet been fully attained.

This paper briefly cites classical procedures for the isolation and fractionation of HSs, and also modern procedures that show promise. It will refer to classical procedures for studies of composition and structure, and focus on modern approaches that are advancing our awareness of these aspects of the humic sciences.

2 ISOLATION OF HUMIC SUBSTANCES FROM SOILS AND WATERS

2.1 Isolation of HSs from Soils

Hayes[16,17] has given detailed accounts of the principles and procedures for the isolation of HSs from soils, and Swift[8] has provided details of the procedures used to isolate the humic standards of the International Humic Substances Society (IHSS). The account that follows is from Hayes.[17]

Solubility is best achieved through solvation of anionic functionalities in the HSs, and their proportions increase as the pH is raised. The behaviour and properties of H^+-exchanged HSs are similar to those of neutral polar macromolecules that are hydrogen bonded to each other. Hydrogen bonding, together with associations of the non-polar moieties through hydrophobic bonding/van der Waals forces, and through charge transfer processes can confer HSs molecules with properties similar to those of high molecular weight (HMW) materials. Only the most polar and the least associated HSs molecules will dissolve in water. At higher pH the most strongly acidic functionalities are first to dissociate, and the conjugate bases (generally carboxylates) then solvate, and eventually the most weakly acidic functionalities (some phenols, enols) dissociate and solvate. Use can be made of these principles to fractionate HSs on the basis of charge density differences.

The dominant exchangeable cations in agricultural soils are di- and trivalent (especially Ca^{2+}, Mg^{2+}, Fe^{3+} and Al^{3+}). These cations are strongly held by the anionic functionalities of the HSs. Such cations form inter- and intramolecular bridges between anionic sites, suppressing repulsion and inhibiting solvation. Cation bridging does not take place when charge balancing metal cations are monovalent. Thus, repulsion between the charged species takes place, the molecules (or the molecular associations) assume expanded conformations, and solvation can occur readily (provided that the ratio of charged to neutral moieties is adequate).

Cation bridging, hydrogen bonding and hydrophobic interactions/van der Waals association effects (which also apply for the H^+-exchanged species) cause HS molecules to assume shrunken (or condensed) conformations, and water is partially excluded from the matrix. These effects can be overcome by replacing the divalent/polyvalent cations with monovalent species (other than H^+). Thus, addition of sodium pyrophosphate ($Na_4P_2O_7$) to the system allows the pyrophosphate to complex the polyvalent cations and the sodium to neutralize the negative charges, and the charged species then solvate in water.

These principles were behind the approach used by Clapp and Hayes,[9] Hayes[18] and Hayes *et al.*[7] for the isolation of HSs from soils. Exhaustive extraction with water isolates the highly charged, and predominantly FA and XAD-4 type acids. Sodium pyrophosphosphate (Pyro, 0.1 M) adjusted with mineral acid (often phosphoric) to pH 7 complexes polyvalent metals and forms the sodium salts of the carboxylates. The next convenient solvent in the sequence is 0.1 M Pyro (pH 10.6). In theory, that solvent system would cause phenols to dissociate and solvate. The final aqueous solvent in the sequence (0.1 M Pyro + 0.1 M NaOH) causes very weak acids (including enols) to dissociate and to solvate.

Clapp and Hayes[9] carried out a final single extraction with dimethyl sulphoxide (DMSO, 94%) and 12M HCl (6%). DMSO is a poor dipolar aprotic solvent for anions but a good solvent for cations,[19] so the metal cations that neutralize the charges on humates would be solvated and the conjugate bases (carboxylates and phenolates) would not. Hence DMSO is a poor solvent for ionized humates. However, when the humates are H$^+$-exchanged they behave essentially as polar molecules. They associate through hydrogen bonding and van der Waals forces.

There are strong interactions between the polar face of DMSO and water, carboxyl, and phenolic groups, and DMSO-water interactions are stronger than the associations between water molecules. Thus DMSO will associate with the phenolic and carboxyl groups to break the intra- and intermolecular hydrogen bonds. The non-polar face of DMSO can associate with the hydrophobic moieties in HSs, and the combination of hydrogen bond breaking and disruption of hydrophobic associations makes DMSO a very effective solvent for HSs. Removal of the solvent is not a problem because the HSs sorb to the XAD-8 resin, and the DMSO and acid wash through.[5,9]

An alternative to exhaustive sequential extraction using a series of solvents would be to extract with the NaOH/Pyro system and subsequently to fractionate by eluting at different pH values from XAD resins. HS extraction with DMSO/HCl would follow exhaustive extraction with base.

The procedure used to isolate the IHSS standards employs a 0.1 M HCl/0.3 M HF treatment to remove finely divided inorganic soil colloids and thereby provide materials with acceptable ash contents.[8] The HCl/HF treatment is repeated as often as necessary to lower the ash content to < 1%. This treatment, followed by dialysis, leads to considerable losses of the humic fractions. Clapp and Hayes[9] and Hayes *et al.*[7] found that filtration through partially clogged 0.45 μm or 0.2 μm filters very effectively lowers the ash contents of HS fractions.

2.2 Isolation of HSs from Waters

Aiken[20] describes a variety of procedures for the isolation of HSs from waters. These include filtration and applications of co-precipitation, ultrafiltration, reverse osmosis, solvent extraction, ion exchange and sorption, including uses of alumina, carbon and non-ionic macroporous and weak anion exchange resins (with secondary amines and their salts providing the active functionalities).

2.2.1 The XAD Resin Procedure. Leenheer[21] refers to six fractions of dissolved organic carbon (DOC). These are the hydrophobic acids, bases and neutrals, and the hydrophilic acids, bases and neutrals. The hydrophobic fractions are retained by XAD-8 resins, and that principle was used to isolate the IHSS Standard aquatic HAs and FAs from the Suwannee River. The procedure is described by Thurman and Malcolm,[22] Thurman,[23] and Aiken.[20]

2.2.2 The Reverse Osmosis (RO) Procedure. The RO process concentrates OM readily from large volumes of water under ambient conditions.[24] Serkiz and Perdue[25] and Sun *et al.*[26] describe instrumentation which processes 150–200 L of water h^{-1} with 90% recovery of organic carbon without exposing the DOM to harsh chemical reagents. All the solutes, including inorganic salts and non-HSs, are concentrated and so resin treatments must be employed for their separation.

3 FRACTIONATION OF HUMIC SUBSTANCES

Swift[8,27] and Stevenson[12] describe the fractionation techniques in current use. Leenheer[21] and Thurman[23] have reviewed the principles and the procedures used to fractionate HSs from water. Techniques for aquatic HSs apply equally well to those from soil.

The classical fractionation of HSs is based on solubility differences in aqueous media at different pH values. In principle, the HA fractions that have weakly dissociable acids should be first to precipitate as the pH is lowered, and the strongest acids would be last. Additions of salt decrease the intra- and intermolecular charge repulsion which results in shrinking of the aggregates by exclusion of water from the matrix. Also the suppression of the electrical double layer allows the molecules to approach more closely and promotes coagulation (expressed as self association or pseudo micelle-type structure formation). These are considered to be salting out effects. Selective precipitation can be achieved with heavy metals.

HSs have low solubilities in most common organic solvents,[16] and solvents such as ethanol, methanol, and acetone can be used to precipitate HAs fractionally from alkaline solutions. Attempts to partition HSs between alkaline solutions and non- (or sparingly) miscible organic solvents can give a concentration of HSs at the water–solvent interface. The hydrophobic components of the HSs are attracted to the organic solvent at the interface, but the polar moieties are not solvated and hence the HSs do not cross the interface. Eberle and Schweer[28] developed a hydrophobic extractable ion pair by dissolving long chain tertiary or quaternary amines in chloroform and extracting HSs and lignosulphonic acids from water at pH 5, then recovering the HSs at pH 10. This gave promise of applying counter current distribution techniques for the fractionation of HSs. Such techniques are tedious and more emphasis has been given to fractionations based on molecular size and charge density differences.

Considerable emphasis has been placed[4,29–31] on the uses of gel chromatography, or gel permeation chromatography (GPC), for the fractionation of HSs on the basis of molecular size differences (MSD). Swift[8] gives a diagrammatic representation of the tedious procedure required to produce fractions with a degree of molecular size homogeneity. The procedure involves reprocessing (through the gel column) the substances eluted between specific volume boundaries from a standardized gel column till all of the material is contained within these volume boundaries. In theory, it should be possible to isolate several homogeneous fractions in that way. However, as the fractions are reprocessed, smaller sized components are released, and that might be interpreted in terms of the breakdown of molecular associations (*vide infra*). At best the reprocessing gives concentrations of components that would appear to have similar molecular sizes.

The work of Piccolo *et al.*[32] suggests that a re-evaluation of the effectiveness of GPC for HS fractionation on the basis of MSD is in order. They subjected two HAs from different sources to GPC and collected, dialysed, and freeze dried the components eluted in 0.02M borate ($Na_2B_4O_7$) in the void volume of a Sephadex G-100 gel. These materials were considered to have a nominal $<M_W>$ > 100 kDa. A stock solution was made by suspending the HAs in water and raising the pH to 11.8 (using 0.5 M KOH), and storing under N_2. When these solutions were eluted in a 0.02 M solution of $Na_2B_4O_7$ through an LKB K 16-70 analytical GPC column packed with Biorad P100 Biogel (molecular range 5–100 kDa) in the same buffer, a large peak was obtained at an elution volume corresponding to the void volume (V_o) of the column, and a smaller peak (eluted at a larger volume) indicated that materials of smaller molecular sizes had separated from the macromolecules (or molecular associations) in the stock solution. The implications are discussed in Section 4.

Other fractionation techniques based on molecular size/shape differences include dialysis and ultrafiltration (uf). Membranes with pores from the nm to the μm sizes are available. Use of a range of such membranes should allow fractions of relatively discrete molecular sizes to be isolated. However, the question of self associations would also apply to applications of uf techniques.

Fractionation based on charge density differences includes electrophoresis, the movement of charged species in solution in response to an applied electrical potential. Duxbury[33] has reviewed applications of the different electrophoretic separation methods, including zone electrophoresis, moving boundary electrophoresis, isotachophoresis, and isoelectric focussing (IEF). Preparative column electrophoresis and continuous flow paper electrophoresis techniques[34] have been used in studies of HSs. Although polysaccharides were separated from coloured HSs, only a Gaussian distribution of colour (and hence no clear cut fractionation of coloured HSs) was achieved.

Ciavatta *et al.*[35] have studied the carrier ampholite (CA) on the applicability of IEF for the preparative fractionation of SOM. They concluded that likely complex formation between the CAs and the HSs makes applications of the technique to preparative fractionations unreliable .

Trubetskoj *et al.*[36] suggest that polyacrylamide gel electrophoresis (PAGE) can fractionate HSs based on M_W differences. HAs subjected to PAGE were fractionated into three zones. Three fractions were collected when the HAs were eluted from Sephadex G-75 in 7M urea. When subjected to PAGE, each of these fractions gave a pattern that corresponded to one of the zones in the electrophoretogram of the unfractionated material.

Klavins *et al.*[37] concluded that the best separation efficiencies for HSs use sorbents that engage in hydrophobic interactions. Good separation was obtained with a sorbent matrix with a weakly basic (1,2,4-aminotriazole) functionality. The eluent was a gradient of 0.1 to 10% dioxane in water. Each of the five fractions isolated had different elemental compositions, capabilities to complex metals, and spectral characteristics.

4 SIZES AND SHAPES OF HUMIC MOLECULES

The classical studies of Cameron *et al.*[31] used an ultracentrifuge to determine the $<M_W>$ of a series of HAs isolated from an organic soil (sapric histosol) and fractionated using gel chromatography and pressure filtration through membranes of known pore sizes. The HAs had $<M_W>$ from *ca.* 2 kDa to 1300 kDa. From frictional ratio data they concluded that the macromolecules had random coil conformations in solution (see also Swift[38]).

The gel permeation studies of Piccolo *et al.*[32] (see Section 3) involved a stock solution of material that appeared to have a high $<M_W>$. However, when methanoic, ethanoic, propanoic or butanoic acids were added to adjust the pH of the stock solution to 2.1, and when this material (at pH 2.1) was applied to the gel column, all of the HA materials were eluted (in the borate solution) in a volume that suggested that the M_W of the materials was < 25 kDa. Benzoic acid did not affect the elution volume. When the pH was readjusted from 2.1 to 3.5, 4.0, 6.0, and 8.5, and the materials were transferred to the gel columns and eluted with the borate buffer, it was evident that reassociations of the humic molecules had taken place at pH 4.5 and above, and most of the substances were eluted in the column void volume.

Simpson[39] added increasing concentrations of ethanoic acid to his HA preparations (in borate). Retention of the HA by Biogel P6 increased with increasing ethanoic acid concentration. This might suggest that the HAs were broken into smaller molecules, but the same result was obtained for methyl red. Perhaps a 'plug' of hydrogen bonded associations was formed between the ethanoic acid, the HAs and the gel, and this 'plug' moved slowly down the column.

Recently the Piccolo group[40] has explored high pressure size exclusion chromato-graphy (HPSEC) of HAs from different sources using dilute organic acids and even aqueous ethanol. Their results clearly show that in these conditions the apparent high $<M_W>$ HAs behave as relatively low $<M_W>$ materials, which suggests that HAs are associations of molecules, and not covalently linked macromolecules.

Silylation of natural organic matter by Hertkorn *et al.*[41] gave materials with relatively low $<M_W>$ that were soluble in organic solvents. Replacement of active hydrogen by the hydrophobic Si-substituents decreases hydrogen bonding and other aggregating inter-actions. The work of Kenworthy and Hayes[42] supports the molecular association concept. They found that the quenching of pyrene fluorescence by bromide in the presence of HS fractions was impeded, and suggested that the HSs provide a hydrophobic environment that shields the pyrene probe from the quenching ion. Treatment of the HSs with ethanoic acid and with base removes the protection, suggesting that the HSs in solution could be associations of relatively low M_W molecular masses held together by hydrophobic bonding.

Wershaw[43] proposed that hydrophobic bonding causes humic molecules to associate in micelle-like aggregates. As the concentration of a surfactant in solution is increased, a break at the *critical micelle concentration* (CMC) occurs in the plot of the specific conductance/g-equivalent of solute (equivalent activity) *vs.* the square root of the solution normality. The free energy of the system increases when the structure of the solvent is disrupted during additions of the surface active material. This increase is minimized when the molecules concentrate at the surface and orientate so that their hydrophobic groups are directed away from the liquid.[44] Alternatively, if the surface active molecules aggregate into clusters or micelles with the hydrophobic groups oriented towards each other and the interior, and the hydrophilic functionalities are on the outside, clusters or micelles are formed and the free energy increase is minimized. The hydrophilic groups in the micelle structures can interact through hydrogen bonding and dipole-dipole interactions with the solvent water. Entropy increases as water structure is disrupted and decreases with micelle formation. There is, however, a favourable increase in energy when roughly spherical micelles are formed (with radius approximately equivalent to the length of the hydrophobic group in linear surfactants) at solution concentration less than ten times the CMC. Non-spherical micelles can form at higher concentrations.

At the lower pH values, inter-micelle associations between the non-dissociated acid groups on the exteriors of the micelles should result in large molecular aggregates that would ultimately precipitate. Piccolo *et al.*[32] suggest that the organic acids penetrate the hydrophobic core of the micelle structures, and thereby give a potentially high charge density molecular structure. Association between the organic acids and the HS structures is feasible because the acids have hydrophobic and hydrophilic character: the hydrophobic face would associate with the hydrophobic inner core. As the pH is raised, the sorbed acids in the inner core of the HAs dissociate. As a result, the aggregate and the micelle structures "blow apart" because of repulsion between the negatively charged groups.

The compositions of humic molecules are, perhaps, too different to form regular micelle structures, and it may be better to consider self association phenomena which could also give rise to pseudo high $<M_W>$ molecules.

The random coil concept for the conformations of humic acids in solution has been useful with regard to interpretations of the reactions and interactions of HSs. However, if HSs are not macromolecular, but associations of molecules that may or may not form micelle-like aggregates, then the random coil model would not hold. The frictional ratio data[31,38] could also be interpreted in terms of an oblate ellipsoid model,[38,45] which could be representative of the types of structures that form through molecular associations.

5 COMPOSITIONS OF HUMIC SUBSTANCES

The compositions of HSs refer to the elemental composition, functional groups, 'building blocks', and actual HS molecules. Analytical data are most meaningful when the samples analysed and compared were subjected to the same isolation and fractionation procedures. A set isolation and fractionation procedure was followed during the isolation of the IHSS Standards. Refer to Huffman and Stuber[46] for analytical determinations of moisture, ash, and the elements of the Standard and Reference IHSS samples.

Swift[8] has updated the wet chemical and some spectroscopy procedures for determinations of functional groups in HSs. There are comprehensive reviews of acidic functionalities,[47] of various spectroscopy procedures such as infrared,[48] of both proton and ^{13}C nuclear magnetic resonance (NMR),[49–51] electron spin resonance (ESR),[52] and of vibrational, electronic, and high energy spectroscopic methods.[53]

The cation exchange capacities (CEC) of HSs increase as the pH of their aqueous medium is raised. Leenheer *et al.*[54] have shown that some of the HS functional groups are strong acids (due to activating substituents α to carboxyl groups) and Perdue[47] has stressed how the same functional groups can have different pK_a values in different local molecular environments.

IR spectra of HSs show broad bands characteristic of functional groups in a variety of environments. Diffuse reflectance Fourier-transform IR (DRIFT) measurements improve resolution and allow analysis of whole soil samples. Niemeyer *et al.*[55] have provided DRIFT spectra of IHSS Reference HAs. These spectra emphasize that IR provides only limited information about the compositions of HSs.

Raman spectroscopy is complementary to IR spectroscopy. Fluorescence from HSs interferes with Raman measurements. Yang and Wang[56] sorbed basic HS solutions on the wall of a truncated NMR tube. Their FT–Raman spectra had two broad bands of different intensities at *ca.* 1300 cm^{-1} and 1600 cm^{-1}, suggesting graphite-like quality for the HSs. However, spectra of the acidic forms were different and downshifts were attributed to sample amorphism.

The peaks in *ultra violet–visible* (UV–vis) spectroscopy have limited applications in the humic sciences. The peaks for humic fractions are broad, and provide little information that is interpretable. E_4/E_6 ratios (absorbance at 465/665 nm) give useful comparisons for humic samples from different sources and environments. The ratio values can have distinct differences, but these are not unambiguously interpretable.[57,58]

Fluorescence spectroscopy is finding applications in the humic sciences, and well defined spectra are obtainable using modern instrumentation. Simpson *et al.*[59] have provided fluorescence spectra for HAs isolated in water, and at pH 7, 10.6, and 12.6 from the Ah horizon of a forested podzol, and for FAs and XAD-4 acids isolated from the same soil at pH 7, 10.6, and 12.6. In general, the HAs had main excitation bands at 465, 480, and 490 nm and these were least intense in the cases of fractions isolated at the lower pH values. The FAs had excitations in the same regions and with intensities that were greater than for the HAs, and with additional excitation bands at 400, 420, and 450 nm that increased in intensity as the pH of the extracting solution increased. The excitation shifted to shorter wavelengths in the cases of the XAD-4 acids isolated in the same solvent systems from the same soil, and new bands appeared at 330 and 370 nm.

It is not yet possible to assign the bands to definite functionalities. Bloom and Leenheer[53] have suggested the presence of two main types of fluorophores in HSs, one at an excitation between 315 and 390 nm, which may be attributable to carboxyphenol, and a second at an excitation between 415 and 470 nm. Studies with model compounds with compositions relevant to HSs will help with functionality assignments for the bands observed. In the meantime it is appropriate to develop the technique as a fingerprint for humic fractions. However, in order to achieve meaningful comparisons it will be important to develop standard sample preparation methods.

Electron spin resonsnce *(ESR)* provides information about free radicals.[52] Although radicals can significantly influence interactions of HSs, they are at relatively low levels in terms of the total HS compositions.

6 NUCLEAR MAGNETIC RESONANCE (NMR)

NMR gives the most useful HS compositional information. Progress was made when techniques such as multiple pulse Fourier transform NMR, dipolar decoupling, cross-polarization nuclear induction spectroscopy, and cross polarization magic-angle spinning (CPMAS) NMR were introduced. These techniques enabled better functional group assignments and allowed a degree of quantification to be achieved.[51,60] However, one dimensional (1-D) proton and carbon spectra of HSs are still relatively ill-defined and yield only limited structural information. The heterogeneous nature of HSs causes overlapping of signals and prevents unambiguous mapping of the H-C-O framework.

Two-dimensional (2-D) NMR methods developed during the past decade have significantly advanced structural determinations of organic compounds and biomolecules. Uses of these techniques to study HS structures are relatively new. Information from liquid state one- (1-D) and two-dimensional (2-D) NMR experiments is listed in Table 1.

A brief outline follows of recent and ongoing work by the author and his colleagues which indicates the usefulness 2-D NMR in HS studies. Figure 1 shows the 1-D ^{1}H NMR spectrum of a FA isolated at pH 12.6 (in the sequential extraction process; see Section 2.1) from the humified root zone of a moss culture in Co. Kerry, Ireland.[61] This spectrum is

Table 1 *Summary of NMR spectroscopy experiments applicable to structural studies of humic substances (from Simpson et al.[61])*

NMR experiment	Structural information
1-D ^{1}H	Chemical shift provides information on the functional groups present
1-D ^{13}C	Chemical shift-functional groups
2-D COSY	^{1}H–^{1}H two and three bond coupling
2-D TOCSY or HOHAHA	^{1}H–^{1}H coupling throughout a complete spin system
2-D ROESY or NOESY	^{1}H–^{1}H dipolar (through space) coupling information which provides inter-proton distances. Helps to refine structure of the molecule
2-D ^{1}H–^{13}C HMQC	Via one bond, provides ^{1}H–^{13}C connectivity information on the specificity of proton-carbon bonding. Supports functional group assignments.
2-D ^{1}H–^{13}C HMBC	Via two and three bonds, provides ^{1}H–^{13}C connectivity information to confirm functional group assignments and substitution patterns

relatively well resolved because the sample had been subjected to extensive fractionation, and a high field strength (600 MHz) spectrometer was used.

The spectrum can be split into three main regions: 0.8–3 ppm, attributable to aliphatic protons;[62] 3–5.5 ppm representing a wide range of protons associated with oxygen-containing functional groups (these protons can be exchangeable and non-exchangeable); and the 6–8.5 ppm region, attributed to aromatic/amide protons.

Table 2 gives the major functional group assignments from the 1-D proton moss FA spectrum. The resonance at 1.5–2.0 ppm is relatively weak, which could indicate protons of functionalities such as R-CH$_2$-C-O-R' (long chain ethers), R$_1$R$_2$CH-R$_3$ (the ethine protons of branched aliphatics), and be used to identify possible moieties. They can

Table 2 *Chemical shift assignments for the 1-D ^{1}H NMR spectrum of the FA fraction isolated at pH 12.6 from humified moss using the sequential extraction process*

Chemical shift (ppm)	Assignment
0.8	Methyl protons
1.2	Methyl protons (likely to be CH$_3$-C-O protons)
1.4	Methine protons
2.4	CH$_3$-CO-OAr or R-CH$_2$-CO-R
2.5	Solvent, DMSOd$_6$
2.6–2.7	CH$_3$-CO-Ar or R-CH$_2$-Ar
3–5.5	Protons associated with oxygen containing functionalities
6–8.5	Aromatic/amide protons

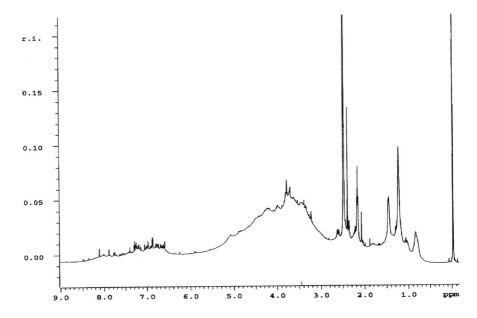

Figure 1 One dimensional ^1H NMR (600 MHz) spectrum of a moss FA (isolated at pH 12.6 in the sequential extraction process) in DMSO d_6 (from Simpson et al.[61])

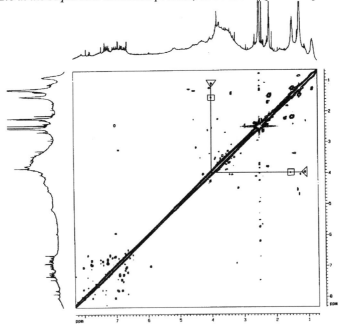

Figure 2 Two dimensional COSY (^1H-^1H; 500 MHz) NMR spectrum of the moss fulvic acid (isolated at pH 12.6) in DMSO d_6. Cross peaks at 4.1 ppm represent R-CH$_2$-O-CO-R protons that couple to methylene (□) and methyl (Δ) protons

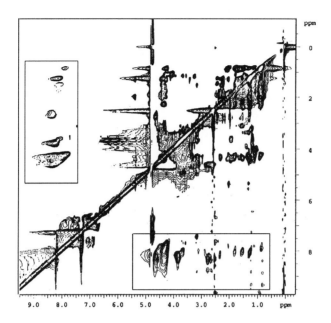

Figure 3 *Two dimensional TOCSY (^{1}H-^{1}H; 600 MHz) NMR spectrum of the IHSS Standard Mollisol soil humic acid in DMSO d_6. Couplings from amino acid moieties are highlighted by the boxes*

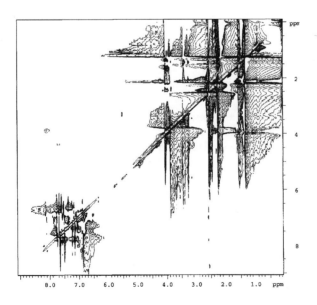

Figure 4 *Two dimensional TOCSY (^{1}H-^{1}H; 600 MHz) NMR spectrum of the fulvic acid (isolated at pH 12.6 from the Bh horizon) of a podzol under ancient oak forest) in DMSO d_6*

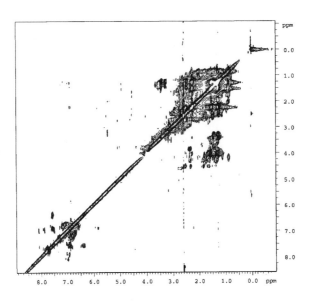

Figure 5 *Two dimensional TOCSY (^1H-^1H; 600 MHz) NMR spectrum of the fulvic acid (isolated at pH 12.6 from the Bh horizon of a podzol) in DMSO d_6 after the addition of D_2O*

indicate which units may be connected to each other. For example, the two peaks at 4.1 ppm indicate protons in R-CH$_2$-O-CO-R units. Their connectivities to the methylene and methyl regions imply the presence of CH$_3$-CH$_2$-O-CO-R and R-CH$_2$-CH$_2$-O-CO-R units.

CH$_3$-C=C and R-CH$_2$-C-C=C protons (associated with certain unsaturated systems) did not contribute significantly to these FA structures. The relatively weak resonances at 2.3 and at 2.7 ppm, representative of CH$_3$-Ar and R-CH$_2$-Ar units, respectively, suggest that alkyl groups may not be major substituents in the aromatic functionalities. The peak intensities may also provide information. For example, in a compound containing long chain aliphatic hydrocarbons, the peak at 1.4 ppm (CH$_2$ protons) would predominate over those at 1.2 and 0.8 ppm (mainly methyl protons), unlike the FA spectrum shown. Long chain aliphatic components are unlikely to be major contributors to the FA, and aliphatic substituents and bridges are more likely to be short, one or two unit structures.

Addition of D$_2$O causes exchange of amide and hydroxyl protons for deuterium, and the protons of these functionalities thus "disappear". D$_2$O addition to the moss FA sample decreased the intensity of the 3–5.5 ppm resonances, indicating the exchange under acidic conditions of amino/amido/hydroxyl protons. The intense water signal at δ = 3.3 ppm dwarfed resonances in other regions that appeared to be less well defined on expansion. A watergate pulse sequence was used to quench the water signal partially.[62]

The COSY spectrum for the moss FA sample has good resolution and connectivities, and couplings are observed through two and three bonds (Figure 2). The broad band from δ = 3–5.5 ppm, caused by many overlapping signals, could not be interpreted from the 1-D spectrum. However, the COSY spectrum indicates a number of coupled hydrogens in the form of assignable cross peaks. For example, the two cross peaks at δ = 4.1 ppm would indicate protons on RCH$_2$-O-CO-R units. Their connectivities to both methylene and methyl regions suggest the presence of CH$_3$CH$_2$-O-CO-R and RCH$_2$CH$_2$-O-CO-R units.

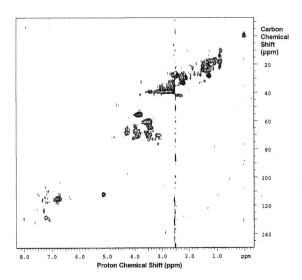

Figure 6 *Two dimensional HMQC (^1H-^{13}C; 600 MHz) NMR spectrum of the moss fulvic acid in DMSO d_6*

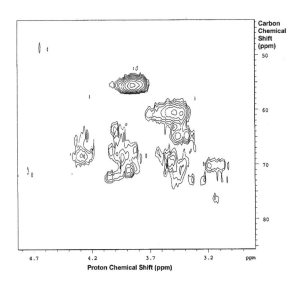

Figure 7 *Two dimensional HMQC (^1H-^{13}C; 600 MHz) NMR spectrum of the moss fulvic acid for the δ = 2.8–4.8 ppm resonance (assigned to protons associated with oxygen-containing functionalities) in DMSO d_6*

TOCSY experiments reveal short and long range couplings of hydrogens. TOCSY data complement COSY information and hint about how units are fitted together. The TOCSY experiment is especially useful for detecting the couplings between the amido protons and protons on the side chains of amino acid residues. Although COSY spectra reveal adjacent couplings that link amino acids (as in peptide-type structures), TOCSY spectra allow long range couplings to be observed and used to identify amino acids present.

The TOCSY spectrum of the moss sample gave little information additional to that from the COSY experiment. However, the spectrum for the IHSS Standard soil HA demonstrates amino acid coupling (Figure 3). (This Standard contains *ca.* 8% amino acids.) Amido couplings can be confirmed by taking advantage of the exchangeable properties of the amido group. The couplings disappeared on the addition of D_2O.

D_2O addition had an enhancing effect on the TOCSY spectral resolution of some samples. For the FA extracted from the B*h* horizon, the TOCSY spectrum was complex, and with large areas of blurring (Figure 4). However, spectral resolution was greatly enhanced after D_2O addition (Figure 5). This is attributed to a decrease in rolling in the baseline due to intense, broad, exchangeable bands. This allows a more refined data phasing and the signals hidden by blurring become apparent.

The heteronuclear HMQC experiment produced significant data for H-C bonds present in the moss FA sample (Figure 6). Little information could be obtained for the δ = 3–5.5 ppm region from from the 1-D and 2-D homonuclear experiments. However, that region is highly resolved when both the carbon and hydrogen chemical shifts are considered and 20–30 different C-H bonds are seen to contribute to the resonance (Figure 7). The cross peak at δ = 3.9 ppm (^1H) and at 55 ppm (^{13}C) indicates methoxyl bound to an aromatic ring, as expected for structures from lignin.

The aromatic region is difficult to interpret because of the diversity of compounds present. More than 40 aromatic/amide proton resonances are seen in the 1-D (^1H) spectrum. Protons attributable to amide structures are marked by arrows (Figure 8), and these are readily identified by D_2O exchange. The COSY spectrum provides comprehensive data on the H-H couplings present in the aromatic systems. Because of the specific nature of the chemical shifts in the aromatic ring it is possible to propose assignments based solely on the chemical shifts of coupled pairs of protons. These assignments can then be further confirmed by the heteronuclear experiments. However, in the cases of some of the lesser abundant aromatic moieties, the weak H-C couplings are not identified by such experiments and identification has to rely on H-H coupling data.

There are 24 cross peaks representing aromatic protons in the COSY spectrum (Figure 9), and the amide protons are not coupled. In aromatic systems, the presence of neighbouring hydrogens gives information about the level and pattern of substitution. With adjacent hydrogens on a ring there cannot be substituents on the 1,3,5 ring positions and rings with more than four substituents are ruled out.

Simpson *et al.*[63] have described in detail the assignments in the aromatic region of a COSY spectrum of a FA isolated at pH 12.6 from the B*h* horizon of a podzol. Chemical shift and coupling data from standard compounds were used to identify possible moieties present. A preliminary search used library spectra and standard compounds from the laboratory to determine the chemical shifts and couplings in over 1000 aromatic moieties, some of which have been identified in the degradation digests of HSs. Others were derivatives of compounds in forms that might be expected in humic structures. From this, over 30 aromatic compounds were found to give similar or identical shift and coupling

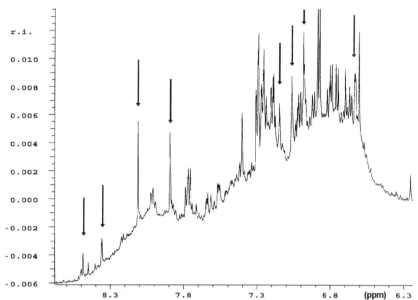

Figure 8 *One dimensional ¹H NMR (600 MHz) spectrum for the aromatic/amide resonance (in DMSO d₆) of a moss fulvic acid isolated in the sequential process at pH 12.6. Arrows indicate resonances from amide protons that disappear on addition of D₂O (from Simpson et al.[61])*

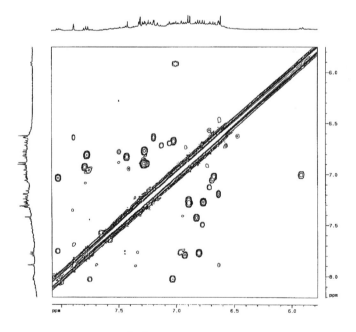

Figure 9 *Two dimensional COSY (¹H-¹H; 500 MHz) NMR spectrum of the aromatic/-amide resonance of the moss FA (isolated at pH 12.6) in DMSO d₆*

data to those observed in the spectra. It was deduced that the aromatic structures fall into 6 zones on the COSY spectra, each zone representing a category, or categories of structure.

From the standard compound search eight compounds were found to fall into zone 1. All of these structures were *para* substituted, with carbonyl on the 1- and oxygen at the 4-position. The data cannot be interpreted as indicating that such standard compounds are present as such within the humic structures. The aromatic proton shifts remain relatively constant when the length of the chain of the 1- or 4- positions is increased, and it can be inferred that cross peaks in zone 1 are consistent with structures of the type described, even though the chain lengths attached to carbonyl or ether are not fully understood. The reader is referred to Simpson *et al.*[63] for details of the technique and assignments.

Cross polarization magic-angle spinning (CPMAS) ^{13}C NMR provides an excellent fingerprint for humic fractions from different sources. The CPMAS ^{13}C NMR data of Watt *et al.*[3] illustrate compositional differences between humic fractions from different pristine waters. The data suggest close similarities in the fractions from watersheds of similar soil types. ^{13}C NMR data also provide information about functionalities.

Malcolm[51] lists chemical shift data that are useful for assignments of functionalities to humic structures. Chemical shift data[17] listed in Figure 10 were obtained using the Chem

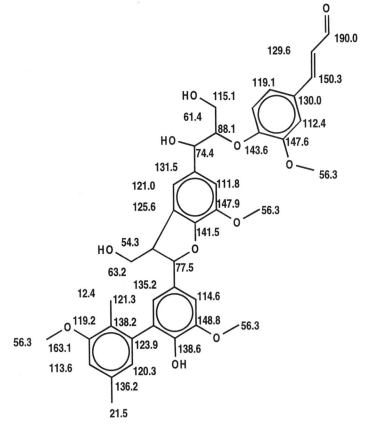

Figure 10 ^{13}C NMR chemical shift assignments for each carbon in a segment of a lignin-like structure, as determined using a computer programme

Windows Program and the attached C-13 Module Version 1 (from SoftShell International) that gives chemical shift data for each carbon of a lignin-type segment. The data show resonances for aromatic carbon ranging from 112 to 148 ppm. Data for HSs showing such a broad spread would suggest that lignin-/tannin-type residues persist.

Figure 11 shows the CPMAS ^{13}C NMR spectra[7,18] of HAs, FAs and XAD-4 acids isolated from the 0–15 cm layer (4.1% organic carbon) of a long term grassland soil from Devon, England. Spectra 1, 4, and 7 are for the HAs, FAs, and XAD-4 acids, respectively, isolated in sodium pyrophosphate (Pyro) at pH 7; spectra 2 and 8 are for the HAs and XAD-4 acids isolated in Pyro at pH 10.6, and spectra 3, 6, and 9 are for the HAs, FAs, and XAD-4 acids, respectively, all isolated in 0.1 M Pyro/0.1 M NaOH at pH 12.6.

The spectra show distinct differences between the same operationally defined fractions isolated at the different pH values, and the data for the integrated areas of the selected resonance bands reflected these differences.[7,18] The HA isolated at pH 7 was the most highly aromatic (spectrum 1), and its aromaticity was almost double that for the HAs isolated at pH 10.6 and 12.6. The integrated areas for the resonances at 160–190 ppm (spectra 2 and 3), representing the carbonyl functionalities of carboxylic acids, esters, and amides, indicate that these functionalities contribute about equally to each of the three fractions. The contribution of amide structures to this resonance is often overlooked, but it was clear from analyses[18] that amino acids (probably in peptide structures) were significant contributors to the compositions of the HAs (the measured contribution of amino acids to the overall composition of the HAs isolated at pH 12.6 was 16.3%).

Although the integrated areas suggest that the HAs isolated at pH 7 have most O-aromatic functionality (hydroxy-, methoxy-, alkoxy-), the spectra show more definite evidence for such functionalities in the HAs isolated at pH 10.6 and 12.6. The resonances in the so-called 'saccharide' region of the spectrum (65–110 ppm) include O-alkyl and C-OH resonances. Only when the anomeric carbon resonance, centered at 105 ppm, has an integrated area of one-fourth to one-fifth of that for the 65–110 ppm can the saccharide composition be validated. The neutral sugar contents of the HAs were about 5–7 per cent, and yet the anomeric carbon resonance is seen to increase as the pH of the extractant is raised. It is likely, therefore, that ether functionalities, or sugar derived/related structures are major contributors to the 65–90 ppm resonance, at least in the case of the HAs isolated at pH 7. Peptide structures can also contribute to the resonance in this region.

The resonance at 65–45 ppm includes methoxyl (a sharp peak at 56 ppm). This peak (especially prominent in lignins) is characteristic of HSs in the process of humification.

The compositions of the HAs that gave spectrum 5 are different from those for spectra 1, 2, and 3 (Figure 11). In the preparation process, dilute soil extracts were adjusted to pH 2 and passed in tandem onto XAD-8 and XAD-4 resin columns. The materials that sorbed to the resins were back eluted in base, and the HAs were precipitated at pH 1 from the eluate from XAD-8. The HAs were then redissolved in base, and as the pH was being adjusted to 2 (in order to desalt with the resin column), a precipitate formed in the pH range 2.5 to 2. That precipitate was found, as spectrum 5 shows, to be different in composition from the material that did not precipitate at that pH (spectrum 1). Aromaticity was low (for the HAs in spectrum 5), and the major components had resonances in the aliphatic hydrocarbon functionality region (10–45 ppm) and in the 65–110 ppm band. Evidence for anomeric carbon was weak, and the sugar content (6.7%) was similar to that for the HAs isolated at pH 7. However, the amino acid content was high (14%), and it is likely that peptide functionalities contributed significantly to the resonances at 160–190 and 65–90 ppm.

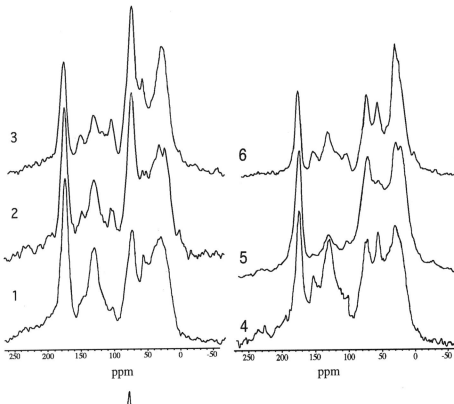

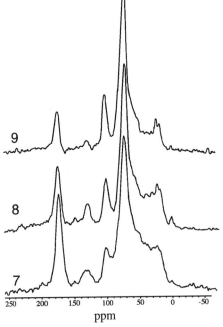

Figure 11 *CPMAS ^{13}C NMR spectra for the Devon soil humic (1, 2 and 3), fulvic (4 and 6) and XAD-4 (7, 8 and 9) acids isolated at pH 7 (1, 4, 7), pH 10.6 (2, 8), and at pH 12.6 (3, 6, 9) and processed following the XAD-8 and XAD-4 resin tandem procedure. Spectrum 5 is of HA isolated at pH 7, subjected to XAD-8 resin treatment, and precipitated at pH 2.0–2.5.*

The spectra for the FAs isolated at pH 7 (spectrum 4) and pH 12.6 (spectrum 6) are also significantly different from each other, and both have features in common with the HAs isolated at the same pH values, which would support the view that HAs and FAs are part of a continuum. Both have significant aromaticity, and the anomeric carbon signal is strong for the FAs isolated at pH 12.6. The sugar and amino acid analyses indicated that these likely are the major contributors to the resonance at 65 to 110 ppm. There are clear indications for methoxyl in both samples, which is unusual for FAs.

The spectra for the XAD-4 acids are different from those for the FAs and HAs, and there are distinct similarities between the spectra of the XAD-4 acids isolated at the different pH values. Aromaticity was low for each sample, and especially in the case of that isolated at pH 12.6. In all cases the 65–110 ppm resonance was strongest, and there is evidence for anomeric carbon. There was not, however, noticeable enrichment in amino acids. Nevertheless it seems likely that these were of microbial rather than plant origins.

Chemical Degradation Procedures. Chemical degradation procedures for studies of the component molecules of HSs are now less favoured than in the past. Arguably, they still provide the best methods for determining the molecules that compose HSs.

It is relatively easy to cleave the components of proteins, polysaccharides, and nucleic acids because the links between the molecules are hydrolysable. However, although HSs are made up of different relatively simple 'building blocks', the units are linked by bonds that are, for the most part, difficult to cleave. The major structural units consist of a 'backbone' whose component molecules are linked by carbon-carbon bonds, ether linkages, and other intractable units. Hydrolysis can cause up to 50% of the masses of soil HAs to be lost[64] as CO_2 and as soluble molecules (*e.g.* sugars, amino acids, small amounts of purine and pyrimidine bases, and phenolic substances).

The amounts of sugars and amino acids detected in the hydrolysates of aquatic humic substances are relatively small percentages of those in the hydrolysates of the similar nominal fractions from soils. However, such comparisons will not be truly valid until analyses are carried out on the hydrolysates of soil HAs and FAs subjected to treatment with XAD-8, or similar acting resin. Although losses of CO_2 during hydrolysis (through decarboxylation of activated carboxyl groups, such as β-keto acids, and of hydroxy-benzenecarboxylic acids) are significant, the total acidities of soil HAs are not decreased by the hydrolysis. This suggests that new acid groups are formed (*e.g.* from esters and lactones).

A boron trifluoride–methanol transesterification procedure has been adapted by Almendros and Sans[65,66] for studies of the compositions of HAs and humins. This mild procedure allows the removal of humic components linked by ester functionalities, and the labile (derivatized) structures can then be identified. Yields of identifiable products were of the order of 30–35% and were similar to those for oxidative degradation procedures. A variety of monobasic straight chain and branched fatty acids, long chain dicarboxylic acids, di-hydroxymonobasic acids, tri-hydroxymonobasic acids, methoxybenzenecarbox-ylic acids, di-, tri-, and tetra-methoxybenzenecarboxylic acids, and a variety of other miscellaneous acids were detected in the digests of humins. It seems likely that these acids were present as esters in the humins, and are typical of depolymerization components of plants, such as cutins and suberins.

More than 100 products have been identified in digests of a variety of oxidative degradations of HSs. This does not indicate the presence of more than 100 'building blocks' because several different products can form from a single precursor in the course of a degradation. The products depend on the type and concentration of reagent, the temperature, the reaction time, and so on. Most of the compounds identified are acids. The aromatic acids range from benzenedi- to benzehexa-carboxylic acids and there is abundant evidence for mono- to tri-hydroxy (or methoxy), and hydrocarbon substituents on the aromatic rings.

The abundance of benzenecarboxylic acids in the digests of alkaline permanganate oxidative degradations of HSs could signal contributions of fused aromatic components to the structures.[4,67,68] However, the same acids are found in the digests of alkaline cupric oxide degradations and such reagents would not degrade fused aromatic structures to benzene polycarboxylic acids. Therefore, it seems that the carboxylic acid substituents on the aromatic nuclei were formed from oxidizable substituents, such as aldehydes, alcohols, olefinic and other structures that cleave under the reaction conditions. Carbonylation reactions could give benzenecarboxylic acids, and this possibility has not been resolved for alkaline degradations of HSs.

Phenols are degraded in alkaline media, but their ether derivatives resist degradation. Therefore it is necessary to methylate the substrates prior to oxidative degradation, and the phenolic hydroxyls persist as methoxy substituents. Because methoxyl substituents on aromatic nuclei as well as phenolic hydroxyls are indigenous to humic substances, we cannot predict whether or not the methoxyl groups detected in methylated digests were present in indigenous structures or were formed from phenolic hydroxyls.

Aliphatic dicarboxylic acids, ranging from ethanedioic to decanedioic, are abundant in the digests of oxidative degradation reactions, and there is evidence also for aliphatic tricarboxylic acids. The carboxyls could come from cleavages of unsaturated groups along aliphatic chains and could be formed from oxidizable functionalities, such as alcohols and aldehydes. Inevitably, some are present as carboxyl groups on the aliphatic structures. Hayes and Swift[4] have discussed mechanisms of release of long chain acids and hydrocarbon compounds from HS structures.

In general, it is unlikely that more than one carboxyl substituent resides on a single aromatic unit in humic molecules. This is evident from the products identified in digests of reductive degradations using sodium amalgam (see reviews by Hayes and Swift[4] and by Stevenson[69]). None of the aromatic structures identified from such digests have more than one carboxyl substituent. There is evidence for the presence of one to three hydroxyl or methoxyl substituents on the aromatic nuclei, and such activating substituents could promote the decarboxylation of di- and poly-hydroxybenzenecarboxylic acids. A variety of non-carboxyl bearing hydroxy/methoxy benzene structures identified in the digests might have resulted from decarboxylation of hydroxy/methoxy substituted benzenecarboxylic acids.

7 CONCLUSIONS

It is evident from Section 5 that the component molecules of HSs are still not known unambiguously, and so far the best indications of their natures can be deduced from identifications of products of degradation reactions, and from an awareness of the mechanisms involved in the degradation processes. Awareness of likely component molecules, and of functionality [provided by techniques such as wet chemical methods, NMR, FTIR, titration data and developments in Raman spectroscopy (which may provide valuable information in the near future)] has not changed dramatically in the last nine years. We knew then, as now, that 25 to 45% of the components of soil HAs are aromatic,[51] and that value may be higher in some instances, *e.g.* for samples from the B*h* horizons of some podzols. HAs from aqueous systems are less aromatic and more highly oxidized than those from soils. The consensus nine years ago regarded the aromaticity as being composed of single ring compounds[49] and with 3 to 5 ring substituents. That view has not changed.

Much data have emerged about sugar and amino acid contents of humic fractions. Aquatic HSs have significantly fewer sugars and amino acids than those in soils, and the contents in the aquatic FAs are especially low. That might suggest that the sugars and amino acids are sterically protected in their humic associations (or physically protected through associations with mineral colloids, and in soil aggregates) from enzymatic attack. Should the associations involve covalent links between the biological molecules and the humic core, then the M_W values of the systems would be increased. It is plausible to consider that conformational changes to the HSs in the soil environment, or the disruption of aggregates, could expose glycosidic and peptide linkages to microbial attack releasing core humic structures, some of which dissolve in the soil solution. The lower M_W values of aquatic HSs (compared with soil HSs) might be explained by the hydrolysis thesis.

The hydrogens in three to five of the (aromatic) ring positions may be replaced by substituent groups, and these consist of hydroxyl and methoxyl, aliphatic hydrocarbon structures, and some of these may be involved in linking aromatic structures. There is also evidence for aldehyde and keto functional groups attached to some of the aromatic nuclei, for phenylpropane (3-carbon chains attached to the aromatic rings) units, and for hydroxyl and methoxyl substituents. The phenylpropane structures, and hydroxy/methoxy substituents in the 3- and 4-, and in the 3-, 4- and 5- ring positions would suggest origins in lignins, whereas the presence of these substituents in the 3- and 5- ring positions suggests origins in the skeletal structures, or the metabolic products of microorganisms.

There is evidence that ether functionalities link aromatic structures, and it is logical to assume that aromatic-aliphatic ethers are also present. It is not certain that all of the aliphatic units are linked to aromatic structures, and long chain hydrocarbons, with origins in plant cuticles and in algal metabolites, could possibly be present as 'impurities' held to the humic structures by van der Waals forces. Some of the hydrocarbon structures are olefinic, as evidenced by the presence of mono- and di- (and sometimes tri-) carboxylic acids in the digests of oxidative degradation reactions (see Section 5).

Fatty acids in degradation digests may be be released from esters of phenols and other hydroxyl groups in the 'backbone' structures. Also, these could arise from waxes and suberins associated with HSs, and could contribute to the hydrophobic properties that HSs display in some instances, and have a role in the self association concept.

Titration data indicate that the acid groups in HSs provide a continuum of dissociable protons representing a range of acid strengths, from the strong to the very weak. The strongest acids are carboxylic, and some are activated by appropriate adjacent groups. Phenolic hydroxyls also contribute to the acidity, and these are most abundant in the HA fraction. It would seem that their contribution to the total acidity is greatest in newly formed humic substances, and especially in those with origins in the lignified components of plants. As oxidation takes place, the phenols are oxidized, and eventually carboxylic acids are formed. Some of the phenolic substituents can have enhanced acidity because of the influences of other substituents on the aromatic structures. Enols and other weakly dissociable groups also contribute to the charge characteristics under alkaline conditions

FAs have many compositional characterisatics similar to those of HAs, but they have differences as well. FAs tend to be less aromatic than than HAs (and can contain as little as 25 per cent aromatic components in some instances), and they appear to be smaller, more polar, and more highly charged. These effects would cause the FAs to be less self associated than HAs, and in terms of the random coil concept, to be more linear than coiled. Fulvic acids are not precipitated under acid conditions, or by low concentrations of divalent metals. They might be expected to be readily removed from soils in drainage waters. That this does not happen is an indication of interactions between fulvic acids and insoluble components of soils. Fulvic acids in mineral soils are held by the inorganic colloids, and by entrapment within (associations with) the more hydrophobic HAs.

The XAD-4 acids would seem to be an association of HSs with carbohydrate and peptide type materials. It will be possible to fractionate these substances further.

Because of their polydispersity, it would be pointless at this stage to attempt to provide exact structures for HS fractions. It is not, however, necessary to know the exact structures in order to have a good understanding of composition, and to be able to predict how the macromolecules will react with other species. The awareness of structure that is emerging allows us a degree of understanding of the processes involved in their binding of metals and of anthropogenic organic chemicals, in their adsorption to clays and to (hydr)-oxides and of the ways in which they respond to chemical treatments of waters.

References

1. G. R. Aiken, D. M. McKnight, R. L. Wershaw and P. MacCarthy, in 'Humic Substances in Soil, Sediment, and Water', G. R. Aiken, P. MacCarthy, R. L. Malcolm and R. S. Swift (eds.), Wiley, New York, 1985, p. 1.
2. G. R. Harvey and D. A. Boran, in 'Humic Substances in Soil, Sediment, and Water', G. R. Aiken, P. MacCarthy, R. L. Malcolm and R. S. Swift (eds.), Wiley, New York, 1985, p. 233.
3. B. E. Watt, R. L. Malcolm, M. H. B. Hayes, N. W. E. Clark and J. K. Chipman, *Water Res.*, 1996, **6**, 1502.
4. M. H. B. Hayes and R. S. Swift, in 'The Chemistry of Soil Constituents', D. J. Greenland and M. H. B. Hayes (eds.), Wiley, Chichester, 1978, p. 179.
5. M. J. Häusler and M. H. B. Hayes, in 'Humic Substances and Organic Matter in Soil and Water Environments', C. E. Clapp, M. H. B. Hayes, N. Senesi and S. M. Griffith (eds.), International Humic Substances Society, University of Minnesota, St. Paul, 1996, p. 25.
6. C. L. Ping, G. J. Michaelson and R. L. Malcolm, in 'Soils and Global Change' R. Lal, J. Kimble, E. Levine and B. A. Stewart (eds.), CRC Lewis Publishers, Boca Raton, FL, 1995, p. 307.
7. T. M. Hayes, M. H. B. Hayes, J. O. Skjemstad, R. S. Swift and R. L. Malcolm, in 'Humic Substances and Organic Matter in Soil and Water Environments', C. E. Clapp, M. H. B. Hayes, N. Senesi and S. M. Griffith (eds.), International Humic Substances Society, University of Minnesota, St. Paul, 1996, p. 13.
8. R. S. Swift, in 'Methods of Soil Analysis. Part 3. Chemical Methods', D. L. Sparks, A. L. Page, P. A. Helmke and R. H. Loeppert (eds.), Soil Science Society of America, Madison, WI., 1996, p. 1011.
9. C. E. Clapp and M. H. B. Hayes, in 'Humic Substances and Organic Matter in Soil and Water Environments', C. E. Clapp, M. H. B. Hayes, N. Senesi and S. M. Griffith (eds.), International Humic Substances Society, University of Minnesota, St. Paul, 1996, p. 3.

10. R. L. Malcolm and P. MacCarthy, *Environ. Int.*, 1992, **18**, 597.
11. M. H. B. Hayes and R. S. Swift, in 'Soil Colloids and their Associations in Aggregates', M. F. DeBoodt, M. H. B. Hayes and A. Herbillon (eds.), Plenum, New York, 1990, p. 245.
12. F. J. Stevenson, 'Humus Chemistry: Genesis, Composition, Reactions', 2nd edn., Wiley, New York, 1994.
13. R. S. Swift, in 'Advances in Soil Organic Matter Research: the Impact on Agriculture and the Environment', W. S. Wilson (ed.), The Royal Society of Chemistry, Cambridge, UK, 1991, p. 153.
14. W. Flaig, in 'Humic Substances, Peats and Sludges: Health and Environmental Aspects', M. H. B. Hayes and W. S. Wilson (eds.), The Royal Society of Chemistry, Cambridge, UK,1997, p. 346.
15. D. L. MacDonald and J. K. Chipman, in 'Humic Substances, Peats and Sludges: Health and Environmental Aspects', M. H. B. Hayes and W. S. Wilson (eds.), The Royal Society of Chemistry, Cambridge, UK, 1997, p. 337.
16. M. H. B. Hayes, in 'Humic Substances in Soil, Sediment, and Water', G. R. Aiken, P. MacCarthy, R. L. Malcolm and R. S. Swift (eds.), Wiley, New York, 1985, p. 329.
17. M. H. B. Hayes, in 'Humic Substances, Peats and Sludges: Health and Environmental Aspects', M. H. B. Hayes and W. S. Wilson (eds.), The Royal Society of Chemistry, Cambridge, UK, 1997, p. 3.
18. T. M. Hayes, Ph.D. Thesis, University of Birmingham, UK, 1996.
19. D. Martin and H. G. Hauthal, 'Dimethyl Sulphoxide' (translated by E. S. Halberstadt), Van Nostrand-Reinhold, New York, 1975.
20. G. R. Aiken, in 'Humic Substances in Soil, Sediment, and Water', G. R. Aiken, P. MacCarthy, R. L. Malcolm and R. S. Swift (eds.), Wiley, New York, 1985, p. 363.
21. J. A. Leenheer, in 'Humic Substances in Soil, Sediment, and Water', G. R. Aiken, P. MacCarthy, R. L. Malcolm and R. S. Swift (eds.), Wiley, New York, 1985, p. 409.
22. E. M. Thurman and R. L. Malcolm, *Environ. Sci. Technol.*, 1981, **15**, 463.
23. E. M. Thurman, 'Organic Geochemistry of Natural Waters', Martinus Nijhoff/Dr W. Junk Publishers, Dordrecht, The Netherlands, 1985.
24. M. Deinzer, R. Melton and D. Mitchell, *Water Res.*, 1975, **9**, 799.
25. S. M. Serkiz and E. M. Perdue, *Water Res.*, 1990, **24**, 911.
26. S. M. Sun, E. M. Perdue and J. F. MacCarthy, *Water Res.*, 1995, **29**, 1471.
27. R. S. Swift, in 'Humic Substances in Soil, Sediment, and Water', G. R. Aiken, P. MacCarthy, R. L. Malcolm and R. S. Swift (eds.), Wiley, New York, 1985, p. 387.
28. S. H. Eberle and K. H. Schweer, *Vom Wasser*, 1974, **41**, 27.
29. R. S. Swift, R. S. and A. M. Posner, *J. Soil Sci.*, 1971, **22**, 237.
30. R. S. Cameron, R. S. Swift, B. K. Thornton and A. M. Posner, *J. Soil Sci.*, 1972, **23**, 342.
31. R. S. Cameron, B. K. Thornton, R. S. Swift and A. M. Posner, *J. Soil Sci.*, 1972, **23**, 394.
32. A. Piccolo, S. Nardi and G. Concheri, *European J. Soil Sci.*, 1996, **47**, 319.
33. J. M. Duxbury, in 'Humic Substances II: In Search of Structure', M. H. B. Hayes, P. MacCarthy, R. L. Malcolm and R. S. Swift (eds.), Wiley, Chichester, 1989, p. 593.

34. J. E. Dawson, C. E. Clapp and M. H. B. Hayes, Special Publication, No. 205, Division of Scientific Publications, Ministry of Agriculture, Bet Dagan, Israel, 1979, p. 278.

35. C. M. Ciavatta, G. Govi, G. Bonoretti, D. Montecchio and C. Gessa, in 'Humic Substances and Organic Matter in Soil and Water Environments', C. E. Clapp, M. H. B. Hayes, N. Senesi and S. M. Griffith (eds.), International Humic Substances Society, University of Minnesota, St. Paul, 1996, p. 53.

36. O. A. Trubetskoj, O. E. Trubetskaya, G. V. Afanasieva and O. I. Reznikova, in 'Humic Substances and Organic Matter in Soil and Water Environments', C. E. Clapp, M. H. B. Hayes, N. Senesi and S. M. Griffith (eds.), International Humic Substances Society, University of Minnesota, St. Paul, 1996, p. 47.

37. M. Klavins, M. Purite and E. Apsite, in 'Humic Substances and Organic Matter in Soil and Water Environments', C. E. Clapp, M. H. B. Hayes, N. Senesi and S. M. Griffith (eds.), International Humic Substances Society, University of Minnesota, St. Paul, 1996, p. 41.

38. R. S. Swift, in 'Humic Substances II: In Search of Structure', M. H. B. Hayes, P. MacCarthy, R. L. Malcolm and R. S. Swift (eds.), Wiley, Chichester, 1989, p. 467.

39. A. J. Simpson, Ph.D. Thesis, University of Birmingham, UK, 1998.

40. P. Conte and A. Piccolo, *Chemosphere,* in press.

41. N. Hertkorn, A. Günzl, C. Wang, D. Freitag and A. Kettrup, in 'The Role of Humic Substances in the Ecosystems and in Environmental Protection', J. Drodzd, S. S. Gonet, N. Senesi and J. Weber (eds.), Polish Chapter of IHSS, Wroclaw, Poland, 1997, p. 139.

42. I. P. Kenworthy and M. H. B. Hayes, in 'Humic Substances, Peats and Sludges: Health and Environmental Aspects', M. H. B. Hayes and W. S. Wilson (eds.), The Royal Society of Chemistry, Cambridge, UK, 1997, p. 39.

43. R. L. Wershaw, *J. Contaminant Hydrology*, 1986, **1**, 29.

44. S. Burchill, M. H. B. Hayes and D. J. Greenland, in 'The Chemistry of Soil Processes', D. J. Greenland and M. H. B. Hayes (eds.), Wiley, Chichester, 1981, p. 221.

45. R. S. Swift, in 'Humic Substances II: In Search of Structure', M. H. B. Hayes, P. MacCarthy, R. L. Malcolm and R. S. Swift (eds.), Wiley, Chichester, 1989, p. 449.

46. E. W. D. Huffman and H. A. Stuber in 'Humic Substances in Soil, Sediment, and Water', G. R. Aiken, P. MacCarthy, R. L. Malcolm and R. S. Swift (eds.), Wiley, New York, 1985, p. 433.

47. E. M Perdue, in 'Humic Substances in Soil, Sediment, and Water', G. R. Aiken, P. MacCarthy, R. L. Malcolm and R. S. Swift (eds.), Wiley, New York, 1985, p. 493.

48. P. MacCarthy and J. A. Rice, in 'Humic Substances in Soil, Sediment, and Water', G. R. Aiken, P. MacCarthy, R. L. Malcolm and R. S. Swift (eds.), Wiley, New York, 1985, p. 527.

49. C. Steelink, R. L. Wershaw, K. A. Thorn and M. A. Wilson, in 'Humic Substances II: In Search of Structure', M. H. B. Hayes, P. MacCarthy, R. L. Malcolm and R. S. Swift (eds.), Wiley, Chichester, 1989, p. 282.

50. M. A. Wilson, in 'Humic Substances II: In Search of Structure', M. H. B. Hayes, P. MacCarthy, R. L. Malcolm and R. S. Swift (eds.), Wiley, Chichester, 1989, p. 309.

51. R. L. Malcolm, in 'Humic Substances II: In Search of Structure', M. H. B. Hayes, P. MacCarthy, R. L. Malcolm and R. S. Swift (eds.), Wiley, Chichester, 1989, p. 340.

52.   N. Senesi and C. Steelink, in 'Humic Substances II: In Search of Structure', M. H. B. Hayes, P. MacCarthy, R. L. Malcolm and R. S. Swift (eds.), Wiley, Chichester, 1989, p. 373.

53.   P. R. Bloom and J. A. Leenheer, in 'Humic Substances II: In Search of Structure', M. H. B. Hayes, P. MacCarthy, R. L. Malcolm and R. S. Swift (eds.), Wiley, Chichester, 1989, p. 409.

54.   J. A. Leenheer, R. L. Wershaw and M. M. Reddy. *Environ. Sci. Technol.*, 1995, **29**, 393.

55.   J. Niemeyer, Y. Chen and J.-M. Bollag, *Soil Sci. Soc. Am. J.*, 1992, **56**, 135.

56.   Y. Yang and T. Wang, *Vibrat. Spectrosc.*, 1997, **14**, 105.

57.   Y. Chen, N. Senesi and M. Schnitzer, *Soil Sci. Soc. Am. J.*, 1977, **41**, 352.

58.   A. J. Simpson, B. E. Watt, C. L. Graham and M. H. B. Hayes, in 'Humic Substances, Peats and Sludges: Health and Environmental Aspects', M. H. B. Hayes and W. S. Wilson (eds.), The Royal Society of Chemistry, Cambridge, UK, 1997, p. 73.

59.   A. J. Simpson, J. Burdon, C. L. Graham and M. H. B. Hayes, 'Humic Substances, Peats and Sludges: Health and Environmental Aspects', M. H. B. Hayes and W. S. Wilson (eds.), The Royal Society of Chemistry, Cambridge, UK, 1997, p. 83.

60.   C. M. Preston, *Soil Sci.*, 1996, **161**, 144.

61.   A. J. Simpson, R. E. Boersma, W. L. Kingery, R. P. Hicks and M. H. B. Hayes, in 'Humic Substances, Peats and Sludges: Health and Environmental Aspects', M. H. B. Hayes and W. S. Wilson (eds.), The Royal Society of Chemistry, Cambridge, UK, 1997, p. 46.

62.   D. G. Gadian, 'Nuclear Magnetic Resonance and Its Application to Living Systems', Oxford University Press, Oxford, 1982.

63.   A. J. Simpson, J. Burdon, M. H. B. Hayes, N. Spencer and W. J. Kingery, submitted to *European J. Soil Sci.*

64.   J. W. Parsons, in 'Humic Substances II: In Search of Structure', M. H. B. Hayes, P. MacCarthy, R. L. Malcolm and R. S. Swift (eds.), Wiley, Chichester, 1989, p. 99.

65.   G. Almendros and J. Sans, *Soil Biol. Biochem.*, 1991, **23**, 1147.

66   G. Almendros and J. Sans, *Geoderma*, 1992, **53**, 79.

67.   S. M. Griffith and M. Schnitzer, in 'Humic Substances II: In Search of Structure', M. H. B. Hayes, P. MacCarthy, R. L. Malcolm and R. S. Swift (eds.), Wiley, Chichester, 1989, p. 69.

68.   R. F. Christman, D. L. Norwood, Y. Seo and F. H. Frimmel, in 'Humic Substances II: In Search of Structure', M. H. B. Hayes, P. MacCarthy, R. L. Malcolm and R. S. Swift (eds.), Wiley, Chichester, 1989, p. 33.

69.   F. J. Stevenson, in 'Humic Substances II: In Search of Structure', M. H. B. Hayes, P. MacCarthy, R. L. Malcolm and R. S. Swift (eds.), Wiley, Chichester, 1989, p. 122.

USE OF ^{13}C NMR AND FTIR FOR ELUCIDATION OF DEGRADATION PATHWAYS DURING NATURAL LITTER DECOMPOSITION AND COMPOSTING. II. CHANGES IN LEAF COMPOSITION AFTER SENESCENCE

Robert L. Wershaw,[1] Kay R. Kennedy[1] and James E. Henrich[2]

[1] U.S. Geological Survey, Denver Federal Center, Denver, CO 80225
[2] Denver Botanic Gardens, 909 York Street, Denver, CO 80206

1 INTRODUCTION

Senescence of leaves is a programmed aging process that culminates in cell death.[1] During leaf senescence of deciduous plants in the Fall, polymeric species such as polysaccharides and proteins are hydrolysed to soluble monosaccharides and amino acids, which are transported to storage organs for subsequent use when the plants reawaken in the spring.[2-4] Leshem[4] pointed out that oxygen-containing free radical species such as superoxide, hydroxyl, peroxyl, alkoxyl, polyunsaturated fatty acid and semiquinone free radicals are active during plant senescence. It is these free radicals that cause the catabolic breakdown of plant tissue components.[5,6] In addition to degradation by active-oxygen species, peptide hydrolases degrade proteins during senescence.[7] The catabolic breakdown of plant components that takes place during senescence is the first step in the humification process that leads to the formation of humus in natural soils and compost in man-made systems.

Further humification is the result of degradation processes catalysed by extracellular enzymes secreted by microorganisms as discussed in part I of this series.[8] These degradation reactions accomplish the same purpose as those that take place during senescence, that is, the conversion of biopolymers and other large biomolecules into smaller chemical species that can be assimilated by living cells for use in the anabolic and catabolic processes that take place within the cells. An additional similarity between senescence and the degradation phase of humification is that both processes, in general, result from the action of active oxygen species.[9]

The purpose of this study is to attempt to follow the large-scale chemical changes that take place during senescence by comparison of the solid-state ^{13}C NMR spectra of senescent leaves with those of nonsenescent leaves. In addition to fallen senescent leaves, yellowed, senescent leaves were also collected prior to dropping from the plants. These leaves presumably had not yet undergone any microbial degradation, and therefore all of the changes can be attributed to senescence. It will be demonstrated that solid-state ^{13}C NMR spectroscopy provides a simple, rapid, nondestructive technique for detecting chemical changes in leaves caused by senescence.

2 MATERIALS AND METHODS

2.1 Leaf Collection and Preparation

Senescent and nonsenescent leaves were collected from different families of plants growing in different environments as indicated below in order to provide a more complete picture of the types of changes brought about by senescence.

Fallen leaves consisting mainly of *Ulmus pumila* (elm) leaves were collected from a residential yard in Lakewood, Colorado in Fall 1995 and stored in a refrigerator until use. Nonsenescent leaves were obtained from elm saplings from the same yard in June 1996.

Attached, green, nonsenescent leaves were collected from four different *Ulmus* species at the Denver Botanic Gardens in late August 1996. Attached and fallen senescent (yellow) leaves were collected from the same trees in November 1996.

Tropical deciduous plants (*Bixa orellana* L. and *Acalypha wilkesiana* Müll. Arg.) that do not undergo senescence of all foliage at one time were also sampled in the conservatory at the Denver Botanic Gardens. Simultaneous sampling of attached senescent and nonsenescent leaves of these plants was carried out in June 1996. The senescent and nonsenescent leaf samples for a given species were collected from a single plant. For this study, the yellowed leaves on a healthy, nonstressed plant were considered senescent.

All of the fresh leaves and the senescent leaves from the Denver Botanic Gardens were vacuum dried at room temperature overnight. The vacuum drying of the Botanic Gardens' samples was started within two hours of collection to reduce the possibility of changes in the leaves. The dry fallen leaves collected from the Lakewood, Colorado residential yard in Fall 1995 were not further dried. At the beginning of this work the dried leaves were ground in a blender and sieved through a 0.4 mm sieve prior to packing into the NMR rotors. However, during the course of this work it was found that grinding and sieving were not necessary for smooth spinning of the rotors, and therefore, the leaves collected in Fall 1996 were crushed and packed into the rotors without grinding.

Elemental analyses (carbon, hydrogen, oxygen, nitrogen) of nonsenescent and senescent leaves were performed by Huffman Laboratories, Inc.

2.2 NMR Measurements

Solid-state cross-polarization magic-angle-spinning (CPMAS) ^{13}C spectra were measured on a 200 MHz Chemagnetics* CMX spectrometer. The samples were contained in 7.5 mm-diameter zirconia rotors. Spinning rates of 4600 or 5000 Hz were used. The acquisition parameters were contact time 1 ms, pulse delay 1s and a pulse width of 4.5 microseconds for the 90° pulse. Adequate signal-to-noise ratios were obtained for all of the samples with 1000 to 2000 scans. Lorentzian line broadenings of 100 Hz or 25 Hz were applied in the Fourier transformation of the free induction decay data. The line broadening of 25 Hz was used for some of the nonsenescent-senescent overlays to provide enhanced resolution for better visualization of the changes brought about by senescence.

The solid state ^{13}C CPMAS NMR spectra of the leaves measured in this study are composed of envelopes of overlapping bands. The positions of the bands are defined as being between pairs of inflection points on the spectral envelopes (Figure 1). The spectral bands in a given envelope were resolved using an iterative curve-fitting routine in the

---

* The use of trade marks in this report is for identification purposes only and does not constitute endorsement by the U.S. Geological Survey.

Chemagnetics Spinsight™ software to find the best combination of band intensities to fit the measured spectral envelopes as explained previously.[8] The spinning sidebands observed above 200 ppm were very weak, indicating that the spectra did not have appreciable intensity in spinning sidebands. This obviated the necessity of applying spinning-sideband corrections to the band intensities.

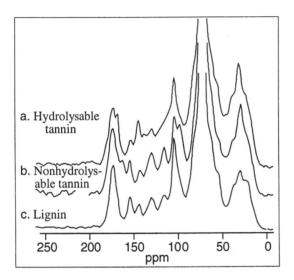

Figure 1 *^{13}C NMR spectral patterns of hydrolysable and non hydrolysable tannins and lignins in nonsenescent leaves from: a. Bixa orellana, b. Malus 'Manbeck Weeper' and c. Zelkova serrata (Thunb.) Mak. Line broadening of 100 Hz was used.*

Interrupted decoupling (dipolar dephasing) experiments were performed as described by Alemany *et al.*[10] and Hatcher.[11] Coupling times (τ) between 30 and 45 msec were tested for dephasing of methine and the methylene carbons; for most of the samples a τ value of 40 msec provided satisfactory results. At this τ value a small residual carbohydrate methine band at 73 ppm was present and the baseline was somewhat distorted; $\tau = 35$ msec provided a better baseline, but the carbohydrate methine band was stronger. A τ of 45 msec totally eliminated the carbohydrate methine band at the expense of baseline distortion. These results are in agreement with those of Opella and Frey[12] who showed that τ values between 40 and 100 msec allow one to differentiate between protonated and nonprotonated carbons. Methyl carbons do not dephase because methyl groups generally rotate freely, thereby reducing dipolar interactions.

3 RESULTS AND DISCUSSION

3.1 NMR Spectra of Nonsenescent Leaves

3.1.1 General Appearance of Spectra. The spectra of the nonsenescent leaves are composed of bands that represent the functional groups of the major chemical components of the leaves (Figure 1). The spectrum of each chemical component consists of a characteristic group of bands that occurs in well-defined spectral regions as indicated in Table 1. We shall demonstrate that the distribution and relative intensities of the bands

Table 1 *^{13}C NMR spectral regions of major leaf components*

Chemical shift regions	Types of carbon atoms	Chemical components
0–50 ppm	Aliphatic carbon atoms	Lipids, polyesters and flavonoids
50–54 ppm	Methyl esters	Pectins and peptides
55–60 ppm	Methyl ethers	Lignin units and peptides
60–90 ppm	Aliphatic alcohols and ethers	Carbohydrates and flavonoids
90–105 ppm	Anomeric carbon atoms	Carbohydrates
95–135 ppm	Aromatic carbons attached to protons or other carbon atoms	Lignin units, tannins and flavonoids
135–160ppm	Aromatic carbons attached to oxygens	Lignin units, tannins and flavonoids
160–180 ppm	Carbonyl carbons	Acids, esters, amino acids and peptides
180–220 ppm	Carbonyl carbons	Quinones, ketones and aldehydes

within each of these spectral regions provides important clues about the identities of the constituent chemical components of a given leaf sample.

3.1.2 Aliphatic Hydrocarbon Bands. Bands at approximately 20, 30 and 40 ppm are present in the region between 0-50 ppm in the spectra of most of the leaf samples. These are nominal chemical shifts that can vary by as much as ±2 or 3 ppm. Additional bands are seen in some of the spectra. The dominant aliphatic band in most of the spectra is that of methylene groups in long-chain structures at 30 ppm.[13] The chemical shift of the terminal C-4 carbon atom (Figure 2) of the alicyclic ring of flavonoid tannins (Table 5) is also near 30 ppm.

A variety of different leaf components contain aliphatic or alicyclic hydrocarbon groups including amino acids, flavonoids, fats and waxes and cutin. Cutin is one of the constituent polymers of the cuticle layer that coats the epidermal cells of leaves and other aerial organs of plants. The cuticle serves as a barrier to water loss from plant tissue. Cutin is a polyester polymer of hydroxy and hydroxy-epoxy fatty acids in which the major monomeric units are C_{16}- and C_{18}- fatty acids;[14] it is readily depolymerized by alkaline hydrolysis. Nip *et al.*[15] found evidence for an aliphatic polymeric component of leaf cuticles that is not susceptible to alkaline hydrolysis. However, they were not able to elucidate the chemical structure of this polymer.

Kögel-Knabner *et al.*[16] found that the solid-state ^{13}C CPMAS NMR spectrum of isolated leaf cuticle had a relatively broad band in the region between 0-50 ppm. They attributed this band to "long-chain and branched aliphatic structures associated with fatty

a. Hydrolysable tannin

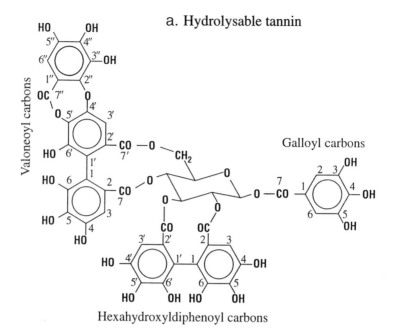

Valoneoyl carbons

Galloyl carbons

Hexahydroxyldiphenyl carbons

b. Nonhydrolysable tannin

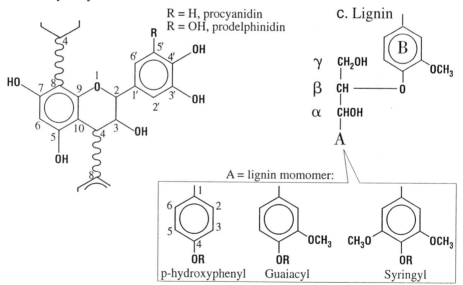

R = H, procyanidin
R = OH, prodelphinidin

c. Lignin

A = lignin momomer:

p-hydroxyphenyl   Guaiacyl   Syringyl

Figure 2 *Structural units of hydrolysable tannins, nonhydrolysable tannins and lignins.*

acids, lipids, waxes and cutin acids." There appear to be at least three bands at approximately 20, 30 and 40 ppm in their published cuticle spectrum; unfortunately, however, the positions of the inflection points defining these bands are obscured by the poor signal-to-noise ratio of the spectrum.

Zlotnik-Mazori and Stark[13] isolated cutin from lime skin and measured the ^{13}C CPMAS NMR spectrum of the isolate. A sharp band at 29 ppm with well-defined shoulders at 25 and 42 ppm is the most prominent peak in the spectrum. Additional aliphatic bands at 64 and 72 ppm were also present in the spectrum. The authors assigned the 29 ppm band to methylene groups in long chains; the 42 ppm shoulder was assigned to methylenes β to ester linkages, the 64 ppm band to methylenes α to ester linkages, and the 72 ppm band to methines α to ester linkages.

The ^{13}C CPMAS NMR spectra of tobacco leaves measured by Wooten[17] contain well-defined bands in the 0-50 ppm region. In addition to bands at 20, 30 and 40 ppm, there were distinct bands between 40 and 50 ppm that were assigned to carbons in salts of citric acid and to the α-carbon atoms of proteins.

The region between 50 and 60 ppm consists mainly of methyl ester and ether bands. The α carbons of proteins and amino acids that cover a range from approximately 40 ppm to 65 ppm can also occur in this region;[18] however, it appears from the work of Wooten[17] that this band is generally centered near 48 ppm. The well-defined band at 56 ppm in the leaf spectra is due most likely to methyl ethers in lignin structures.

As pointed out above, methyl carbons do not dephase in the dipolar dephasing experiment because dipolar interactions are greatly reduced in freely rotating methyl groups. A well-defined band at about 55 ppm in the dipolar-dephased *Malus* spectrum (Figure 3) provides confirmation for the presence of methoxyl groups in the *Malus* leaves.

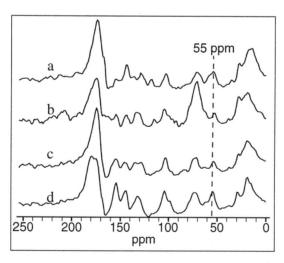

Figure 3 *Interrupted, decoupled (dipolar, dephased) spectra of a. nonsenescent Acer campestre L. leaves, $\tau = 40$ μsec; b. nonsenescent Malus 'Manbeck Weeper, $\tau = 30$ μsec; c. Zelkova serrata (Thunb.) Mak., $\tau = 40$ μsec; d. senescent Zelkova serrata (Thunb.) Mak, $\tau = 40$ μsec. Line broadening of 100 Hz was used throughout.*

The very strong 73 ppm band was not completely dephased with $\tau = 30$ msec as used for the spectrum in Figure 3b; a τ of 40 msec provided more complete dephasing. Dipolar-dephased spectra were also measured on nonsenescent *Acer campestre L.* and nonsenescent and senescent *Zelkova serrata* (Thunb.) Mak. leaves. The dipolar-dephased spectra of the leaves of both of these trees (Figures 3a and 3c) also had bands near 55 ppm indicating that lignin is present in these leaves.

3.1.3 Aliphatic Alcohol and Ether Bands. The bands in the 60-90 ppm region of plant tissue spectra are due mainly to carbohydrates. The spectra of most leaf samples in this region consist of a single broad band centered at about 73 ppm with one or two shoulders on each side of the band (Figure 1). The 73 ppm band is due to the C-2, C-3 and C-5 atoms of cellulose, hemicelluloses and pectins.[19,20] The bands for the C-6 atoms are generally between 63 and 65 ppm for all carbohydrates. Crystalline cellulose is distinguished by a C-4 band at 89 ppm; the C-4 band of noncrystalline cellulose, hemicelluloses, and pectins generally occurs near 84 ppm. These bands generally are not well resolved in the leaf spectra. Aliphatic carbon atoms attached to oxygen atoms in lignin and nonhydrolysable tannins also occur in the 60-90 ppm region, but these bands are masked by the much more intense carbohydrate bands.

The chemical shifts of the anomeric carbon atoms of carbohydrates occur in the region between 90 and 105 ppm. All of the leaf spectra have a band at about 105 ppm, which represents the anomeric carbon of cellulose. In leaf and wood samples the 105 ppm band appears to be the only anomeric carbon band that is present.[17,19,21] Protonated carbon atoms of some phenolic functional groups also have bands between 90 and 105 ppm. These bands, along with the other phenolic bands, provide important clues about the identities of the types of phenolic compounds present in leaves in concentrations high enough to be detected by solid-state ^{13}C NMR.

3.1.4 Aromatic Bands. Three patterns of aromatic carbon bands have been observed in the NMR spectra of the leaf samples examined in this work (Figure 1). As we shall show below, these patterns are characteristic of hydrolysable tannins, nonhydrolysable tannins, and lignins, which are the major aromatic components of plant tissue. The nonhydrolysable tannins are polyflavonoids, which are more properly called proantho-cyanidins.[22] The monomeric units of these three types of polymers are given in Figure 2. The hydrolysable tannins are generally gallic acid esters (gallotannins) or hexahydroxy-diphenoyl esters (ellagitannins) as shown in Figure 2a.[22] Czochanska *et al.*[23] have shown that the proantho-cyanidins in plants generally exist as oligomers or polymers of procyanidin or prodelphinidin units in linear chains (Figure 2b). The flavan-3-ol monomeric units in these polymers are linked by C-4 to C-8 bonds (shown in Figure 2b) or C-4 to C-6 bonds.[24] Lignin is formed by the enzymatically mediated dehydrogenation polymerization of the three monolignols as shown for the A ring in Figure 2c.[25,26] The most common linkages formed in the polymerization are β-O-4 (Figure 2c); however, as Saake *et al.*[26] pointed out other linkages such as α-O-4, β–β, and β-5 are also found in lignins.

The chemical shifts for the three major types of aromatic species in leaves are given in Tables 2-7. The number of structural units for each of these types of species is very limited; however, differences in distribution of structural units and patterns of bonding of these units lead to a wide assortment of species within each group. For this reason, the chemical shifts given for the tannins in the tables are those that have been measured for a particular chemical species. The lignin chemical shifts have been measured on simple lignin monomeric or dimeric units. Major differences in tannin and lignin composition exist from one plant species to another, and even within a given species more than one type of tannin or lignin can be found. The values given in the tables below should be considered as nominal values because published values for some types of carbon atoms may vary by 3 ppm.

Table 2 *Chemical shifts of galloyl carbon atoms of Praecoxin C, a hydrolysable tannin (data from Hatano et al.[27])*

Chemical shifts (ppm)	Carbon atoms
110	C-2, C-6
120	C-1
140	C-4
146	C-3, C-5
165	C-7

Table 3 *Chemical shifts of hexahydroxyldiphenoyl carbon atoms of Praecoxin C, a hydrolysable tannin (data from Hatano et al.[27])*

Chemical shifts (ppm)	Carbon atoms
107	C-3, C-3'
114–115	C-1, C-1'
126	C-2, C-2'
139	C-5, C-5'
144	C-6, C-6'
145	C-4, C-4'
168–169	C-7, C-7'

The proanthocyanidin units differ from the other aromatic species by having bands between 95 and 101 ppm (Table 5). The presence of a band in this region in a leaf spectrum provides evidence for nonhydrolysable tannins in the leaf. As pointed out above, the anomeric carbon bands of carbohydrates occur in the region between 90 and 105 ppm. In most of the leaves we have examined only a single band at about 104 ppm is present in this region, which we have attributed to the anomeric carbon atoms of carbohydrates.

Table 4 *Chemical shifts of valoneoyl carbon atoms of Praecoxin C, a hydrolysable tannin (data from Hatano et al.[27])*

Chemical shifts (ppm)	Carbon atoms
107	C-3
110–111	C-1", C-3', C-6"
115	C-1
122	C-1'
125	C-2
133	C-2'
136–137	C-5, C-5', C-3"
142–144	C-2", C-4", C-5"
145–148	C-4, C-6, C-6'
152	C-4'
163–168	C-7, C-7', C-7"

However, in the *Malus* spectrum (Figure 1b) a second band at about 98 ppm is superimposed on the 104 ppm band. The interrupted decoupled (dipolar dephased) spectrum of the *Malus* leaves (Figure 3b) has a quaternary carbon band at about 103 ppm. The 98 ppm band and the quaternary band at 103 ppm provide evidence for the presence of polyflavonoid structures in the *Malus* leaves.[30] The 98 ppm band most likely represents terminal C-6 or C-8 carbons and the 103 ppm band C-10 carbons (Table 5).

Table 5 *Chemical shifts of polyflavonoid (nonhydrolysable) tannins (data from Czochanska et al.,[23] Pizzi and Stephanou[24] and De Bruyne et al.[28])[a]*

Chemical shifts (ppm)	Carbon atoms
27–28	C-4 terminal
38	C-4 joined to C-6 or C-8
67–68	C-3
81–82	C-2
95–96	C-8 terminal
96–98	C-6 terminal
101–107	C-10
108	C-2' and C-6' in PD
110	C-6 or C-8 joined to C-4
115–117	C-2' and C-5' in PC
120–121	C-6' in PC
131–133	C-1'
145–146	C-3'
145–146	C-4' in PC
156–157	C-5, C-7, C-9

[a] Proanthocyanidin units are designated as PC and prodelphinidin units designated as PD; where chemical shifts for both PC and PD are the same, no designation is given.

The assignment of the 103 ppm band is open to some question because the dipolar dephased spectra of *Zelkova* and *Acer* leaves also have bands between 103-105 ppm (Figure 3a,c,d). Bands near 98 ppm, however, are absent from the CPMAS spectra of these leaves. One possible explanation for these 103-105 ppm bands is that they represent the anomeric carbons of ketoses such as fructose in sucrose or a related oligosaccharide.

Table 6 *Chemical shifts of aromatic carbons in guaiacylpropanoid lignin monomeric units (data from Hassi et al.[29])*

C-4 Phenol		C-4 Ether	
Chem. shifts (ppm)	Carbon atoms	Chem. shifts(ppm)	Carbon atoms
111	C-2	111	C-2
115	C-5	113	C-5
118	C-6	121	C-6
133	C-1	135	C-1
146	C-4	148	C-4
147	C-3	149	C-3

Opella and Frey[12] found that the dipolar dephased spectrum of sucrose has a band near 103 ppm. Another possible explanation is that the 95 ppm bands are relatively weak in the *Zelkova* and *Acer* spectra and are obscured by the cellulose anomeric carbon band. It is also possible that the 103-105 ppm bands in all of the dipolar dephased spectra represent both C-9 carbons of polyflavonoids and anomeric carbons of saccharides.

Table 7 *Chemical shifts of aromatic carbons in syringylpropanoid lignin monomeric units (data from Hassi et al.[29])*

C-4 Phenol		C-4 Ether	
Chem. shifts (ppm)	*Carbon atoms*	*Chem. shifts (ppm)*	*Carbon atoms*
104	C-2, C-6	104	C-2, C-6
132	C-1	138	C-1
135	C-4	136	C-4
147	C-3, C-5	153	C-3, C-5

3.1.5 Carbonyl Bands. There is considerable overlap of the chemical shifts of the carbonyl-containing structures of the chemical components of plant tissue. The chemical shifts of the carbonyls of carboxylic acids or carboxylate salts, esters and peptides can all fall within the range between about 160 and 180 ppm. Aldehydes, ketones, and quinones generally have chemical shifts between about 180 and 220 ppm.

The NMR spectra of all of the leaves have a broad band centered at about 175 ppm (Figure 1). This band generally has been attributed in the NMR spectra of plant tissue to amide carbonyl groups in peptides and carboxyl groups of lipids, pectins, and cutin.[20,21,31] An estimate of the contribution of peptide carbonyls to the 175 ppm band may be calculated from elemental composition data by assuming that all of the nitrogen is present in amide groups. This assumption is based on the finding by Knicker and Lüdemann[32] that in rye grass and wheat more than 80% of the nitrogen is present in amide structures. These amide structures are most likely in peptides. For each amide nitrogen atom there will be a carbonyl carbon atom, and the total mass of the amide carbonyl carbons will be 12/14 (0.86) the mass of nitrogen. Thus the fraction of the total carbon that is amide carbonyl carbon, C_{amide}, in a sample will be given by the equation:

$$C_{amide} = 0.86 \, M_N / M_C$$

where M_N is the mass of nitrogen in the sample, and M_C is the mass of carbon in the sample determined by elemental analysis. Comparison of C_{amide} with the fraction of the total spectral area under the 175 ppm band, A_{175}, provides an estimate of the contribution of amide (peptide) carbonyls to the 175 ppm band. The variability of the peptide 175 ppm contribution, % pep., is shown in Table 8 for a representative set of leaf samples.

A sharp band or shoulder at about 168 ppm is observed in spectra of some of the leaves (Figure 1a). This band is most likely due to ester linkages in hydrolysable tannins (see Tables 2-4). Tannins are common in leaves and other organs of many dicotyledonous plants.[22] The 168 ppm band is probably the most diagnostic band for the presence of hydrolysable tannins in leaves because the other bands of hydrolysable tannins overlap the positions of some of the lignin and nonhydrolysable tannin bands.

Table 8 *Carbon and nitrogen analyses of leaves and calculated peptide contribution to 175 ppm band for green and senescent leaves*

Genus and species	C %	N %	C_{amide}	A_{175}	% pep.
Acalypha wilkesiana Müll. Arg.-green	41.09%	2.49%	0.052	0.112	46
Acalypha wilkesiana Müll. Arg.-senesent	36.38	0.99	0.023	0.064	36
Bixa orellana L.-green	45.35	2.09	0.040	0.070	57
Bixa orellana L.-senescent	43.48	1.33	0.026	0.026	100
Dombeya wallichii (Lindl.) Benth.-green	44.69	3.32	0.064	0.085	75
Dombeya wallichii (Lindl.) Benth.-senesent	41.88	1.02	0.020	0.043	47
Malus 'Manbeck Weeper'-green	48.91	2.46	0.043	0.088	49
Malus 'Manbeck Weeper'-senescent	48.51	0.81	0.014	0.063	22

3.2 NMR Spectra of Senescent Leaves

3.2.1 General Appearance of Spectra. The spectral changes brought about by senescence will be considered within each of the spectral regions outlined in Table 1. The types of changes observed within these regions for a given species may be classified into four different categories: (1) the bands in a given spectral region of the senescent leaves are the same bands as those for the nonsenescent leaves; however, the relative intensities of the bands are different; (2) the senescent leaves have bands that are not present in the nonsenescent leaves in the same region; (3) the senescent leaves are missing some bands within a region that are present in the nonsenescent leaves; and (4) the senescent leaves differ from the nonsenescent leaves by presence of some additional bands and the absence of other bands within the same region. These differences provide insight into the chemical changes that are brought about by senescence.

The differences may be most easily observed by the overlaying the spectra of the senescent and nonsenescent leaves. In these overlays the spectra have been normalized such that the intensities of the 73 ppm bands of all of the spectra are the same. Quantitative measures of the differences may be obtained by deconvolution of the spectra. The spectral overlays indicate that for most regions the observed differences mainly involve changes in relative intensities of the bands. Examples of the types of changes within each of the spectral regions revealed by the overlays will be discussed below.

3.2.2 Aliphatic Hydrocarbon Region. The *Acer campestre L.* and *Acalypha wilkesiana* overlays (Figures 4 and 5) show the types of changes brought about by senescence that can be observed by ^{13}C NMR. The aliphatic regions of the two *Acer campestre L.* spectra each have a broad band centered at about 30 ppm. At least four shoulders are evident on the sides of this band in the nonsenescent spectrum, indicating that it is composed of several overlapping bands. The most prominent shoulder in the nonsenescent spectrum is between about 38-39 ppm. This shoulder is absent or very weak in the senescent spectrum. More profound changes are seen in the *Acalypha wilkesiana* overlay. Well-defined bands at 25 and 30 ppm in the nonsenescent spectrum have been reduced to shoulders in the senescent spectrum, and a broad new band between 40 and 50 ppm appears in the

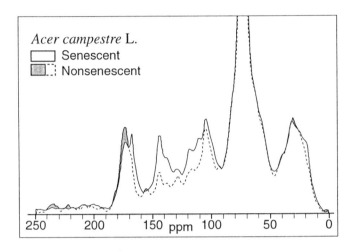

Figure 4 *Overlay of the ¹³C NMR spectra of senescent and nonsenescent Acer campestre L. leaves. Line broadening of 100 Hz was used.*

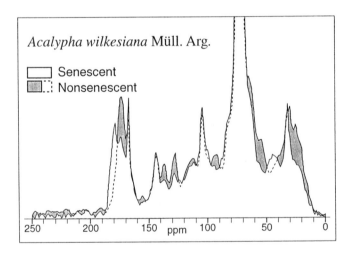

Figure 5 *Overlay of the ¹³C NMR spectra of senescent and nonsenescent Acalypha wilkesiana Müll. Arg. leaves. Line broadening of 25 Hz was used.*

senescent spectrum. In both the *Acer* and *Acalypha* overlays the intensities of the shoulders at about 55 ppm are reduced in the senescent spectra. These shoulders most likely represent methoxyl groups in lignin structures. Apparently, these methoxyl groups undergo partial hydrolysis during senescence. Evidence of this partial hydrolysis has been observed in the NMR spectra of all the nonsenescent-senescent pairs of leaves that we have examined.

The spectral changes below 50 ppm cannot be interpreted solely from the NMR data collected in this study. They probably result mainly from oxidation of leaf lipid and cutin components. Definitive identification of the changes in the lipid and cutin components will require that the components be isolated and identified by chemical and spectroscopic means before and after senescence. Typically, during senescence lipids undergo enzymatic oxidation reactions that cleave the lipids into smaller fragments. The four major mechanisms of oxidative degradation of lipids in plants are α-oxidation, β-oxidation, ω-oxidation and hydroperoxidation catalysed by lipoxygenase.[33] In α-oxidation C_n fatty acids are converted to C_{n-1} aldehydes, which then may be further oxidized to C_{n-1} carboxylic acids; in β-oxidation, C_n acids are converted to C_{n-2} acids.[34] In ω-oxidation, dicarboxylic acids are produced by oxidation of the terminal methyl group of a fatty acid to a carboxyl group.[35]

3.2.3 Aliphatic Alcohol and Ether Bands. Senescence causes less apparent change in the carbohydrate region of the spectra than in the other regions. Careful examinations of senescent-nonsenescent pairs of spectra indicate that there are changes in the anomeric carbon bands and the shoulders of the 73 ppm band. These changes will be discussed in part III of this series where comparison with the leachate spectra will provide more insight into the mechanisms of carbohydrate degradation. The relative stability of the 73 ppm band has allowed us to normalize all of the spectra to a constant intensity of the 73 ppm band. This normalization readily allows one to assess relative changes in the concentrations of the major structural components of the leaves before and after senescence.

3.2.4 Aromatic Bands. The simplest of the three different aromatic patterns is the lignin pattern. The spectrum of the *Zelkova serrata* leaves provide a good example of the lignin pattern. A series of spectra measured on leaves collected from the Summer through the Fall of 1996 is shown in Figure 6. The ratio of the intensity of the 153 ppm band to that of the 147 ppm band (153/147 ratio) is reduced in the senescent leaf sample in comparison to nonsenescent leaf samples (Figure 6). A similar reduction in 153/147 ratio was observed in *Ulmus pumila* senescent leaves compared to nonsenescent leaves collected at the same time (Figure 7). These reductions in 153/147 ratios are most likely due to hydrolysis of C-4 ether linkages of syringylpropanoid units of lignin. As shown in Table 7, hydrolysis of the C-4 ether linkage causes the chemical shifts of the C-3 and C-5 carbon atoms to change from 153 ppm to 147 ppm. Therefore, if some of these ether linkages are hydrolysed in the leaves there will be a reduction in the 153 ppm band and a concomitant increase in the 147 ppm band. Hemmingson and Dekker[36] and Haw and Maciel[37] found that steam hydrolysis of wood causes reduction of 153/147 ratios. Hemmingson and Dekker[36] proposed that steam hydrolysis depolymerizes lignin by cleaving β-O-4 linkages. The reduction of the 153/147 ratios in the *Zelkova* and *Ulmus* leaves indicates that the hydrolytic and oxidative processes of senescence are also causing lignin depolymerization.

The NMR spectrum of *Acer campestre L.* leaves exhibits the galloyl bands (Table 2) of the hydrolysable tannin pattern (Figure 4). The intensities of the tannin bands with respect to the aliphatic alcohol and ether bands are larger in the senescent leaves than in

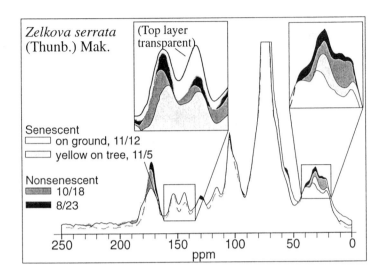

Figure 6 *Multiple overlay of the ^{13}C NMR spectra of senescent and nonsenescent Zelkova serrata (Thunb.) Mak. leaves collected at different times of the year. Line broadening of 100 Hz was used.*

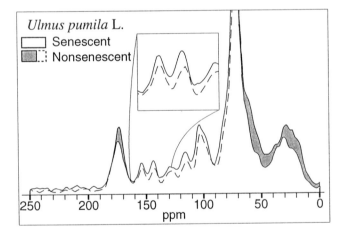

Figure 7 *Overlay of the ^{13}C NMR spectra of senescent and nonsenescent Ulmus pumila L. leaves. Line broadening of 100 Hz was used.*

the nonsenescent leaves. This indicates that there has been a loss of carbohydrates from the senescent leaves. These leaves were collected from the ground beneath the tree and had undergone some leaching by melting snow prior to collection. During the leaching, carbohydrates and tannins would both be removed from the leaves, but apparently more carbohydrates than tannins were lost. The hydrolysable tannin pattern in the senescent leaves is similar to that in the nonsenescent leaves, differing only in the relative intensities of some of the bands. The intensity variations suggest that some structural alterations may have taken place.

In addition to the galloyl bands, bands at 130 and 153 ppm are present in the aromatic region of the *Acer* spectra. These bands may indicate that the tannins contain valoneoyl groups (Table 4) or that lignins are also present in the *Acer* leaves. The shoulders at 55 ppm which are characteristic of the methyl ethers of lignins indicate that the 130 and 153 ppm bands are at least partially due to lignin.

The aromatic regions of the *Malus* 'Manbeck Weeper' nonsenescent and senescent spectra (Figure 8) are more complex than the *Acer* and *Zelkova* spectra. In addition to the aromatic bands seen in these spectra, there is a band at about 98 ppm. As pointed out above, a band in this region is diagnostic of polyflavonoid (nonhydrolysable) tannins (Table 5). Bands at 163 and 167 ppm are also present in the spectra; these bands most likely are indicative of hydrolysable tannins. A similar pattern is present in the NMR spectrum of *Lespedeza cuneata* leaves.[20] Gamble *et al.*[20] isolated nonhydrolysable tannins from these leaves. Benner *et al.*[38] reported that tannins constitute the majority of the aromatic components of mangrove leaves.

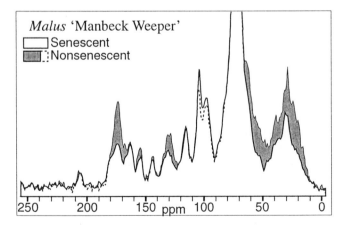

Figure 8 *Overlay of the ^{13}C NMR spectra of senescent and nonsenescent Malus 'Manbeck Weeper' leaves. Line broadening of 25 Hz was used.*

The most prominent change in the aromatic region of the spectrum of the senescent *Malus* leaves compared with the spectrum of the nonsenescent leaves is a marked reduction in the relative intensity of the 131 ppm band (Figure 8). This may indicate scission of C-C bonds in tannin structures.

3.2.5 Carbonyl Bands. The bands between about 162 and 170 ppm in the leaf spectra are most likely due to ester linkages. As pointed out above, both aromatic and aliphatic esters are abundant in leaves. Aliphatic esters in general would be expected to have chemical shifts near 170 ppm (Wooten[17] assigned a band at 171 ppm to pectins), and those of aromatic esters would be expected to be between 163 and 169 ppm (Tables 2-4). We assigned the band at 167 ppm in the *Acer campestre L.* spectrum to hydrolysable tannins both because of its position and because the relative intensity of the band increases along with the intensities of the aromatic bands in the spectrum of senescent *Acer campestre L.* leaves compared with the spectrum of the nonsenescent leaves (Figure 3).

The data in Table 8 show that the reductions in the areas of the 175 ppm bands (A_{175} values) in senescent leaves are associated with reduced C_{amide} values in the senescent leaves. These results provide spectral evidence for the removal of peptides from leaves during senescence.

A new sharp band at 179 ppm is present in the spectrum of the senescent *Acalypha wilkesiana* leaves (Fig. 5). The most likely possibilities are that this new band represents either a low molecular weight aliphatic acid or a quinone. Wooten[17] found that the chemical shifts of the carboxylate bands of potassium salts of citric and malic acid are between 178 and 182 ppm. Endogenous polyphenol oxidases such as those that are thought to be active during senescence can convert *o*-diphenols to *o*-quinones.[39] The chemical shifts of *o*-quinones are near 180 ppm.[18]

4 CONCLUSIONS

The biochemical processes that are active during senescence of leaves cause changes in the chemical structures and relative concentrations of the major components of the leaves that can be detected by solid-state ^{13}C NMR spectroscopy. The NMR spectra of the leaves may be divided into spectral regions that are diagnostic of different types of functional groups. The combinations of functional groups indicated by the spectra provide information about the general types of compounds present in the leaves. In particular, it is possible to distinguish spectral patterns that are characteristic of lipids, carbohydrates, lignins, hydrolysable and nonhydrolysable tannins and peptides. Three distinct patterns of aromatic-carbon bands characteristic of either (1) hydrolysable tannins, (2) nonhydro-lysable tannins or (3) lignins were observed in the NMR spectra of the leaf samples.

Changes in the spectral patterns brought about by senescence allow one to assess changes in chemical-component concentrations and structures. The observed changes in concentration and structure indicate that the following types of reactions take place during senescence: oxidation of lipids and cutins, hydrolysis of the methyl ether groups of lignins, lignin depolymerization and hydrolysis of peptides. The senescence-induced hydrolysis of methyl ethers and lignin depolymerization in leaves have not previously been reported.

Leaves from different species exhibit different types of senescence-induced changes. These differences are apparently the result of species-specific enzyme systems. After the leaves fall to the ground and are incorporated into the litter layer and ultimately into the soil, one would expect a mixing of enzymes from different species. This means that one would expect that all of the fallen leaves in a given area should eventually be subjected to degradation by a mixture of enzymes that will be characteristic of the types of plants present. In addition to these endogenous enzymes, exogenous enzymes secreted by soil microorganisms will also degrade the leaves.

References

1. S. Gan and R. M. Amasino, *Science*, 1995, **270**, 1986.
2. R. Aerts, *J. Ecology*, 1996, **84**, 597.
3. K. T. Killingbeck, *Ecology*, 1996, **77**, 1727.
4. Y. Y. Leshem, *Free Radical Biology & Medicine*, 1988, **5**, 39.
5. S. Philosoph-Hadas, S. Meir, B. Akiri and J. Kanner, *J. Agric. Food Chem.*, 1994, **42**, 2376.
6. S. Strother, *Gerontology*, 1988, **34**, 151.
7. L. D. Noodén and A. C. Leopold, 'Senescence and Aging in Plants', Academic Press, San Diego, 1988.
8. R. L. Wershaw, J. A. Leenheer, K. R. Kennedy and T. I. Noyes, *Soil Sci.*, 1996, **161**, 667.
9. J. S. Valentine, C. S. Foote, A. Greenberg and J. F. Liebman, 'Active Oxygen in Biochemistry', Blackie Academic & Professional, London, 1995.
10. L. A. Alemany, D. M. Grant, T. D. Alger and R. J. Alemany, *J. Am. Chem. Soc.*, 1982, **105**, 6697.
11. P. G. Hatcher, *Org. Geochem.*, 1987, **11**, 31.
12. S. J. Opella and M. H. Frey, *J. Amer. Chem. Soc.*, 1979, **101**, 5854.
13. T. Zlotnik-Mazori and R. E. Stark, *Macromolecules*, 1988, **21**, 2412.
14. P. E. Kolattukudy and K. E. Espelie, in 'Biosynthesis and Biodegradation of Wood Components', T. Higuchi, (ed.), Academic Press, Orlando, FL, 1985, Chap. 8, p. 161.
15. M. Nip, E. W. Tegelaar, J. W. de Leeuw, P. A. Schenck and P. J. Holloway, *Naturwissenschaften*, 1986, **73**, 579.
16. I. Kögel-Knabner, J. W. De Leeuw, E. W. Tegelaar, P. G. Hatcher and H. Kerp, *Org. Geochem.*, 1994, **21**, 1219.
17. J. B. Wooten, *J. Agric. Food Chem.*, 1995, **43**, 2858.
18. E. Breitmaier and W. Voelter, 'Carbon-13 NMR Spectroscopy', VCH Verlagsgesellschaft, Weinheim, Germany, 1987.
19. G. R. Gamble, A. Sethuraman, D. E. Akin and K-E. L. Eriksson, *App. Environ. Microbiol.*, 1994, **60**, 3138.
20. G. R. Gamble, D. E. Akin, H. P. S. Makkar and K. Becker, *Appl. Environ. Microbiol.*, 1996, **62**, 3600.
21. W. Kolokziejski, J. S. Frye and G. E. Maciel, *Anal. Chem.*, 1982, **54**, 1419.
22. E. Haslam, in 'The Biochemistry of Plants', E. E. Conn, (ed.), Academic Press, New York, 1981, Vol. 7, Chap. 18, p. 527.
23. Z. Czochanska, L. Y. Foo, R. H. Newman and L. J. Porter, *J. Chem. Soc. Perkin Trans. 1*, 1980, 2278.
24. A. Pizzi and A. Stephanou, *J. Applied Polymer Sci.*, 1993, **50**, 2105.
25. C. J. Douglas, *Trends in Plant Sci.*, 1996, **1**, 171.

26. B. Saake, D. S. Argyropoulos, O. Beinhoff and O. Faix, *Phytochem.*, 1996, **43**, 499.
27. T. Hatano, K. Yazaki, A. Okonogi and T. Okuda, *Chem. Pharm. Bull.*, 1991, **37**, 1689.
28. T. De Bruyne, L. A. C. Pieters, R. A. Dommisse, H. Kolodziej, V. Wray, T. Domke and A. J. Vletinck, *Phytochem.*, 1996, **43**, 265.
29. H. Y. Hassi, M. Aoyama, D. Tai, C.-L. Chen and J. S. Gratzl, *J. Wood Chem. Tech.*, 1987, **7**, 555.
30. M. A. Wilson and P. G. Hatcher, *Org. Geochem.*, 1988, **12**, 539.
31. G. R. Gamble, G. N. Ramaswamy, B. S. Baldwin and D. E. Akin, *J. Sci. Food Agric.*, 1996, **72**, 1.
32. H. Knicker and H.-D. Lüdemann, *Org. Geochem.*, 1995, **23**, 329.
33. B. A. Vick and D. C. Zimmerman. 'The Biochemistry of Plants. Vol. 9: Lipids: Structure and Function', P. K. Stumpf, (ed.), Academic Press, Orlando, FL, 1987, Chap. 3, p. 53.
34. T. Galliard, 'The Biochemistry of Plants. Vol. 4 Lipids: Structure and Function', P. K. Stumpf, (ed.), Academic Press. New York, 1980, Chap. 3, p. 85.
35. A. White, P. Handler and E. L. Smith, 'Principles of Biochemistry', McGraw-Hill, New York, 1964, pp. 444-445.
36. J. A. Hemmingson and R. F. H. Dekker, *J. Wood Chem. Technol.*, 1987, **7**, 229.
37. J. F. Haw and G. E. Maciel, *Holzforschung*, 1984, **38**, 327.
38. R. Benner, P. G. Hatcher and J. I. Hedges, *Geochim. Cosmochim. Acta*, 1990, **54**, 2003.
39. K. B. Hicks, R. M. Haines, C. B. S. Tong, G. M. Sapers, Y. El-Atawy, P. L. Irwin and P. A. Seib, *J. Agric. Food Chem.*, 1996, **44**, 2591.

USE OF ^{13}C NMR AND FTIR FOR ELUCIDATION OF DEGRADATION PATHWAYS DURING NATURAL LITTER DECOMPOSITION AND COMPOSTING. III. CHARACTERIZATION OF LEACHATE FROM DIFFERENT TYPES OF LEAVES

Robert L. Wershaw, Jerry A. Leenheer and Kay R. Kennedy

U.S. Geological Survey, Denver Federal Center, Denver, CO 80225

1 INTRODUCTION

Decomposing leaves are an important imput for the formation of natural dissolved organic carbon compounds (DOC) and soil organic matter or humus.[1] The purpose of this series of papers is to develop a detailed understanding of the reactions that lead to the formation of DOC and humus from decomposing leaf litter.

In the first part of the series[2] we demonstrated that the early stages of the oxidative degradation of the lignin components of leaves follow the sequence of O-demethylation, and hydroxylation followed by ring-fission, chain-shortening and oxidative removal of substituents. Oxidative ring-fission leads to the formation of carboxylic acid groups on the cleaved ends of the rings, and in the process transforms aromatic structures into aliphatic compounds (see Barz and Weltring[3] for a discussion of mechanisms). The carbohydrate components are broken down into aliphatic hydroxy acids and aliphatic alcohols. These conclusions were reached by ^{13}C NMR and IR spectrometric characterization of the water-soluble products of senescent leaves. A much more extensive fractionation procedure than has been used previously on DOC was developed for the separation of groups of compounds of different polarities from the water extract.

In the second part of the series[4] we demonstrated that distinctive patterns are observed in the ^{13}C NMR spectra of leaves of different species. The spectra of the nonsenescent leaves are composed of bands that represent the functional groups of the major chemical components of the leaves. The spectrum of each chemical component consists of a characteristic group of bands that occur in well-defined spectral regions. Of particular interest are the bands in the aromatic regions of the spectra because major differences were observed in the spectra of the leaves of different species in this region. For a given species, one of three aromatic carbon patterns characteristic of either hydrolysable tannins, nonhydrolysable tannins or lignins was observed.

In this part of the series we shall demonstrate that the leachates from different types of leaves have distinctive ^{13}C NMR spectral signatures, which should be useful in following the diagenesis of plant-derived organic material to DOC species in natural water systems. As we have shown previously,[2] ^{13}C NMR spectroscopy provides a unique tool for the characterization of whole DOC samples. Changes in spectral patterns of leachate DOC from leaves of a particular species as the DOC moves from litter layer to soil water to groundwater and to surface water should provide a means of describing the changes that take place in each compartment of a hydrologic system. Ultimately, this should lead to a better understanding of the processes that give rise to DOC in natural systems.

2 MATERIALS AND METHODS

2.1 Leaf Collection and Leachate Preparation

Leachates were prepared from senescent *Acer campestre L.* (maple), *Malus* 'Manbech Weeper' (crab apple), and *Zelkova serrata* (Thunb.) Mak. leaves, collected from the Denver Botanic Gardens as described previously,[4] by soaking for one hour in distilled water. After separation from the leaves, each leachate sample was poured through a plug of glass wool in a funnel and then pressure filtered through a prewashed Gelman Supor[*] 0.45 micron-pore-size polysulfone membrane filter and freeze-dried.

In addition to the one hour leaching, *Acer campestre L.* leaves were subjected to another leaching experiment. Unleached leaves were first soaked for 19 hours in distilled water and then the leached leaves were soaked a second time for 72 hours. These two leachates were fractionated into fractions of different polarities by sequential adsorption on XAD macroreticular resin.[2] Briefly, the leachate was first pumped onto a 150 mL XAD-8 column without pH adjustment and the adsorbed material (fraction 1) was eluted with 100 mL of 75%/25% (v/v) acetonitrile/water solution. The solvents were removed in a rotary vacuum evaporator and the DOC was freeze-dried from water. The pH of the leachate that passed through the XAD-8 column in the first step of the procedure was lowered to 2 and the leachate was pumped through the XAD-8 column again. The adsorbed fraction 2 was eluted with 100 mL of 75%/25% (v/v) acetonitrile/water solution and freeze-dried. The pH of the leachate from the second step was adjusted to 5 and its volume was reduced to 500 mL by rotary evaporation. After volume reduction, the pH was adjusted to 2 and the leachate was pumped onto the XAD-8 column for a third time. The adsorbed fraction 3 was eluted with 75%/25% (v/v) acetonitrile-water and freeze-dried.

The pH of the leachate that passed through the XAD-8 column (fourth step) was adjusted to 4 and it was then taken to dryness on a rotary evaporator. The residue in the evaporation flask was dissolved in 100 mL of 1 N HCl, filtered through a glass wool plug to remove silica, and pumped through a 100 mL XAD-4 column. The column was eluted with 100 mL of 75%/25% (v/v) acetonitrile/water and vacuum evaporated to 20 mL. Residual HCl and water were removed by repeated additions and evaporations of anhydrous acetonitrile. The dried sample (fraction 4) was taken up in 50 mL of water and freeze-dried. Fraction 5 was not isolated because previous studies have shown it to be a very minor fraction.[2]

The leachate that passed through the XAD-4 was deionized on 80 mL columns of MSC-1H cation exchange resin and Duolite A-7 anion exchange resin in series. The volume of effluent was reduced by rotary evaporation and the hydrophilic neutral material remaining in the effluent was freeze-dried (fraction 6). The organic acids sorbed on the A-

[*]The use of trademarks in this report is for identification purposes only and does not constitute endorsement by the U.S. Geological Survey.

7 resin (fraction 7) were not isolated because previous work has shown that this fraction is very small.

The fractions from the second *Acer campestre L.* leachate are identified by a "B" appended to each of the fraction numbers. During the fractionation of the second leachate it was noted that after elution of fraction 1 some color remained on the column. A second 100 mL aliquot of 75%/25% (v/v) acetonitrile–water was used to elute the remaining colored material from the column. The resulting fraction was designated 1B'.

2.2 NMR Spectroscopy

Solid-state cross-polarization magic-angle-spinning (CPMAS) ^{13}C NMR spectra were measured at 50.298 MHz on a 200 MHz (proton frequency) Chemagnetics CMX spectrometer with a 7.5 mm-diameter probe. The spinning rate was 5000 Hz. The acquisition parameters for the freeze-dried samples were contact time of 1 ms, pulse delay of 1s and a pulse width of 4.5 microseconds for the 90° pulse. A line broadening of 100 Hz was applied in the Fourier transformation of the free induction decay data.

The liquid-state ^{13}C NMR spectra were measured in a solution of approximately 200 mg/mL of the sample dissolved in D_2O in 10-mm diameter tube on a Varian Inova 400 spectrometer at 100.618 MHz. Quantitative spectra were obtained using inverse gated-decoupling in which the proton decoupler was on only during the acquisition of the free induction decay (FID) curve. An 8-second delay time and a 45 degree tip angle were used. The sweep width was 30,000 Hz.

2.3 Infrared Spectroscopy

Pellets for infrared analysis were prepared by mixing approximately 5 mg of sample with 250 mg of KBr, grinding the ingredients together with a mortar and pestle, and pressing the mixture in a die. The spectra of the pellets were measured on a Perkin-Elmer 2000 Fourier transform spectrometer.

3 RESULTS AND DISCUSSION

3.1 Spectra of Unfractionated Leachates

3.1.1 General Appearance of Spectra. Distinctive patterns of bands are generally observed in the solid-state ^{13}C NMR spectra of unfractionated leachates derived from leaves of different species. These patterns are similar to the patterns that have been observed in the leaf spectra.[4] The spectral band patterns may be most readily understood by dividing the spectra into spectral regions characteristic of different types of chemical compounds.[4]

In Figure 1 the spectra of three pairs of leaves and leachates are shown. Comparison of the band pattern of a region of a leachate spectrum with the band pattern of the same region of the spectrum of the leaves from which the leachate was derived provides information about the compositional differences between the leaves and the leachates. These differences provide insight into the humification process.

3.1.2 Aliphatic Hydrocarbon Bands. In all three pairs the leachate bands in the aliphatic region are different than those in the unextracted leaves. In the *Acer campestre*

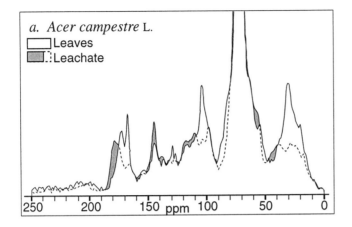

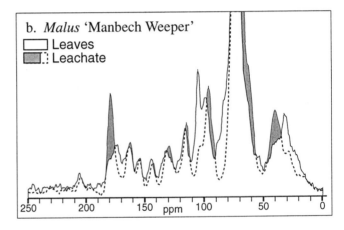

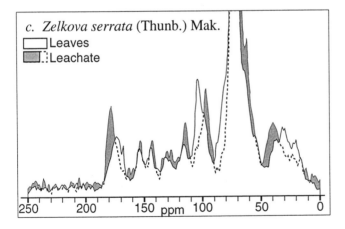

Figure 1 *^{13}C NMR spectra of senescent leaves and leachates of: a. Acer campestre L., b. Malus 'Manbech Weeper' and c. Zelkova serrata (Thunb.) Mak.*

L. spectra the well-defined band at 31 ppm in the leaves has been replaced by two weaker bands at 29 and 39 ppm. The *Malus* 'Manbeck Weeper' leaves have what appears to be a doublet centered at about 30 ppm flanked by two shoulders, one at about 22 ppm and one at about 40 ppm. The *Malus* leachate on the other hand, has a strong band at 41 ppm and weaker band at 29 ppm. The band pattern of the *Zelkova serrata* (Thunb.) Mak. leaves is similar to that of the *Malus* with a band at about 30 ppm flanked by two lesser bands at 21 and 38 ppm. The *Zelkova* leachate has a strong band at 40 ppm and a broad shoulder from about 10 to 30 ppm. Computer simulation using the ChemWindow program (SoftShell International) indicates that the strong band near 40 ppm in the leachate NMR spectra could result from substituted 1,4-butanedioic acid groups resulting from the oxidative fission of aromatic rings.[3] The 40 ppm band would represent a carbon atom α to one of the carboxylate groups that has an aromatic or aliphatic substituent on it. Such a structure could arise from the fission of one aromatic ring in a polymer. For example, Kirk[5] pointed out that an aromatic ring in a lignin polymeric structure can be degraded while remaining attached to the rest of the polymer.

3.1.3 Aliphatic Alcohol and Ether Bands. The shoulder at 55 ppm is reduced in all of the leachate spectra compared with the leaf spectra. This reduction in intensity most likely indicates a loss of methyl ether groups in lignin-like structures. The *Acer campestre L.* pair provides an example of the differences between a leaf spectrum and a leachate spectrum. In the *Acer* spectra a well-resolved band at 55 ppm in the leaf spectrum has been replaced by a shoulder centered at about 57 ppm. This shoulder may represent residual methoxyl groups or it may be due to carbon atoms α to amino groups of amino acids or peptides, which are in the range of 40 to 65 ppm.[6] As will be demonstrated below, the spectra of the fractions of the *Acer* leachate allow one to distinguish between these two possibilities.

All the leaf and leachate spectra have strong bands between 72 and 73 ppm. The C-2, C-3, and C-5 carbons of polysaccharides occur in this region. Wooten[7] has pointed out that the C-4 and C-6 carbon atoms of crystalline cellulose microfibrils have chemical shifts at 89.1 and 65.5 ppm respectively. In disordered cellulose these bands are at 84.7 and 63.0 ppm. In addition to cellulose, leaves contain hemicellulose, starch, and pectin. Wooten's data indicate that the C-4 bands of hemicellulose and pectin are near 85 ppm and are broader and less well-resolved than the C-4 band of crystalline cellulose. The C-6 band of hemicellulose appears at 63.0 ppm; the C-6 carbohydrate band is absent in pectic polysaccharides because it has been replaced by a carbonyl carbon of an acid or methyl ester. Wooten[7] has pointed out that the spectrum of starch should be very similar to that of hemicellulose. The poorly-resolved shoulders observed in the leaf spectra between 60 and 66 ppm, and between 80 and 90 ppm are indicative of the mixed character of leaf polysaccharides.

The chemical shift of the anomeric carbon (C-1) of cellulose is at about 105 ppm; the C-1 chemical shifts of hemicellulose, pectin and starch are generally between 100 and 103 ppm.[7–9] The 105 ppm cellulose band is well-resolved in all of the leaf spectra; in most of the leaf spectra we have examined a broad shoulder between 100 and 103 ppm attributable to hemicelluloses, pectins and starches that is superimposed on the 105 ppm band (see the *Acer* and *Zelkova* spectra).

The *Malus* 'Manbeck Weeper' spectrum is exceptional in that there are two distinct, well-resolved bands in the anomeric region, one at 105 ppm and one at 99 ppm. The deconvolution routine described previously[2] was used to separate the two bands. The integrated areas of the these two deconvoluted bands and the other carbohydrate bands are given in Table 1. If the deconvoluted bands at 105 and 99 ppm represent only anomeric carbons in carbohydrates, then the ratio of cellulose to the other carbohydrate components in a given leaf sample will be equal to the ratio of areas of the two bands. Since most of the carbohydrates in leaves are hexoses, the sum of the areas of the two anomeric bands should be equal to one-fifth of the combined areas of the bands between 60 and 90 ppm. The ratio of the sum of the areas of the two *Malus* anomeric bands to the sum of the other carbohydrate bands calculated from the data in Table 1 is 0.243. Since this ratio is greater than one fifth, there is probably a contribution from nonhydrolysable tannins which have a band near 98 ppm.

Table 1 *Integrated areas of deconvoluted carbohydrate NMR bands in Malus 'Manbeck Weeper' spectrum*

Band	Area	Sum of areas
99	0.056	
105	0.073	
		0.129
84	0.010	
73	0.507	
63	0.013	
		0.530

Comparison of the leaf and leachate spectra of all the species show that there is a partial to complete loss of the shoulders of the 72 ppm band between 80 and 90 ppm and an almost complete disappearance of the anomeric carbon (C-1) band at 105 ppm in the leachate spectra. These differences are consistent with the conclusion that the carbohydrates in the leachates are noncellulosic components. Data presented below on the *Acer campestre L.* fractions indicate that, at least for the *Acer* leachate, some of the leachate carbohydrates are conjugated to hydrolysable tannins by ester linkages. In hydrolysable tannins the chemical shift of an esterified C-4 is between 69 and 71 ppm, and the anomeric carbon is between 91 and 98 ppm.[10]

3.1.4 Aromatic Bands. The spectral patterns of the aromatic bands of the unfractionated leachates are similar to those of the patterns of the senescent leaves from which they are derived. For each of the spectral pairs, differences in the relative intensities of some of the bands are seen when one compares the leachate spectrum to the corresponding leaf spectrum. However, these differences are relatively minor; they will not be discussed here because much more information about the changes in the aromatic components of the leaves can be derived from the spectra of the leachate fractions which will be described in detail below.

3.1.5 Carbonyl bands. In all the spectral pairs the carbonyl bands are shifted from about 175 ppm in the leaves to between 178 to 180 ppm in the leachates. A number of different types of acids, some peptides and flavones have carbonyl chemical shifts near 180 ppm. As will be demonstrated below in the discussion of the leachate fractions, combining NMR analysis with infrared analysis indicates that both acid and flavonoid types of structures are present in the leachates.

3.2 Spectra of Leachate Fractions

Further insight into the degradation processes leading to the formation of humic substances from senescent leaves may be obtained by fractionating leaf leachates into fractions of increasing polarity.[2] The NMR and IR spectra (Figures 2-7) of the fractions isolated from solutions resulting from two successive leaching of senescent *Acer campestre L.* leaves illustrate the types of information that can be obtained from the fractions of increasing polarity. The first set of fractions was isolated from a solution that resulted from a 19-hour leaching of fresh senescent leaves in distilled water. The second set was isolated after an additional 72-hour leaching of the previously leached leaves. The masses of the fractions are given in Table 2.

Table 2 *Masses in grams of fractions isolated from 1-L samples of 19-hour and 72-hour leachates. Mass of leaves extracted was 33 g*

19-Hour leachate			*72-Hour leachate*		
Fraction	*Mass*	*Percent*	*Fraction*	*Mass*	*Percent*
			1B'	0.0149	1.7
1	0.8673	35.1	1B	0.5689	64.1
2	0.3898	15.8	2B	0.1621	18.3
3	0.0996	4.0	3B	0.0190	2.1
4	0.2779	11.3	4B	0.0602	6.8
6	0.8357	33.8	6B	0.0631	7.1
Total	2.4703		Total	0.8882	

In the first paper in this series[2] it was proposed that the fractions of increasing polarity that were isolated represent different stages in the oxidative degradation of lignin components of the leaves. Fraction 1 contained the least degraded lignin fragments and fraction 4 the most degraded fragments. In this study we shall demonstrate that degradation products of other types of aromatic compounds such as tannins also can be identified in the leachate fractions.

The NMR spectra of the fractions from the two leachates are given in Figures 2-5. Table 3 contains a compilation of the chemical shifts of the bands of the fractions. The data in Table 3 demonstrate that there is an approximate repetition of chemical shifts in the NMR spectra of all of the fractions. That is to say, the NMR bands of all of the fractions fall into a single set. The NMR spectra of some of the fractions contain more of the bands in the set than others. The existence of a single set of bands implies that there is a corresponding set of functional groups from which all of the fractions are composed. Some of the fractions contain more of these functional groups than others. The aliphatic and aromatic regions of the spectra will be discussed separately.

The differences between the aliphatic regions of the *Acer campestre L.* leaves and the unfractionated leachate spectra and the most probable band assignments have been discussed above. A question was raised about whether the shoulder at about 57 ppm in the unfractionated leachate spectrum should be assigned to carbon atoms in methoxyl groups (*e.g.* phenolic methyl ethers) or to carbon atoms α to amino groups of amino acids or peptides. In most of the leachate fractions a well-defined band at about 55 ppm is observed in the same region (Figures 2-5). Dipolar dephased spectra were measured on fractions 1, 2, 1B, 1B', 2B' (see Figure 6 for fractions 1 and 2); a well-defined band at

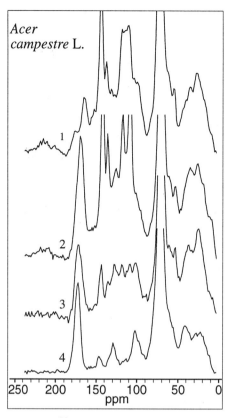

Figure 2 ^{13}C *NMR spectra of fractions 1-4 of Acer campestre L. leachate.*

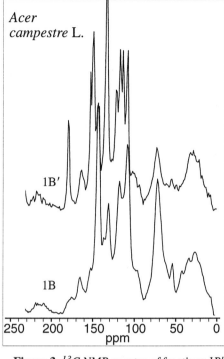

Figure 3 ^{13}C *NMR spectra of fractions 1B' and 1B of Acer campestre L. leachate.*

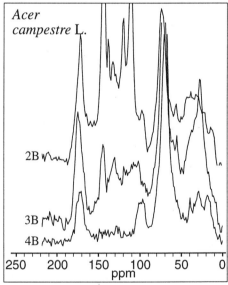

Figure 4 ^{13}C *NMR spectra of fractions 2B-4B of Acer campestre L. leachate.*

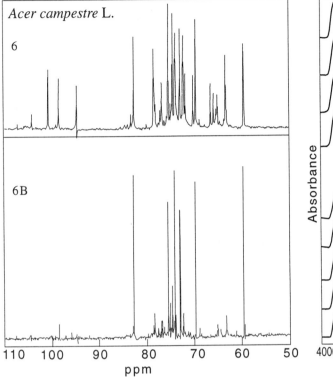

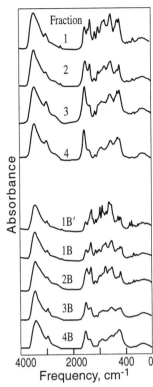

Figure 5 *¹³C NMR spectra of fractions 6 and 6B of Acer campestre L. leachate.*

Figure 7 *FTIR spectra of fractions 1-4 and 1B'-4B of Acer campestre L. leachate.*

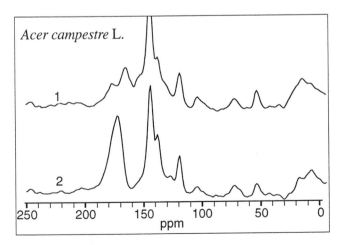

Figure 6 *Dipolar, dephased spectra of fractions 1 and 2 of Acer campestre L. leachate.*

about 55 ppm was observed in each of these spectra, indicating the presence of freely-rotating methyl ether groups in each of the fractions.

Fraction 1B' is the most nonpolar of all of the fractions. Two soakings were required for this fraction to be released from the leaves. Wershaw *et al.*[2] have proposed that the fractions they isolated from the leachate of composted leaves using a fractionation scheme similar to one used here represent different stages in the oxidative degradation of lignin, and that fraction 1 contains the least degraded lignin fragments and fraction 4 the most degraded. Following the same line of reasoning in the present study, fraction 1B' would represent the material that is closest in chemical structure to the parent material from which the other fractions were derived. Further evidence for this hypothesis is provided by the fact that the NMR spectrum of fraction 1B' contains most of the aromatic bands that are present in the other fractions (Table 3). The aromatic NMR bands of fraction 1B' are sharper and better resolved than the aromatic bands of the other fractions (Figure 3), indicating that there is less heterogeneity within each type of aromatic functional group. Partial degradation would increase heterogeneity by producing mixtures of precursor fragments in various stages of degradation. A detailed examination of the NMR spectrum of fraction 1B' therefore should provide important clues about the identity of the major types of aromatic functional groups that are present in the other fractions. Additional information can be obtained from the IR spectrum of fraction 1B' which is also better resolved than the spectra of the other fractions.

Table 3 *Chemical shifts of* ^{13}C *NMR bands between 95 and 200 ppm of Acer Campestre L. leachate fractions*

1	2	3	4	1B'	1B	2B	3B	4B
101[a]			98[a]	98[a]	98[a]	97.2	98[a]	101[a]
104	102[a]	104.2	103.0	103.0			105[a]	
	111.2	111.2		108.9	110.0	111.2		
				114.7				
119.4	119.4			118.2	119.4	119.4	119[a]	
		120.5		121.7				
	127.5	129.9	129.9			128.7		
				133.4	132.2	132.2	131.1	
138.1	138.1	136.9			138.1	138.1		
145.1	143.9	145.1		143.9	143.9	143.9	145.1	
			147.4	149.7				
153[a]				153.3	153.3			
166.1				163.8	166.1			
176.6	171.9	174.3	173.1		176.6	171.9	175.4	174[a]
				180.1				

[a] Broad or poorly-resolved band

The NMR spectrum of fraction 1B' between 95 and 200 ppm contains bands charac-teristic of lignin phenylpropanoid monomeric units and of hydrolysable tannins.[11–13] The pattern of bands that represent the nonphenolic (nonhydroxylated) carbons provides important clues about the structural components of the fraction. The most intense band is at 133 ppm. Carbons 1, 2 and 6 of para-hydroxyphenyl lignin monomers and the C-1 carbons of 3,4-dihydroxy- or 3,4,5-trihydroxy-phenols and phenolic ethers have chemical shifts between 130 and 135 ppm. The fact that the 133 ppm band is the strongest band in the 1B' spectrum suggests that para-hydroxyphenyl structures, which have three carbons with chemical shifts near 133 ppm, are major components in this fraction. Comparison of the integrated area of the 133 ppm band of the dipolar-dephased spectrum of 1B' with that of the CPMAS spectrum indicates that 40 to 50 percent of the carbon atoms represented by the 133 ppm band are not protonated. If all of the structures contributing to this band were phenylpropanoid units, then 33 percent of the 133 ppm carbons would be substituted. The additional 10 to 15 percent probably represents the C-1 carbons of 3,4-dihydroxy- or 3,4,5-trihydroxy-phenols. The probable assignments of the other nonphenolic bands in the spectrum are given in Table 4.

Table 4 *Probable assignments of nonphenolic bands in fraction 1B' NMR spectrum*

Band	Most probable assignment
114.7 ppm	C-3 and C-5 of p-hydroxyphenyls
118.2	C-3 and C-5 of *p*-hydroxyphenyls (C-4 methyl ether); C-6 of guaiacylpropanoids (C-4 phenol)
121.7	C-6 of guaiacylpropanoids (C-4 methyl ether)
133.4	C-1, C-2, C-6 of *p*-hydroxyphenyl monomers; C-1 of 3,4-dihydroxy- or 3,4,5-trihydroxy-phenyls

The increased nonpolar character of fraction 1B' compared with fraction 1B results from the presence of fewer hydroxyl groups in fraction 1B'. The intensity of the 133 ppm band is greatly reduced in fraction 1B compared with fraction 1B', indicating that p-hydroxyphenyl groups that possess only one hydroxyl are in lower concentration in 1B than in 1B'. Fraction 1B is, therefore, composed mainly of phenols with more than one hydroxyl group.

The infrared spectrum (Figure 7) of fraction 1B' also provides evidence of para-substitution, with a prominent band at 1515 cm^{-1}. Colthup *et al.*[14] pointed out that the frequency of the ring stretching vibration of a para-substituted phenyl ring is 1515 cm^{-1}. Gallic acids and gallotannins have a strong infrared band near 1026 cm^{-1}.[15] A strong band in this region is present in the infrared spectra of fractions 1-3 and 1B'-2B (Figure 7). This band provides further evidence for the presence of 3,4,5-trihydroxyphenyl groups in these fractions.

The sharp band at 180 ppm in the 1B' NMR spectrum is displaced from carbonyl bands of all of the other fractions, which are near 175 ppm; we have attributed these bands to carboxylic acids. In addition, the 180 ppm band is sharper than the 175 ppm bands in the other fractions. The carbonyl carbons of flavones generally have chemical shifts near 180 ppm.[6,16] Torssell[17] has pointed out that many flavones have a phloroglucinol A-ring structure (alternating hydroxyl substitution). The chemical shifts of the intervening carbon atoms in the alternating hydroxyl substitution pattern have chemical shifts between 95 and 98 ppm; these bands are absent in the 1B' spectrum. Flavones that do not have the phloroglucinol A-ring do not have the 95 and 98 ppm bands.[16] Therefore, phloroglucinol structures are apparently absent from fraction 1B'. The chemical shifts of the carbonyls of orthoquinones are also near 180 ppm.[6] Orthoquinone structures are readily formed by the oxidation of catechol or galloyl groups, and therefore, orthoquinone structures could be present in these fractions.

The small, relatively broad peak at 164 ppm in the 1B' NMR spectrum most likely represents aromatic esters such as those present in gallotannins. The 1716 cm^{-1} band in the infrared spectrum (Table 5) provides further evidence for this assignment as demonstrated by the work of Hatano *et al.*,[13] who reported carbonyl stretching frequencies between 1710 and 1730 cm^{-1} for a set of gallotannins in which most of the carbonyl groups are associated with ester linkages. Fraction 1B and fraction 1 also have NMR bands near 165 ppm. The carbonyl frequencies of these two fractions are lower than that of 1B', apparently reflecting the presence of more gallic acid groups (carbonyl stretching frequency of 1701 cm^{-1} as measured by Pouchert[15]) in these fractions.

The 165 ppm band is absent from the NMR spectra of fractions 2, 3, and 4, and fractions 2B, 3B, and 4B. In the spectrum of each of these fractions it is replaced by a band between 171 and 176 ppm. The infrared spectra of these fractions provide further insight into the types of groups represented by these bands. Monomeric carboxylic acids (in which carboxylic acid groups are not hydrogen bonded to one to another) absorb between 1740 and 1800 cm^{-1} and dimeric acids (carboxylic acid groups hydrogen bonded one to another) between 1680 and 1720 cm^{-1}.[18] In both instances, aromatic acids occupy the lower frequency ends of the ranges and aliphatic acids the higher frequency ends. Hydrogen bonding of carboxylic acid groups to hydroxyl groups and other hydrogen-bonding groups also reduce the carbonyl stretching frequency. In phenolic compounds such as these it is likely that most of the carboxylate groups are hydrogen bonded either to hydroxyl or other carboxylates. Therefore, the bands near 1720 cm^{-1} most likely represent aliphatic acids and those near 1700 cm^{-1} represent aromatic acids. The bands between 1730 cm^{-1} and 1735 cm^{-1} are probably due to esters, and the 1780 cm^{-1} band to lactones.[19]

Table 5 *Positions of carbonyl bands of carboxylic acids and esters for the different fractions. Shoulders are designated (sh)*

Fraction	Position (cm^{-1})	Fraction	Position (cm^{-1})
		1B'	1715
1	1714	1B	1708
2	1706	2B	1698, 1728 (sh)
3	1720	3B	1722, 1785 (sh)
4	1735	4B	1731, 1780 (sh)

The relative intensities of the 73 ppm bands are greater in fractions 4 and 4B than in any of the other fractions (Figures 2 and 4) indicating that these fractions consist mainly of carbohydrates or carbohydrate derivatives. The spectrum of fraction 4B, which contains only very weak aromatic bands between 110 and 160 ppm, is the best spectrum for the study of carbohydrates in the leachate that are not associated with tannins. Deconvolution of this spectrum allows one to obtain information about types of carbohydrates in the sample. The areas of the deconvoluted bands are given in Table 6. As was pointed out above, the ratio of the area of the anomeric carbon band (101 ppm) to the other carbohydrate bands (58.8, 72.1 and 80.9 ppm) for hexose sugars should be 0.2. In fraction 4B the ratio is 0.13, indicating that about one third of the carbohydrates do not have anomeric carbons. The strong band at 173 ppm suggests that much of the carbohydrate material is present as sugar acids.[6] Aldonic and aldaric acids do not have anomeric carbons, and could therefore account for the carbohydrates without anomeric carbons in the sample. The presence of a broad shoulder centered at about 1780 cm^{-1} in the IR spectrum of fraction 4B provides additional evidence for the presence of sugar acids in the sample. Aldonic and uronic acids commonly form internal esters (lactones). The carbonyl stretching frequency of six-membered-ring lactones is between 1735 and 1750 cm^{-1} and that of five-membered rings between 1760 and 1795 cm^{-1}.[19]

Table 6 *Integrated areas of deconvoluted bands of fraction 4B*

Band (ppm)	Area
19.9	0.070
32.1	0.111
41.8	0.016
58.8	0.054
72.1	0.558
80.9	0.004
101.0	0.080
173.2	0.108

As in our previous study,[2] fraction 6 was a syrup, which required that a liquid-state NMR spectrum be measured. The liquid-state spectra of fractions 6 and 6B are shown in Figure 5. This spectrum is very similar to that of fraction 6 of the overnight-leaching sample of the previous study.[2] Integration of the intensities of the anomeric bands in the NMR spectrum of fraction 6 indicates that approximately half of the sample consists of saccharides and half of sugar alcohols or inositols. In contrast to fraction 6, fraction 6B is a solid. Anomeric carbon lines are practically completely absent from the liquid-state spectrum of fraction 6B (Figure 5); a weak anomeric band is present in the solid state spectrum. The liquid-state spectrum is consistent with that of a mixture of inositols with perhaps some sugar alcohols also present. Inositols are produced during senescence from the phosphoinositides of disintegrating cell membranes.[20,21] The weak anomeric band in the solid-state spectrum probably indicates that there was some saccharide material in the sample that did not dissolve when the liquid-state spectrum was measured.

4 CONCLUSIONS

The unfractionated leachates derived from leaves of different species of trees have distinctive ^{13}C NMR spectral signatures that should be useful in following the diagenesis of plant-derived organic material to dissolved organic carbon species (DOC) in natural water systems. Comparisons of the ^{13}C NMR spectral patterns of the unfractionated leachates with those of the senescent leaves from which they were derived provides evidence of the types of degradation processes that the different leaf constituents have undergone. The appearance in all the leachate spectra of a strong band near 40 ppm probably indicates the presence of substituted 1,4-butanedioic acids arising from ring fission of phenols and β-oxidation of lipids. The reduction in intensity of the 55 ppm band in the leachate spectra indicates a loss of methyl ether groups in lignin-like phenolic structures.

Noncellulosic components such as pectins, starches and hemicelluloses are removed from the leaves during the leaching process. The NMR and FTIR spectra of some of the

Acer campestre L. fractions indicate that some of the carbohydrates in these fractions are conjugated to hydrolysable tannins by ester linkages. It is reasonable to assume that such linkages exist in most leaf leachates that contain hydrolysable tannins.

There is a replication of band-positions in the aromatic and carbonyl regions of the NMR spectra of all of the fractions. This replication corresponds to the replication of phenolic-group substitution patterns in hydrolysable tannins, nonhydrolysable tannins (proanthocyanidins) and lignins. The NMR spectra indicate that some of the fractions are composed of the mixtures of two or more of these categories of components, whereas other fractions consist mainly of one type of component.

References

1. J. B. Wallace, S. L. Eggert, J. L. Meyer and J. R. Webster, *Science*, 1997, **277**, 102.
2. R. L. Wershaw, J. A. Leenheer, K. R. Kennedy and T. I. Noyes, *Soil Sci.,* 1996, **161**, 667.
3. W. Barz and K. -M. Weltring, 'Biosynthesis and Biodegradation of Wood Components', T. Higuchi, (ed.), Academic Press, Orlando, FL, 1985, Chap. 22, p. 607.
4. R. L. Wershaw, K. R. Kennedy and J. E. Henrich, Part II, this volume.
5. T. K. Kirk, *Les Colloques de l'INRA*, 1987, No. 40, p.51.
6. E. Breitmaier and W. Voelter, 'Carbon-13 NMR Spectroscopy', VCH Verlagsgesellschaft, Weinheim, Germany, 1987.
7. J. B. Wooten, *J. Agric. Food Chem.*, 1995, **43**, 2858.
8. M. C. Jarvis, *Carbohydrate Research*, 1990, **201**, 327.
9. M. C. Jarvis and D. C. Apperley, *Plant Physiol.*, 1990, **92**, 61.
10. T. Hatano, T. Yoshida, T. Shingu and T. Okuda, *Chem. Pharm. Bull.*, 1988, **36**, 2925.
11. H. Y. Hassi, M. Aoyama, D. Tai, C.-L. Chen and J. S. Gratzl, *J. Wood Chem. Tech.*, 1987, **7**, 555.
12. P. G. Hatcher, *Org. Geochem.*, 1987, **11**, 31.
13. T. Hatano, K. Yazaki, A. Okonogi and T. Okuda, *Chem. Pharm. Bull.*, 1991, **37**, 1689.
14. N. B. Colthup, L. H. Daly and S. E. Wiberley, 'Introduction to Infrared and Raman Spectroscopy', Academic Press, San Diego, 1990, p. 263.
15. C. J. Pouchert, 'The Aldrich Library of FTIR Spectra', Aldrich Chemical Co. Milwaukee, Wisconsin, 1985, Vol. 2, p. 263.
16. D. W. Aksnes, A. Standnes and Ø. M. Anderson, *Mag. Reson. Chem.*, 1996, **34**, 820.
17. K. B. G. Torssell, 'Natural Product Chemistry : A Mechanistic and Biosynthetic Approach to Secondary Metabolism', Wiley, Chichester, England, 1983.
18. N. B. Colthup, L. H. Daly and S. E. Wiberley, 'Introduction to Infrared and Raman Spectroscopy', Academic Press, San Diego, 1990, p. 314.
19. N. B. Colthup, L. H. Daly and S. E. Wiberley, 'Introduction to Infrared and Raman Spectroscopy', Academic Press, San Diego, 1990, pp. 289-319.
20. J. Bücher and R. Guderian, *J. Plant Physiol.*, 1994, **144**, 121.
21. R. C. Fialho and J. Bücher, *Can. J. Botany*, 1996, **74**, 965.

USE OF ^{13}C NMR AND FTIR FOR ELUCIDATION OF DEGRADATION PATHWAYS DURING NATURAL LITTER DECOMPOSITION AND COMPOSTING. IV. CHARACTERIZATION OF HUMIC AND FULVIC ACIDS EXTRACTED FROM SENESCENT LEAVES

Robert L. Wershaw and Kay R. Kennedy

U.S. Geological Survey, Denver Federal Center, Denver, CO 80225

1 INTRODUCTION

The composition and origin of soil organic matter has been the subject of considerable controversy and confusion. Most workers have attempted to make a distinction between the humified components of soil organic matter that have been called humus, or humic substances, and the nonhumified components of soil organic matter. Unfortunately, however, there are no universally accepted definitions of humus or humic substances. Stevenson,[1] for example, defines humus as "Total of the organic compounds in soil exclusive of undecayed plant and animal tissues, their 'partial decomposition' products, and the soil biomass". Included within the humus category are humic substances, which Stevenson[1] defines as "A series of relatively high-molecular weight, yellow to black colored substances formed by secondary synthesis reactions." In contrast to Stevenson,[1] Tate[2] includes within his definition of humic substances both plant degradation products and compounds formed by "*de novo* synthesis."

In practice, the distinction that Stevenson[1] makes between "secondary synthesis products" and plant "partial decomposition" products is totally unworkable. The commonly used procedures for the isolation and fractionation of soil humic substances are based entirely on solubility in aqueous solutions of different pH and not on origin of the compounds isolated. Thus, soil humic substances are normally fractionated into three components: humic acid, fulvic acid and humin, which are operationally defined by their solubility in acidic and basic solutions. In the most commonly used extraction procedure, humic and fulvic acids are extracted from a soil or a sediment with 0.1 to 0.5 N sodium hydroxide solution. The unextracted fraction of the organic matter that is insoluble in the sodium hydroxide is called humin. The fraction of the extract that precipitates when the pH is lowered to 1 or 2 is called humic acid, and the fraction that remains in solution is called fulvic acid.

Recently, a number of workers have proposed that humic substances result from the partial degradation of plant polymers.[3-5] Studies of soil isolates of different sizes indicate that plant degradation products are stabilized by association with soil minerals.[6] The work of Skjemstad[7] and Baldock *et al.*[3] suggests that there are at least two different types of mineral-organic matter associations in soils. In the clay-size (< 2 mm) soil isolates it appears that the humic substances form coatings on the mineral surfaces, whereas in the coarser size fractions (> 20 mm and 2–20 mm) fragments of plant tissue are stabilized by encapsulation in aggregates of mineral grains. Baldock *et al.*[3] have measured the solid-state ^{13}C NMR spectra of different particle sizes isolated from five mineral soils. They have proposed "...that the extent of decomposition of organic materials in soils follows a continuum from fresh plant residues in the large particle size fractions (> 20 mm diameter) through partially degraded residues in intermediate fractions (2–20 mm diameter) to degraded residues in the finest fractions (< 2 mm diameter)." The NMR spectra of the >20 mm fractions isolated from some of the soils are very similar in general appearance to the spectra of senescent leaves,[8] while the NMR spectra of the 2–20 mm fractions show depletion of carbohydrate material compared with the leaves.

It will be demonstrated here that the plant fragments in soils, and in particular leaf fragments, constitute a possible source of humic and fulvic acids in soils. We have obtained organic isolates which meet criteria for humic and fulvic acids from freshly fallen senescent leaves that had previously been extracted with distilled water to simulate leaching by rainwater and soilwater. Comparison of the solid-state ^{13}C NMR spectra of these isolates with the published spectra of humic and fulvic acids isolated from soils provide important clues about the diagenesis of humic substances in soils.

2 MATERIALS AND METHODS

The leaves used in this study were senescent *Acer campestre L.* leaves that had been collected and extracted twice with distilled water as described previously.[8] An 82.2 g sample of the extracted leaves was placed in a polyethylene bottle, and 1900 mL of 0.1 N NaOH that had been purged with nitrogen for 1 hour to remove oxygen was added to the leaves. The headspace above the NaOH solution was displaced with nitrogen, and the bottle was tightly capped. At the end of 20 hours the solution was decanted from the extracted leaves and immediately neutralized to pH 7 with HCl. The solution was then filtered through a glass wool plug, and the pH was lowered to 1. The solution was refrigerated overnight, and the precipitated humic acid was isolated by centrifugation. The humic acid was washed twice with distilled water and freeze-dried.

The fulvic acid fraction was isolated from the supernatant by sorption on XAD-8[*]. The supernatant (pH adjusted to 2) was divided into three aliquots of approximately 600 mL each, and each aliquot was pumped onto a 147 mL XAD-8 column. The column was then washed with 375 mL of water, and then eluted with 250 mL of acetonitrile–water solution (3:1 v/v). The eluates from the three aliquots were combined, and the acetonitrile–water was removed in a rotary evaporator. The fulvic acid was dissolved in water and freeze-dried.

Solid-state cross-polarization magic-angle-spinning (CPMAS) ^{13}C NMR spectra were measured on a 200 MHz Chemagnetics CMX spectrometer with a 7.5 mm-diameter probe. The spinning rate was 5000 Hz. The acquisition parameters were contact time of 1 ms, pulse delay of 1 s and a pulse width of 4.5 microseconds for the 90° pulse. A line broadening of 25 Hz was applied in the fourier transformation of the free induction decay data.

---

[*] The use of trademarks in this report is for identification purposes only and does not constitute endorsement by the U.S. Geological Survey.

3 RESULTS AND DISCUSSION

The solid-state ^{13}C NMR spectra of the fulvic and humic acids extracted from senescent *Acer campestre L.* leaves are shown in Figure 1. These spectra consist of a number of sharp, well-defined bands in the aromatic region (100–160 ppm). These bands are in the same positions as the aromatic bands observed in the water extracts of the leaves; they represent different phenolic carbon atoms as discussed in Wershaw *et al.*[9] The spectral band pattern of the fulvic acid is very similar to that of fraction 1 (Figure 2) of the water extract with two exceptions; the carbonyl band is shifted from 166 ppm to 170 ppm in the fulvic acid, and the relative intensity of the carbohydrate band at 74 ppm is greater in fraction 1 than in the fulvic acid. The humic acid band pattern does not match the band patterns of any of the water extract fractions.

Wershaw *et al.*[9] concluded that fraction 1 is composed mainly of gallotannins conjugated to carbohydrates. The shift in the carbonyl band to 170 ppm in the fulvic acid most likely indicates hydrolysis of many of the ester linkages to acids. Further evidence of this hydrolysis is provided by the lower carbohydrate concentration in the fulvic acid. Carbohydrates are normally bound to gallotannins by ester linkages.

The infrared spectra of the fulvic and humic acids (Figure 3) provide further evidence for the presence of gallotannin groups in these isolates. The carboxylic-acid carbonyl stretching band is at 1702 cm^{-1} in the fulvic acid IR spectrum; the humic acid spectrum has three overlapping carbonyl bands centered at about 1711 cm^{-1}. The fulvic acid carbonyl band is most likely indicative of aromatic acids; the humic acid bands probably represent both aromatic and aliphatic acids. Pouchert[10] showed that the carbonyl stretching frequency of gallic acid groups is 1701 cm^{-1}. Additional evidence for gallo-tannin structures is provided by the prominent bands near 1515 cm^{-1} in both spectra; these bands are evidence of para-substituted phenyl rings such as those in gallotannins.[11] The strong bands at 1196 and 1078 cm^{-1} in the fulvic acid IR spectrum represent C-O stretching modes of carbohydrates. The 955 and 792 cm^{-1} bands are probably also due to carbohydrates.[11] Aliphatic structures in the humic acid are represented by the methyl and methylene stretching bands between 2970 and 2830 cm^{-1}.

Deconvolution of the humic and fulvic acid NMR spectra into their constituent bands provides additional clues about the composition of these two isolates. The positions and relative areas of the bands (Table 1) may be used to infer the types of functional groups present in the isolates. The bands in the aliphatic region at 17, 29 and 39 ppm correspond to terminal methyl groups, methylene groups in long chains, and methylene groups near points of chain branching, respectively.[12] Bands in the region between 40 and 50 ppm have been assigned by Wooten[13] in the NMR spectra of tobacco leaves to malate and citrate groups and some amino acids. The higher concentration of long-chain methylene groups (29 ppm band) in the humic acid than in the fulvic acid most likely is indicative of higher concentrations of lipid- and cutin-derived material in the humic acid than in the fulvic acid. Cutin that forms the outer coating of leaves is a polyester polymer composed of C_{16} and C_{18} monomeric units.[14]

The methoxyl band at 56 ppm is diagnostic of lignin structures. The integrated area of the methoxyl band is greater in the humic acid spectrum than in the fulvic acid spectrum, indicating a higher lignin concentration in the humic acid. Further evidence of this is provided by the larger integrated area of the 154 ppm band in the humic acid. This band most likely represents the C_3 and C_5 carbon atoms of syringylpropanoid units of

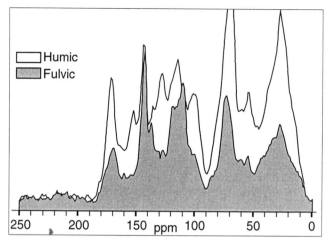

Figure 1 *¹³C NMR spectra of humic and fulvic acids extracted from Acer campestre L. leaves.*

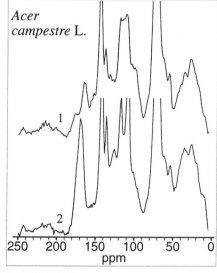

Figure 2 *¹³C NMR spectra of fractions 1 and 2 of water extract from Acer campestre L. leaves.*

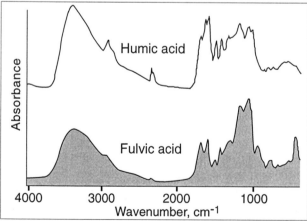

Figure 3 *Infrared spectra of humic and fulvic acids extracted from Acer campestre L. leaves*

lignin structures in which the C_4 carbon is methoxylated.[8] If C_4 is not methoxylated then the chemical shift of the C_3 and C_5 atoms is 147 ppm.

The bands between 62 and 83 ppm indicate that saccharide structures are present in the fulvic and humic acids. In addition to the 62 and 83 ppm bands, saccharides have anomeric carbon bands near 100 ppm. The nonsubstituted carbons of phloroglucinol structures, such as those in polyflavonoid tannins, also have bands in the 100 ppm region.[8] The integrated areas of the sugar band allows one to determine whether the bands near 100 ppm in the fulvic and humic acid spectra are due only to saccharides or if aromatic structures also contribute to this band. In hexose sugars the area of the anomeric carbon band should be one-fifth (0.2) of the sum of the areas of the other sugar bands. In the fulvic acid the ratio of the 101–102 ppm band to the combined 62–83 ppm bands is 0.17; however, in the humic acid, this ratio is 0.35, indicating that about 40% of the 101–102 ppm band represents nonanomeric carbon atoms. These nonanomeric carbons are most likely in polyflavonoid structures.[15]

Table 1 *Integrated areas of deconvoluted bands of fulvic and humic acid spectra*

Band	Integrated area of band (% of total area)	
	Fulvic acid	Humic acid
17–18 ppm	2.7	2.6
29	11.9	18.2
39–40	10.2	5.4
50		2.1
56	1.6	4.2
62	1.7	1.1
71	6.5	11.3
76–77	10.2	6.5
83		1.0
101–102	3.2	6.9
111–112	13.7	4.2
116		4.1
120–121	9.2	4.1
130–132	5.1	7.7
137–138	2.4	1.6
144–145	11.9	6.1
154–155	1.3	4.2
161–164	0.4	1.7
170–173	7.0	7.1
177	1.1	

As pointed out above, the position of the carbonyl band in the fulvic acid sample indicates that the NaOH extraction has hydrolysed some ester linkages. Hydrolysis of lignin units linked to hemicelluloses by ester bonds could result in solubilization of the lignin units.[16] In addition, certain types of α-aryl and β-aryl linkages in lignin structures are susceptible to alkaline hydrolytic cleavage.[17] The presence of lignin fragments in the humic and fulvic acids indicates that the NaOH treatment results in partial depolymerization of lignin structures in the leaves and release of molecular fragments that would not otherwise be susceptible to solubilization. Polyphenolic structures other than lignin may also require NaOH extraction for release. For example, Hillis[18] has pointed out that some tannins can only be extracted with bases.

The results of this study indicate that partially degraded plant leaf fragments, which are plentiful in many soils,[19] can serve as sources for fulvic and humic acids extracted from soils. Fragments of other organs such as roots and stems should also serve as sources of fulvic and humic acids. These fragments are subjected to intermittent leaching by soil water solutions; however, as we have demonstrated above, basic solutions can bring about depolymerization and release of molecular structures that are not susceptible to solution in soil waters.

Published soil humic and fulvic acid solid-state ^{13}C NMR spectra generally consist of much less well-resolved bands in the aromatic region than the spectra in Figure 1. The most prominent aromatic bands in these spectra are between 125 to 130 ppm.[20,21] Bands in this region most likely represent unsubstituted carbon atoms in alkylsubstituted phenyl structures,[12] unsubstituted carbon atoms in the 1-, 2- and 6-positions of para-hydroxyphenyl structures,[22] or unsubstituted carbons in benzene monocarboxylic acid structures. Skjemstad *et al.*[23] have assigned bands centered at 128 ppm in the ^{13}C NMR spectra of organic isolates from an Australian podzolic soil to substituted 1,2-benzenedicarboxylic acid structures. However, these types of structures generally do not have bands near 128 ppm.[12] In some of the published humic acid spectra[5] bands between 140 and 160 ppm are also present. These bands most likely represent di- and tri-hydroxyphenyl structures.[8,9]

A possible reason for the differences between the spectra of leaf-derived fulvic and humic acids and those of soil fulvic and humic acids is that the soil humic substances are derived from more than one source. In addition to extracting humic substances from plant fragments, the basic extractants that are used to remove humic substances from soils will also strip organic coatings from the mineral surfaces. These coatings are formed by adsorption of organic acids from soil waters onto mineral surfaces.[24] Humic substances that have precipitated during evaporation of soil solution will also be dissolved by basic extractants. Soil fulvic and humic acids, therefore, are mixtures of components derived from several different sources. Partially degraded plant tissue constitutes the least altered source of the humic substances, and therefore, the basic extracts of this source will be less oxidized than the extracts from mineral coatings and precipitates.

The aerobic microbial oxidation of aromatic compounds generally proceeds by ring fission to produce low molecular weight acids that can be utilized in catabolism. Fission of aromatic rings is brought about by mono-and di-oxygenases.[5,25] Mono-oxygenases add a second hydroxyl group to a monohydroxyphenol to produce a dihydroxyphenol. Dihydroxyphenols are cleaved by di-oxygenases to produce carboxylic acids and aldehydes. In aerobic environments, one would expect, therefore, that nonsubstituted or alkyl-substituted aromatic structures would be more stable than monophenols, which in turn would be more stable than dihydroxy and trihydroxyphenols. The structures mentioned above that are more resistant to oxidation generally have more carbon atoms with NMR bands in the 125–130 ppm region than those that are less resistant to oxidation. Therefore, strong bands in the 125–130 ppm region of soil fulvic and humic acids indicate preferential preservation of resistant aromatic structures in material that has undergone oxidation.

4 CONCLUSIONS

The fulvic acid from senescent *Acer campestre L.* leaves is composed mainly of gallotannins in which some of the carboxylate groups are conjugated to carbohydrates. The carbonyl band at 170 ppm in the fulvic acid indicates, however, that most of the carboxyl groups are present as acids. These carboxylic acid groups probably arise from hydrolysis of ester bonds linking the gallotannin groups to carbohydrate moieties. The partial hydrolysis of the ester linkages brought about by the sodium hydroxide extraction was one of the factors that led to the release of the fulvic acid isolate from the leaves.

The ^{13}C NMR spectra of the humic acid isolated from senescent *Acer campestre L.* leaves indicate that much of this isolate consists of lignin units and polyflavonoid structures linked to carbohydrates. Long-chain polymethylene structures derived from lipids and cutins are also important components of the humic acid. The lignin fragments in the humic acid, and to a lesser extent in the fulvic acid, were probably solubilized by partial depolymerization of α-aryl and β-aryl linkages brought about by the sodium hydroxide extractant.

The differences between the NMR spectra of soil fulvic and humic acids and those of the leaf-derived fulvic and humic acids provide insight into the diagenesis of organic matter in soils. Published ^{13}C NMR spectra of soil humic and fulvic acids generally consist of much less well-resolved bands in the aromatic region than the spectra in Figure 1. The most prominent aromatic bands in these spectra are between 125 to 130 ppm. These bands are indicative of the preferential preservation of resistant aromatic structures such as unsubstituted phenyl groups, monophenols, and benzene monocarboxylic acid structures in soil fulvic and humic acids. The less well-resolved bands in the NMR spectra of soil fulvic and humic acids compared with those of the leaf fulvic and humic acids indicate that the soil extracts are more complex than the leaf extracts because they are derived from plant material, precipitated organic matter, and coatings on mineral grains.

References

1. F. J. Stevenson, 'Humus Chemistry: Genesis, Composition, Reactions', 2nd Edn., Wiley, New York, 1994, p. 33.
2. R. L. Tate III, 'Soil Organic Matter: Biological and Ecological Effects', Wiley, New York, 1987.
3. J. A. Baldock, J. M. Oades, A. G. Waters, X. Peng, A. M. Vassallo and M. A. Wilson, *Biogeochem.*, 1992, **16**, 1.
4. P. G. Hatcher and E. C. Spiker, in 'Humic Substances and Their Role in The Environment', F. H. Frimmel and R. F. Christman, (eds.), Wiley, Chichester, 1988, p. 59.
5. R. L. Wershaw, 'Membrane-Micelle Model for Humus in Soils and Sediments and Its Relation to Humification', U.S. Geological Survey Water-Supply Paper 2410, 1994.
6. J. M. Oades, 'Environmental Impact of Soil Component Intertactions: Natural and Anthropogenic Organics', P. M. Huang, J. Berthelin, J.-M. Bollag, W. B. McGill and A. L. Page, (eds.), Lewis Publishers, Boca Raton, FL, 1995, p. 119.

7.   J. O. Skjemstad, L. J. Janik, M. J. Head and S. G. McClure, *J. Soil Sci.*, 1993, **44**, 485.
8.   R. L. Wershaw, K. R. Kennedy and J. E. Henrich, Part II, this volume.
9.   R. L. Wershaw, J. A. Leenheer and K. R. Kennedy, Part III, this volume.
10.   C. J. Pouchert, 'The Aldrich library of FT-IR spectra', Aldrich Chemical Co. Milwaukee, Wisconsin, 1985, Vol. 2, p. 232.
11.   N. B. Colthup, L. H. Daly and S. E. Wiberley, 'Introduction to Infrared and Raman Spectroscopy', Academic Press, San Diego, 1990.
12.   E. Breitmaier and W. Voelter, 'Carbon-13 NMR spectroscopy', VCH Verlagsgesellschaft, Weinheim, Germany, 1987.
13.   J. B. Wooten, *J. Agric. Food Chem.*, 1995, **43**, 2858.
14.   P. E. Kolattukudy and K. E. Espelie, in 'Biosynthesis and Biodegradation of Wood Components', T. Higuchi, (ed.), Academic Press, Orlando, FL, 1985, p. 162.
15.   L. Y. Foo, *Phytochem.*, 1987, **26**, 2825.
16.   M. L. Fidalgo, M. C. Terrón, A. T. Martínez, A. E. González, F. J. González-Vila and G. C. Galletti, *J. Agric. Food Chem.*, 1993, **41**, 1621.
17.   R. W. Thring, *Biomass and Bioenergy*, 1994, **7**, 125.
18.   W. E. Hillis, in 'Biosynthesis and Biodegradation of Wood Components', T. Higuchi, (ed.), Academic Press, Orlando, FL, 1985, p. 209.
19.   E. A. Fitzpatrick, 'Soil Microscopy and Micromorphology', Wiley, Chichester, 1993.
20.   R. L. Malcolm, in 'Humic Substances in Soil and Crop Sciences: Selected Readings', P. MacCarthy, C. E. Clapp, R. L. Malcolm and P. R. Bloom., (eds.), American Society of Agronomy, Madison, Wisconsin, 1990, p. 13.
21.   M. Schnitzer, in 'Humic Substances in Soil and Crop Sciences: Selected Readings', P. MacCarthy, C. E. Clapp, R. L. Malcolm and P. R. Bloom, (eds.), American Society of Agronomy, Madison, Wisconsin, 1990, p. 65.
22.   P. G. Hatcher, *Org. Geochem.*, 1987, **11**, 31.
23.   J. O. Skjemstad, A. G. Waters, J. V. Hanna and J. M. Oades., *Aust. J. Soil Res.*, 1992, **30**, 667.
24.   R. L. Wershaw, E. C. Llaguno and J. A. Leenheer, *Colloids and Surfaces A*, 1996, **96**, 213.
25.   W. Barz and K.-M. Weltring, in 'Biosynthesis and Biodegradation of Wood Components', T. Higuchi, (ed.), Academic Press, Orlando, FL, 1985, p. 607.

CHARACTERIZATION AND PROPERTIES OF HUMIC SUBSTANCES ORIGINATING FROM AN ACTIVATED SLUDGE WASTEWATER TREATMENT PLANT

Benny Chefetz,[1] Jorge Tarchitzky,[1] Naama Benny,[1] Patrick G. Hatcher,[2] Jacqueline Bortiatynski[3] and Yona Chen[1]*

[1] Department of Soil and Water Sciences, Faculty of Agricultural, Food and Environmental Quality Sciences, The Hebrew University of Jerusalem, Rehovot 76100, Israel
[2] Department of Chemistry, The Ohio State University, Columbus, OH 43210, USA
[3] Department of Chemistry, The Pennsylvania State University, University Park, PA 16802, USA

1 INTRODUCTION

The use of treated sewage water for crop irrigation as an alternative for effluent water disposal is common in water-scarce countries worldwide. In arid and semi-arid regions, the use of treated water for agriculture means that a larger amount of fresh water is available for domestic use. One of the major concerns of treated effluent water use is its organic matter (OM) load. The total C level (expressed as chemical oxygen demand, COD) in raw sewage and in sewage effluents after the initial and secondary treatments have been reported to be 250–1000, 150–750 and 30–160 mg L^{-1}, respectively.[1] The common sewage treatment methodology employed in Israel is the activated sludge process. In this process, wastewater is fed continuously into an aerated tank where microorganisms metabolize and biologically flocculate the OM. Microorganisms and OM (*i.e.* activated sludge) settle on the bottom under quiescent conditions in the final clarifier and are then returned to the aeration tank. Clear supernatant from the final settling tank is the plant's effluent. Excessive activated sludge is removed from the system to maintain the proper C, N and energy supply to the microorganisms.[2]

Although the use of wastewater is increasing, little research has been performed on the properties of the OM and of the humic substances (HSs) present in this water. Sachedev *et al.*[3] reported that 58–62% of the dissolved organic carbon (DOC) present in wastewater had a molecular weight (M$_W$) < 0.7 kDa whereas the M$_W$ > 5 kDa fraction content was only 20–29%. Ishiwatari *et al.*[4] reported that 40% of the DOC originating from sewage-contaminated river waters was adsorbed by a hydrophobic resin. Only 10% of the adsorbed compounds exhibited M$_W$ > 10 kDa and were described as HSs. Rebhun and Manka[5] defined 42% of the DOC present in treated wastewater as HSs, 1.7% as tannins, 22.4% as proteins, 11.5% as carbohydrates, 13.9% as anionic detergents and 8.3% as ether extractable. Comparison of HSs extracted from stream water, wastewater and soil indicated that the water-originated compounds were similar to each other and that both were easily distinguishable from soil HSs. The water-derived humic acids (HAs) exhibited higher H/C and C/N ratios (*i.e.* higher aliphaticity and biodegradability, respectively) than the HAs isolated from soil. In addition, the water-derived HAs exhibited amide bonds which did not appear in the soil HAs' FTIR spectra.[6]

As compared with Leonardite HA, sludge HAs had lower C/H, O/H and C/N ratios, increased N and H content, decreased total acidity, carboxylic acid and phenolic hydroxyl content and lower E_4/E_6 ratios. The sludge HAs appeared to be more aliphatic, N-rich polymers than the Leonardite HA. The type of digestion process (aerobic or anaerobic) that the sludge underwent did not influence the properties of sludge HA to any great extent.[7]

Israel's climate is semi-arid to arid and its water resources are limited. On the other hand, with a steady population growth and increase in industrial activity there has been a corresponding increase in the demand for fresh water. Other semi-arid regions (*e.g.* in the Middle East, North Africa and California) face similar problems. In order to overcome water shortage, treated wastewater is being used in agriculture (63% of the treated wastewater produced in Israel in 1994 was used in agriculture). Another product of wastewater treatment facilities is sewage sludge, which is generated at a rate of 120,000 ton per year in Israel.

Sewage sludge (SS) poses a major environmental problem worldwide because ocean disposal is becoming illegal. Therefore, utilization of municipal and animal wastes in agriculture is an important alternative. The continuous use of treated wastewater in irrigation and the application of high levels of SS also has potentially adverse effects on soils and it can change the quality and quantity of the soil OM. Thus, the objective of this study was to characterize the HSs present in treated wastewater and in the SS from an activated sludge system.

2 MATERIALS AND METHODS

2.1 Treated Sewage Water and Sludge

Treated sewage water and air-dried, anaerobically digested SS were sampled from the aerobic sewage-treatment facility in Netanya, Israel. The treated sewage water was filtered through a 0.45 μm membrane filter (Supor-450, Gelman Sciences). The concentration of dissolved organic carbon (DOC) was measured immediately after sampling or during the HoA extraction using a total carbon monitor (TCM 480, Carlo Erba Instruments).

2.1.1 Hydrophobic Acid (HoA) Extraction. We decided to focus our study on the HoA fraction in treated sewage water because this fraction comprises about 65% of the DOC. In step 1, 200 L of the filtered sewage was acidified to pH 2.0 with 6 M HCl and then pumped with a peristaltic pump at a flow rate of 10–15 pore volumes per hour through a glass column containing 350 mL of Amberlite XAD-8 resin (Sigma). Following the sample, 2 L of distilled water were passed through the column. In step 2, the HoA fraction was displaced from the XAD-8 resin with a 0.5 pore volume of 0.1 M NaOH followed by 2 pore volumes of distilled water. In step 3, the HoA was desalted by passing it through a protonated cation exchanger (Amberlite 15, Sigma) glass column. The HoA fraction was freeze-dried for further analysis.[8,9]

2.2 Humic Acid (HA) and Fulvic Acid (FA) Extraction

For HA and FA studies, 20 g of anaerobic digested sludge was extracted with 200 mL 0.1 M NaOH under N_2. The supernatant was collected following centrifugation at 15,000 g for 30 min and the residue was resuspended in 0.1 M NaOH. This procedure was repeated 3 times. The combined supernatants were acidified with 6 M HCl to pH~1 and left at room temperature for 24 h. The supernatant containing the extracted fulvic fraction (SSFA) was separated from the humic acid (SSHA) by centrifugation (2500 g for 30 min). The SSFA was further purified by adsorption to Amberlite XAD-8 resin with the method used to purify the HoA fraction. Suspended particles were removed from the SSHA sample by dissolving it in a minimum volume of 0.1 M KOH/0.3 M KCl under N_2. After 4 h of shaking, the suspended solids were removed by centrifugation. The clear HA solution was acidified to pH 1 and the HA allowed to flocculate. The HA solid was separated by centrifugation and shaken repeatedly for 24 h at room temperature with 1 L of dilute HCl + HF (5 mL concentrated HCl + 5 mL 52% HF + 990 mL distilled water) solution until the ash content was < 1.0%. Following this treatment, the HA was dialysed against distilled water until the dialysate was free of chloride and exhibited a neutral pH, and it was then freeze-dried to give solid SSHA.

2.3 Elemental and Functional Group Analysis

Elemental analyses of SSHA, SSFA and HoA were performed on triplicate samples using an EA 1108 Elemental Analyser (Fisons Instruments). Functional group analysis was based on acid-base titration and calculation of total acidity corresponding to phenolic and carboxylic functional groups as described by Inbar *et al.*[10]

2.4 Fourier Transform Infrared (FTIR) Spectroscopy

The FTIR spectra of SSHA, SSFA and HoA samples were obtained over the wavenumber range 4000 to 400 cm^{-1} with a Nicolet 550 Magna-IR^{TM} spectrometer. Samples of SSHA, SSFA and HoA were oven-dried at 65 °C for 48 h and finely ground prior to analysis. KBr disks were prepared by mixing 98 mg of KBr with 2 mg of the respective solid. Forty scans were collected to obtain FTIR spectra. A linear baseline correction was applied using 4000 cm^{-1}, 2000 cm^{-1} and 860 cm^{-1} as zero absorbance points. The major-peak data (intensity and wavenumber) were found with OMNIC® software (Nicolet Instruments).

2.5 ^{13}C-Nuclear Magnetic Resonance (NMR) Spectroscopy

Solid-state ^{13}C-NMR spectra with CPMAS (cross polarization magic angle spinning) were obtained with a Chemagnetics M-100 NMR spectrometer operated at a ^{1}H frequency of 100 MHz and a ^{13}C frequency of 25 MHz. Pertinent experimental parameters were contact time 1 ms, recycle delay time 0.8 s, sweep width 14 kHz (562.5 ppm) and line broadening 30 Hz. Dried solid SSHA, SSFA and HoA samples (48 h at 65 °C) were placed in a rotor and spun at a frequency of 3.5 kHz at the magic angle (54.7° to the magnetic field).

The ^{13}C-NMR spectra were assigned to five regions:[11] region I (0–50 ppm) to aliphatic carbon or carbons bound only to other carbons; region II (50–112 ppm) to C-O and C-N bonds as in carbohydrates, alcohols, esters and amines; region III (112–163 ppm) to aromatic and phenolic carbons; region IV (163–190 ppm) to carboxyl, ester and amide carbon groups; and region V (190–215 ppm) to the carbonyl carbons. The 0–112 ppm region was calculated as aliphatic carbon, 60–112 ppm as polysaccharide carbon and 112–163 ppm as aromatic carbon. Total aromaticity was calculated by expressing aromatic C (112–163 ppm) as a percentage of the aliphatic C (0–112 ppm) plus aromatic C.[12]

3 RESULTS

3.1 Hydrophobic Acid (HoA) Extraction

The DOC concentration of the treated sewage varied between 17 and 25 mg L^{-1}. After acidification and loading the treated effluent water onto a XAD-8 column, the outlet DOC concentration dropped to 8–10 mg L^{-1}. The hydrophobic fractions that adsorb on the XAD-8 resin (*i.e.* hydrophobic acid and hydrophobic neutrals),[8] represent 65% of the total DOC. The hydrophilic fractions that were not adsorbed to the XAD-8 resin (such as low-molecular weight polysaccharides, carbohydrates, proteins, peptides and amino acids) were mostly degraded during the sewage treatment, resulting in increasing relative concentrations of aromatic and aliphatic compounds.

3.2 Solid-state NMR

CPMAS ^{13}C-NMR spectra of the SSHA, SSFA and HoA samples are presented in Figure 1. The major peaks were at 20–25 ppm (methylene C from proteins, methyl C of microbial deoxysugars, or other methyl C), 30 ppm (methylene C in aliphatic polymethylenic components), 40 ppm (methine C), 56 ppm (methoxyl C or N-substituted aliphatic C), 72 ppm (carbohydrates or aliphatic alcohols), 105 ppm (anomeric C of polysaccharides), 115 ppm (protonated aromatic C), 130 ppm (C-substituted aromatic C), 145–150 ppm (O-substituted aromatic C) and 175 ppm (carboxyl C).[13-15]

Distinct differences were observed between the SSHA, SSFA and HoA spectra as follows: (*i*) the aromatic region peaks (130 and 150 ppm) were intense in the SSFA and HoA spectra, whereas in the SSHA spectrum the O-substituted aromatic C (150 ppm) peak was extremely small and the aromatic peak was less intense; (*ii*) the 105 ppm peak appeared only in the SSFA spectrum and the 72 ppm carbohydrate peak was most prominent in the SSFA spectrum; (*iii*) the methoxy peak was absent in the HoA spectrum; and (*iv*) the SSHA spectra exhibited predominantly aliphatic C containing groups compared with the SSFA and HoA. The 41 ppm peak exhibited in the HoA spectrum has been attributed to aliphatic acid structures. The HoA spectrum exhibited large peaks at about 118 ppm that are possibly due to quinones and hydroxy quinone structures. The dominant peaks at 35 and 27 ppm shown in the SSHA spectrum represent methylene C of branched alkyl chains.[16]

The relative changes in the C containing group levels are presented in Table 1. The following changes should be noted: (*i*) alkyl C represented 47% of the total C in the SSHA sample *vs.* only 30 and 36% in the SSFA and HoA samples, respectively; (*ii*) the C-containing groups of polysaccharide alcohols and esters (50–112 ppm) represented 29% of the total C in the SSFA spectrum compared with only 20% in the SSHA and 16% in the HoA spectra; (*iii*) the aromatic components, including aromatic and phenolic C, represented only 15 and 19% in the SSHA and SSFA, respectively compared with 34% of the total C in the HoA spectrum.

The calculated aromaticity of the HoA sample was 39% but only 18 and 24% for the SSHA and SSFA samples, respectively. In contrast, the calculated aliphaticity was 81, 75 and 60% for the SSHA, SSFA and HoA, respectively.

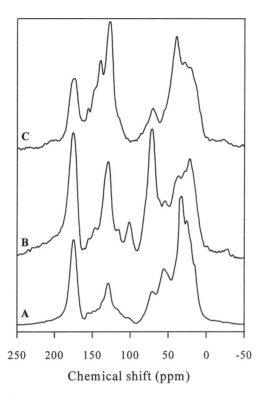

Figure 1 *CPMAS ^{13}C-NMR spectra of (a) sewage sludge humic acid (SSHA); (b) sewage sludge fulvic acid (SSFA); and (c) hydrophobic acid fraction (HoA) obtained from treated sewage water.*

Table 1 *Distribution of C-containing groups as determined by CPMAS ^{13}C-NMR spectra (percent of total carbon).*

Sample		SSHA	SSFA	HoA
Chemical shift (ppm)	C-containing group		% of total carbon	
0–50	Alkyl	47	30	36
50–112	O-Alkyl	20	29	16
112–145	Aromatic	12	16	26
145–163	Phenolic	3	3	8
163–190	Carboxyl	17	17	12
Aromaticity		18	24	39
Aliphaticity		81	75	60

3.3 FTIR Spectra

Fourier transformed infrared (FTIR) spectra of the SSHA, SSFA and HoA samples are shown in Figure 2. The main peaks observed are a broad band at 3300–3400 cm^{-1} (H-bonds, OH groups), two distinct peaks at 2930 and 2850 cm^{-1} (C-H asymmetric, C-H stretch of –CH$_2$), a sharp peak at around 1716 cm^{-1} (C=O of COOH), a peak at 1650 cm^{-1}

(C=C in aromatic structures, COO⁻, H-bonded C=O), a peak at 1540 cm⁻¹ (amide II bonds), a peak at 1525-1518 cm⁻¹ (C=C of aromatic rings), a peak at 1450 cm⁻¹ (C-H deformation of CH_2 or CH_3 groups), a peak at 1400 cm⁻¹ (CH_2, COO⁻), a small peak at 1375 cm⁻¹ (CH_3, COO⁻), a pronounced peak at 1221 cm⁻¹ (aromatic C, C-O stretch), a peak at 1189 cm⁻¹ (C-O stretch of aliphatic OH), a peak at 1126 cm⁻¹ (-C-O stretch and OH deformation of –COOH) and two peaks at 1080 and 1045 cm⁻¹ (C-O stretch of polysaccharides).[10,17–19]

In accordance with the NMR spectra, the SSHA exhibited FTIR evidence for high levels of aliphatic C structures. This spectrum differs from those of the SSFA and HoA. The former exhibited two sharp peaks of the –CH_2– bonds while these peaks were much less intense in SSFA and HoA. In addition, the carboxylic peak at 1720 cm⁻¹ did not appear in the SSHA spectrum, suggesting the presence of metal carboxylate complexes. All spectra showed an amide peak at 1540 cm⁻¹; its intensity was greatest in the SSHA spectrum, intermediate in the SSFA spectrum and lowest in the HoA spectrum. The HoA spectrum exhibited a sharp peak at 1400 cm⁻¹, possibly reflecting the presence of carboxylate ions, whereas this peak is very weak in the SSHA and the SSFA spectra. The polysaccharides peaks at 1000–1100 cm⁻¹ are clearly exhibited in all spectra.

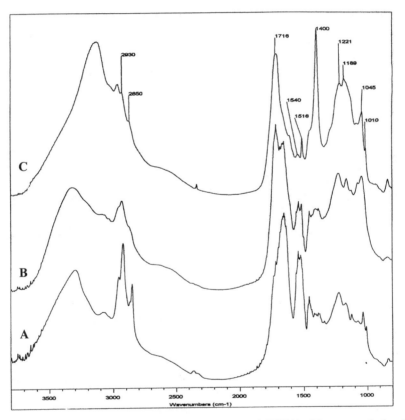

Figure 2 *FTIR spectra of (a) sewage sludge humic acid (SSHA); (b) sewage sludge fulvic acid (SSFA); and (c) hydrophobic acid fraction (HoA) obtained from treated sewage water.*

3.4 Elemental and Functional Group Analyses

The elemental and functional group composition provides essential information about the chemical nature of HSs and can therefore be used to compare sewage-originated HSs with soil, compost and "model" HAs and FAs. Data are summarized in Table 2. The C content of the SSHA was 56.3% whereas that of SSFA and HoA was less than 50%. The O content exhibited the opposite trend. The highest C/N ratio was measured for the SSFA sample.

The highest total acidity and E_4/E_6 ratio were measured for the HoA (Table 2). The carboxyl group level was similar in both HoA and SSFA samples (3.9 mol kg^{-1}) and it was much lower in the SSHA sample (2.3 mol kg^{-1}).

Table 2 *Elemental analysis, atomic ratio and functional group content of sewage sludge humic acid (SSHA), sewage sludge fulvic acid (SSFA) and the hydrophobic acid fraction (HoA) obtained from treated sewage water*

Constituent	SSHA	SSFA	HoA
Elemental analysis (%)			
C	56.3	48.5	43.0
H	7.7	5.5	5.5
N	8.8	5.3	5.6
O	27.2	59.3	54.1
S	1.7	2.1	3.2
Atomic ratio			
H/C	1.6	1.3	1.5
O/C	0.36	0.91	0.94
C/N	6.4	9.1	7.6
Functional group (mol kg^{-1})			
Total acidity	3.7	5.4	7.6
CO$_2$H	2.3	3.9	3.9
Phenolic OH	1.4	1.4	3.7
E_4/E_6 ratio	5.0	9.2	14.6

4 DISCUSSION

The high percentage of the hydrophobic fractions (hydrophobic acids and neutrals) adsorbed by XAD-8 resin suggests that the hydrophilic fractions (such as low-molecular-weight polysaccharides, carbohydrates, proteins, peptides and amino acids, which are widespread in waste water effluent) have been degraded during the aerobic treatment, resulting in a relatively elevated level of aromatic and aliphatic compounds.

The elemental analysis of the SSHA exhibited lower O and higher N content as compared with model HAs.[20] Moreover, the functional group contents were lower than those suggested for model HAs. The same phenomenon was exhibited for the elemental analysis of SSFA compared with model FAs.[20] The HoA fraction exhibited the highest total acidity and E_4/E_6 ratio, suggesting a very active, highly charged low-molecular-weight FA fraction, as expected for DOC.

^{13}C-NMR and FTIR spectra of SSHA suggest that this material contains a high level of aliphatic structures, mainly long alkyl chains. The ^{13}C-NMR spectrum of SSHA is in general similar to that found for SS, SSHA and marine HA obtained from an area contaminated by sewage.[21–23] The prominent aliphatic peaks exhibited in the FTIR spectra of the SSHA and its high H/C ratio are similar to data for other HAs extracted from SS,[24] but the aliphatic proportions calculated from the ^{13}C-NMR spectrum were higher than in HA extracted from uncomposted aerobic SS.[25] The ^{13}C-NMR spectrum of the SSHA exhibited a high content of proteins, in accordance with the FTIR data (1540 cm^{-1}) and the high N content data from elemental analysis (8.8%). The FTIR spectrum of SSHA exhibited features similar to those of other young type III HAs extracted from soil,[26] uncomposted separated cattle manure and municipal solid waste (MSW).[9,27–29] All of these materials showed a strong aliphatic character with sharp peptide bond peaks along with relatively low carboxylic acid content. As with the FTIR spectra, the ^{13}C-NMR data for the SSHA were similar to those of HA derived from uncomposted MSW.

The physico-chemical analysis of SSFA indicates low molecular weight, high carboxylic acid, carbohydrate and aromatic content, which are similar to data for soil FAs. The ^{13}C-NMR data indicate a high proportion of O-alkyl structures, which explains the extremely high O content. In addition, the FTIR spectrum exhibits features similar to those of other young type III soil FAs.[19] The aromatic content of the SSFA is in the range recorded for marine, stream and soil FA. However, the SSFA aromaticity is higher than that for the SSHA, in contrast to HAs and FAs originating from streams and soils.[16]

The elemental and functional group analyses of the HoA resulted in values closer to model FA data than the present SSFA data, although the HoA exhibited a higher C/O ratio and N content. This is in contrast to the similarity between the SSFA and SSHA to FA and HA derived from uncomposted MSW, respectively. The HoA fraction differed significantly from HoA originating from composted MSW.[9,29] The HoA studied exhibited a higher proportion of aromatic and phenolic C-containing groups (34% *vs.* 11–16% in HoA derived from composted MSW). Moreover, the level of O-alkyl compounds in the HoA sample in this study is less than 50% of the level in the HoA originating from composted or uncomposted MSW. The FTIR spectrum of the HoA exhibits peaks similar to those observed for a HoA fraction extracted from a river water sample, with pronounced aliphatic (1400–1450 cm^{-1}) and carboxylic (1720 cm^{-1}) peaks.[8]

5 CONCLUSIONS

Treated wastewater is commonly used in Israel and other semi-arid counties as irrigation water for agriculture. As a result of the properties of the dissolved organic matter present in treated wastewater (solubility, and high aromaticity and functional group content), this OM can interact with aromatic compounds that are common in the environment (herbicides and pesticides) and with heavy metals present in soil. This interaction could cause migration of these contaminants to groundwater along with HSs transport (in particular, the highly soluble HoA fraction added with the irrigation water). The physico-chemical properties of the HSs originating from SS suggest that these compounds are recently-formed humic macromolecules compared with soil HSs. The high concentration of aliphatic and carbohydrate components present in the SSHA and SSFA studied suggest that these materials are relatively unstable and that they can undergo extensive biodegradation. Thus, we suggest that as usually (but not always) done, the raw SS be further stabilized (*e.g.* by composting) before utilization in agriculture.

ACKNOWLEDGEMENT

The authors wish to thank the Israeli Ministry of Science for providing financial support to this research.

References

1. A. Feigin, I. Ravina and J. Shalhevet, 'Irrigation with Treated Sewage Effluent', Springer-Verlag, Berlin, 1991.
2. W. Viessman and M. J. Hammer, 'Water Supply and Pollution Control', Harper Collins College Publishers, New York, 1992.
3. D. R. Sachedev, J. J. Ferris and N. L. Clesceri, *WPCF*, 1976, **48**, 570.
4. R. Ishiwatari, H. Hamana and T. Machiara, *Water Res.*, 1980, **14**, 1257.
5. M. Rebhun and J. Manka, *Environ. Sci. Technol.*, 1971, **5**, 606.
6. G. Peschel and T. Wildt, *Water Res.*, 1988, **22**, 105.
7. T. Hernandez, J. I. Moreno and F. Costa, *Biolog. Wastes*, 1988, **26**, 167.
8. J. A. Leenheer, *Environ. Sci. Tech.*, 1981, **15**, 578.
9. B. Chefetz, Y. Hadar and Y. Chen, *Soil Sci. Soc. Am. J.*, 1998, **62**, 326.
10. Y. Inbar, Y. Chen and Y. Hadar, *Soil Sci. Soc. Amer. J.*, 1990, **54**, 1316.
11. P. G. Hatcher, I. A. Breger, L. W. Dennis and G. E. Maciel, 'Aquatic and Terrestrial Humic Materials', Ann Arbor Science, Ann Arbor, Michigan, 1983, p. 37.
12. P. G. Hatcher, M. Schnitzer, L. W. Dennis and G. E. Maciel, *Soil Sci. Soc. Am. J.*, 1981, **45**, 1089.
13. A. Gutierrez, P. Bocchini, G. C. Galletti and A. T. Martinez, *Appl. Environ. Microbiol.*, 1996, **62**, 1928.
14. I. Kögel-Knabner, P. G. Hatcher and W. Zech, *Soil Sci. Soc. Am. J.*, 1991, **55**, 241.
15. R. L. Wilson, *Soil Sci.*, 1981, **32**, 167.
16. R. L. Malcolm, *Anal. Chim. Acta*, 1990, **232**, 19.
17. A. U. Baes and P. R. Bloom, *Soil Sci. Soc. Am. J.*, 1989, **53**, 695.
18. J. Niemeyer, Y. Chen and J. M. Bollag, *Soil Sci. Soc. Am. J.*, 1992, **56**, 135.
19. F. J. Stevenson, 'Humus Chemistry: Genesis, Composition and Reactions', 2nd Edn., Wiley, New York, 1994.
20. M. Schnitzer, 'Humic Substances: Chemistry, Reactions', Elsevier, Amsterdam, 1978, p. 1.
21. E. G. Piotrowski, K. M. Valentine and P. E. Pfeffer, *Soil Sci.*, 1984, **137**, 194.
22. W. V. Gerasimowicz and D. M. Byler, *Soil Sci.*, 1985, **139**, 270.
23. P. G. Hatcher, PhD Thesis, University of Maryland, 1980.
24. S. Deiana, C. Gessa, B. Manunza, R. Rausa and R. Seeber, *Soil Sci.*, 1990, **150**, 419.
25. C. Garcia, T. Hernandez and F. Costa, *Biores. Technol.*, 1992, **41**, 53.
26. F. J. Stevenson and K. M. Goh, *Geochim. Cosmochim. Acta*, 1971, **35**, 471.
27. Y. Inbar, Y. Chen and Y. Hadar, *Soil Sci. Soc. Am. J.*, 1989, **53**, 1695.

28.   Y. Chen, B. Chefetz and Y. Hadar, in 'The Science of Composting', M. de Bertoldi, P. Sequi, B. Lemmes and T. Papi (eds.), Blackie Academic and Professional, London, 1996, p. 382.
29.   B. Chefetz, Y. Hadar and Y. Chen, *Acta Hydrochim. Hydrobiol.*, 1998, **26**, 172.

STRUCTURE AND ELEMENTAL COMPOSITION OF HUMIC ACIDS: COMPARISON OF SOLID-STATE ^{13}C NMR CALCULATIONS AND CHEMICAL ANALYSES

J. Mao,[1] W. Hu,[2] K. Schmidt-Rohr,[2] G. Davies,[3] E. A. Ghabbour[3] and B. Xing[1]

[1] Department of Plant and Soil Sciences, Stockbridge Hall, University of Massachusetts, Amherst, MA 01003, USA

[2] Department of Polymer Science and Engineering, University of Massachusetts, Amherst, MA 01003, USA

[3] Chemistry Department and the Barnett Institute, Northeastern University, Boston, MA 02115, USA

1 INTRODUCTION

Humic acids (HAs) are found in soils, sediments, waters and some plants.[1-4] HAs are complex polymeric mixtures. Structural and functional information is critical in determining their reactivity with heavy metals and organic contaminants.[5-10] The best way to fingerprint HA structures is to examine HAs without any pretreatment. In this respect, solid state ^{13}C NMR is a powerful tool.[11] Cross Polarization Magic Angle Spinning (CPMAS) ^{13}C NMR has been extensively used in HAs studies.[12] There are, however, several problems with this technique, which can lead to nonquantitative spectra.[12-15] The first problem is the reduced cross-polarization (CP) efficiency for unprotonated carbons, mobile components, or regions with short $T_{1\rho}^{H}$ (proton rotating-frame spin-lattice relaxation time). CP efficiency relies on the distance between ^{13}C carbons and protons. Thus, ^{13}C carbons without directly bonded protons have a reduced CP efficiency. Molecular mobility, which reduces the dipolar interaction, can also significantly decrease the CP efficiency. The shortening of $T_{1\rho}^{H}$ by paramagnetic species in HAs can cause protons to lose magnetization too quickly to transfer to carbon (typical $T_{1\rho}^{H}$s are less than 10 ms in HAs[16]). Ramp sequence CPMAS is a relatively new technique which was first developed by the Smith group at Yale University.[17-19] Cook *et al.*[20,21] applied this technique in the study of humic substances. The applicability of this new technique still requires further examination because the relatively long contact time used (10 ms) may render some carbon species with short $T_{1\rho}^{H}$ invisible.[19]

The second quantification problem arises from spinning sidebands, which reduce the intensity of the centerband, leading to the loss of intensity and distortion of peak areas. For a MAS spectrum, if the spinning speed is less than the chemical shift anisotropy, significant spinning sidebands occur. They are especially significant for aromatic and carbonyl groups which have large chemical shift anisotropy.[22] For instance, at a spinning speed of 8 kHz in a 7 Tesla field, the sidebands of aromatic carbons contain 15% of the total area. The sidebands can be decreased by spinning at a higher rate, but fast MAS can reduce the CP efficiency.[23,24]

An alternative to CPMAS is Direct Polarization Magic Angle Spinning (DPMAS). This technique has not been widely used because it requires that the recycle delays between scans be longer than five times the longest T_1^C (the ^{13}C spin-lattice relaxation time) to generate a quantitative spectrum. In most glassy or crystalline materials, the longest T_1^C is of the order of 5 seconds to > 10,000 seconds, which requires forbiddingly long recycle delays. For some HAs, the longest T_1^C could be of the order of tens of seconds.[15] Thus, it is rather impractical to use DPMAS alone to obtain quantitative spectra, though it avoids CP problems and permits high spinning speeds to reduce sidebands. Nevertheless, DPMAS combined with the T_1 correction obtained from CP/T_1-TOSS (TOSS is TOtal Sideband Suppression) spectra is a good approach for HAs with $T_1^C < 5$ s.

The objective of this study was to explore this combined method to assess the reliability of this technique. Its results were compared with traditional chemical analyses.

2 MATERIALS AND METHODS

2.1 Sample Origins and Preparation

Nine HAs of various origins and locations were used: five soil HAs (Amherst, German, Irish, New Hampshire, and New York), three commercial HAs (Aldrich, ARC and IHSS), and one plant HA (*Pilayella*). The extraction and purification procedures are described elsewhere.[1,25] Aldrich HA was purified in our laboratory before use. The locations and some analytical data for the HAs are shown in Table 1.

Table 1 *Origins and some analytical data for nine HAs*[a]

HA	Origin	%C	%H	%O	%N	%Ash	%Fe
Amherst	a peat in Amherst, Massachusetts, USA	52.9	4.49	40.0	2.57	<0.1	NA[b]
German	a peat from Bad Pyrmont, Germany	50.5	5.32	42.5	1.71	1.5	0.03
Irish	a peat from Turf Board Company, Cork, Ireland	50.5	5.56	41.9	2.06	0.89	0.025
New Hampshire	a bog soil in Rumney, New Hampshire, USA	52.9	5.40	39.7	2.00	0.25	0.024
New York	an alluvial farm soil from New York, USA	53.8	5.08	37.3	3.89	1.2	0.3
Aldrich	a product of Aldrich Chemical Company	60.0	4.47	34.5	0.96	<0.1	NA
ARC	a product of Arctech, Inc., Chantilly, Virginia	56.0	3.97	38.7	1.29	2.6	0.258
IHSS[c]	a standard (Leonardite) from IHSS (International Humic Substances Society)	59.2	4.08	35.6	1.12	2.3	NA
Pilayella	an HA isolated from the brown alga *Pilayella littoralis*	46.6	5.68	42.1	5.68	2.4	0.154

[a] The elemental compositions were calculated on the assumption that the sum of C, H, O and N were 100%; [b] not available; [c] considering water absorption during storage, IHSS elemental compositions were measured again in our laboratory.

2.2 NMR Spectroscopy

NMR techniques of DPMAS, CP-TOSS and CP/T$_1$-TOSS were used. For DPMAS, samples were run at a ^{13}C frequency of 75 MHz in a Bruker DSX-300 spectrometer. HAs were packed in a 4-mm-diameter zirconia rotor with a Kel-F cap. Single pulse excitation and a Hahn echo before detection with a spinning speed of 13 kHz were used. The 90° pulse length was 3 μs. One rotation period (77 μs) was employed as the pre-echo delay in a Hahn echo to avoid baseline distortions due to dead time. The 180°-pulse length was 6 μs. The number of scans (NS) varied from 8192 to 32768, depending on the specific sample. Recycle delays were 6 s or less.

For CP-TOSS, samples were run at 75 MHz in a Bruker MSL-300 spectrometer. HAs were packed in a 7-mm-diameter zirconia rotor with a Kel-F cap. The spinning speed was 4.5 kHz. A ^{1}H 90° pulse was followed by a contact time of 500 μs, and then a TOSS sequence was used to remove sidebands.[22] The 90° pulse length was 3.4 μs and the 180° pulse was 6.4 μs. The recycle delay was 1 s with NS 4096.

CP/T$_1$-TOSS was a modified pulse sequence of CP-TOSS. After the contact time and before the π-pulse train, a +z/-z filter was applied so that the signal decayed from full intensity to zero as a result of T_1^C relaxation. This experiment yielded the T$_1$ corrections for the various peaks in the DPMAS spectra as discussed in more detail below.

2.3 Procedures for Evaluation of the DPMAS-CP/T$_1$-TOSS Technique for HA Quantification

In this study, DPMAS with CP/T$_1$-TOSS correction was used to quantify HAs. Briefly, in order to examine the reliability of this technique, elemental compositions of %C, %H and %(O + N) were calculated from the DPMAS NMR spectra corrected by CP/T$_1$-TOSS, and then were compared with the results from traditional chemical analyses. Good agreement between the two methods would confirm that the new NMR technique is reliable.

2.3.1 *Measure a DPMAS Spectrum.* The spectra were measured under the conditions described in Section 2.2. With recycle delay times of 6 s or less, some samples were fully relaxed but others were not.

2.3.2 *Deconvolute Spectra.* The HA spectra from solid-state NMR measurements are composed of overlapping peaks. Deconvolution (Bruker Xedplot 2.2.0) was used to separate a spectrum into individual bands.[26,27] The positions of the fitted bands were determined from the obvious peaks in the spectrum. However, if one peak could not be fitted with a single band, another band was used to improve the fit. The bands were fitted with 100% Gaussian line shapes because the Lorentzian line shapes have long tails and lead to excessive overlap of neighboring bands.

2.3.3 *Run CP/T$_1$-TOSS Spectra at Two Different T_1^C Filter Times.* The first spectrum had a T_1^C filter time τ equal to 500 μs. With this short filter time there was almost no relaxation along the z axis. The second spectrum had a filter time τ which was equal to the recycle delay used in the DPMAS experiment. For example, the DPMAS spectrum of ARC HA was obtained at recycle delay 5 s and the second T_1^C filter time of its CP/T$_1$-TOSS was also set at τ = 5 s.

There are two steps here. The first is to examine whether the DPMAS spectrum resulted from complete ^{13}C relaxation. If the CP/T$_1$-TOSS spectrum at τ equal to the DPMAS recycle delay relaxed to zero, the DPMAS spectrum was the result of full relaxation. Otherwise, the DPMAS recycle delay was not long enough to obtain a quantitative spectrum. As the second step, if the DPMAS spectrum was not the result of complete relaxation, the two CP/T_1-TOSS spectra were used to correct the DPMAS spectrum.

2.3.4 *Correct Deconvolution Bands with CP/T_1-TOSS for Incompletely Relaxed Spectra.* The following relationship was used for spectrum correction, where $S(t)$ is the

$$S(\infty) = \frac{S(t) \times h(0)}{h(0) - h(t)} \tag{1}$$

DPMAS peak area obtained at time t which is not long enough to obtain a full-relaxation spectrum, $S(\infty)$ is the DPMAS peak area of the spectrum when fully relaxed, $h(0)$ is the CP/T_1-TOSS peak height at time $\tau_1 = 500$ μs, and $h(t)$ is the CP/T_1-TOSS peak height at time τ_1 equal to the DPMAS recycle delay t. The non-fully relaxed spectrum can be corrected with eq. 1.

2.3.5. *Calculate the Sideband Percentage for the Functional Groups with Large Chemical Shift Anisotropy.* Generally, aromatic and carbonyl groups have the greatest chemical shift anisotropy. The higher the spinning rate, the smaller the sidebands. At 13 kHz, the sidebands of a typical aromatic group are 7% of its centerband and 4% in the case of carbonyl groups.

2.3.6. *Calculation of %C, %H and %(O + N).* According to the chemical shifts,[21,28–30] the elemental numbers of C, H and O of different deconvoluted bands can be determined. For each peak or chemical shift range, the elemental number (H or O+N) was based on and normalized by C (*i.e.* the C number is 1). For the chemical ranges with several overlapping functional groups, the numbers of each element were generally obtained based on the average of the elemental numbers in those functional groups and, as a reasonable first-order approximation, we assume equal CO$^-$ and COH proportions. For example, for the 145–162 ppm region the main functional groups are CO$^-$ and COH. Thus, the C number is 1, the O number is 1, and the average of H is 1/2. Hence, the elemental number distribution of this chemical shift range is COH$_{1/2}$.

As a reasonable approximation, data in the 162–190 ppm range are assigned to groups -*C*OOH (50%), *C*OO$^-$ (25%) and C-O-*C*=O in ester groups (25%). In esters, one of the O atoms on *C* is shared with another C atom, so this O atom has elemental number 1/2. With the above proportions of functional groups for this chemical shift range we have 175 O and 50 H atoms associated with every 100 *C* atoms and thus the elemental numbers are CO$_{1.75}$H$_{1/2}$ (Table 2). The data at 96–108 ppm are assigned to anomeric carbon centers *C* in C-O-*C*H-O-C units. Since the two O atoms are joined to other C atoms, they both have elemental number 1/2. The elemental numbers for the atoms associated with anomeric *C* centers in HA samples are thus CHO (Table 2).

For the fully relaxed spectra, only sideband correction was needed. For the non-fully relaxed spectra, sideband and CP/T_1-TOSS corrections both were used. After the sideband and/or CP/T_1-TOSS corrections, the percentages of all the bands were added up and each individual percentage was divided by the sum (*i.e.* the bands were normalized to 100%). The deconvoluted bands were grouped according to eleven peaks or ranges (Table 2). Then, eqs. 2-4 were used to calculate the %C, %H and %(O + N),

$$\%C = \frac{1200}{16 \times \sum_{i=1}^{11} n_i^{O+N} p_i + 1200 + \sum_{i=1}^{11} n_i^{H} p_i} \tag{2}$$

$$\%H = \frac{\sum_{i=1}^{11} n_i^{H} p_i}{16 \times \sum_{i=1}^{11} n_i^{O+N} p_i + 1200 + \sum_{i=1}^{11} n_i^{H} p_i} \tag{3}$$

$$\%(O+N) = \frac{16 \times \sum_{i=1}^{11} n_i^{O+N} p_i}{16 \times \sum_{i=1}^{11} n_i^{O+N} p_i + 1200 + \sum_{i=1}^{11} n_i^{H} p_i} \tag{4}$$

where n_i^{H} and n_i^{O+N} are the elemental numbers of H and (O + N), respectively, and p_i is the percentage for individual ranges or peaks. Because (1) the O- and N-containing functional groups were generally overlapped, (2) the percentage of nitrogen is relatively low in HAs, and (3) the atomic weight of nitrogen is close to that of oxygen, the O and N were calculated together and the calculation weight for (N + O) was chosen as 16, with 12 and 1 for C and H, respectively. Because the sum of carbon p_i was 100 and the atomic weight is 12, carbon contributes 1200 in eqs. 2–4.

Table 2 *Elemental numbers of different chemical shift ranges or peaks*

Ranges/peaks	Chemical shift, ppm	Functional groups	Elemental numbers	
1	190–220	C=O, HC=O[a]	$COH_{1/2}$	
2	162–190	COO$^-$, COOH, C-O-C=O[b]	$CO_{1.75}H_{1/2}$	
3	145–162	C-O$^-$, C-OH[a]	$COH_{1/2}$	
4	130	CH, C[a]	$CH_{1/2}$	
5	115	CH	CH	
6	96–108	O-CH-O		COH
7	60–96	CHOH, CH$_2$OH[a]	$CH_{2.5}O$	
8	50–60	CH$_3$O-, -CH$_2$-O-, CH, C[a]	$CH_{3/2}O_{1/2}$	
9	40	CH$_2$, CH, C[a]	CH	
10	30	CH$_2$	CH_2	
11	20	CH$_3$	CH_3	

[a] Equal proportions of each type are assumed; [b] see text.

2.4 Ratios of sp³-C to sp²-C

The ratios of sp³-C to sp²-C were calculated from the corrected DPMAS spectrum area with eq. 5:

$$sp^3 / sp^2 = \frac{area(0-90ppm)}{area(90-220ppm)} \cdot \tag{5}$$

3 RESULTS AND DISCUSSION

3.1 Comparison of Functional Groups among the Nine HAs

CP-TOSS spectra were used for comparison (Figure 1). DPMAS spectra could not be used for comparison due to the incomplete relaxation of some HAs. Although CP-TOSS spectra could not be used for quantitation,[24] they could be compared because all the HAs were run at the same conditions where fully relaxed spectra were obtained. We assume that the same functional group had a similar CP/TOSS efficiency in HAs. Functional groups were assigned as follows:[21,30] 0–50 ppm for aliphatic-C, 50–60 ppm methoxy-C, 60–96 ppm carbohydrate-C, 96–108 ppm O-C-O, 108–145 ppm aromatic-C, 145–162 ppm phenolic-C, 162–190 ppm carboxylic-C, and 190–220 ppm carbonyl-C. Five soil HA spectra were quite similar (Figure 1). They had distinct peaks at 31 ppm, 75 ppm, 130 ppm, and 175 ppm. New Hampshire, Amherst and Irish HAs exhibited sharp peaks at 55 ppm. Their peaks between 90 ppm and 140 ppm were also very similar. German HA showed a very small peak at 55 ppm and New York HA had a different peak pattern between 90–160 ppm from the other soil HAs. Compared with soil HAs, the *Pilayella* HA had a much higher percentage of sp^3-C and a lower percentage of sp^2-C. Moreover, the *Pilayella* HA spectrum had a more intense peak at 175 ppm and much lower aromatic- and phenolic-C.

The commercial Aldrich and ARC HA spectra were relatively simple, differing significantly from those extracted from soil. They had distinct peaks at 31 ppm and 130 ppm. They also contained some carboxylic-C. An interesting finding was that ARC HA was almost the same as IHSS HA. All the commercial HAs lacked carbohydrate-C.

The significant difference between commercial HAs and soil HAs suggests that commercial HAs cannot be used as surrogates for soil HAs in structural elucidation and environmental research.

3.2 Explanation of Elemental Composition Calculation from NMR Spectra

Based on CP/T$_1$-TOSS and DPMAS results, only IHSS and ARC HAs were fully relaxed. Amherst, New York and Aldrich HAs were relaxed to more than half their respective equilibrium magnetization while German, Irish, New Hampshire and *Pilayella* HAs were relaxed to less than half their respective equilibrium magnetization. Based on the explanation in Section 2.3.4, the DPMAS spectra of ARC and IHSS HAs were quantitative whereas those of Amherst, German, Irish, New York, New Hampshire, Aldrich and *Pilayella* HAs were not because their T$_1$^{C}s were long (*i.e.* the DPMAS recycle delays used in the study were not long enough). Therefore, CP/T$_1$-TOSS correction was used. In principle, we could set their recycle delays long enough to obtain quantitative spectra in DPMAS, but this is not practical due to instrument availability and productivity. In the following, New York HA is used as an example to explain the calculation of elemental compositions from the NMR spectra. The deconvolution of the New York HA DPMAS spectrum is shown in Figure 2.

The whole spectrum was deconvoluted into 12 separate bands. The percentages of different deconvoluted bands are shown in Table 3. The DPMAS spectrum of New York HA was not the result of full relaxation. The correct percentages of each band were obtained by using eq. 1 and correcting for the sidebands of aromatic- and carbonyl-C. Then the corrected percentages were normalized to 100 %. The next step was to group normalized percentages according to the peaks or chemical shift ranges in Table 2.

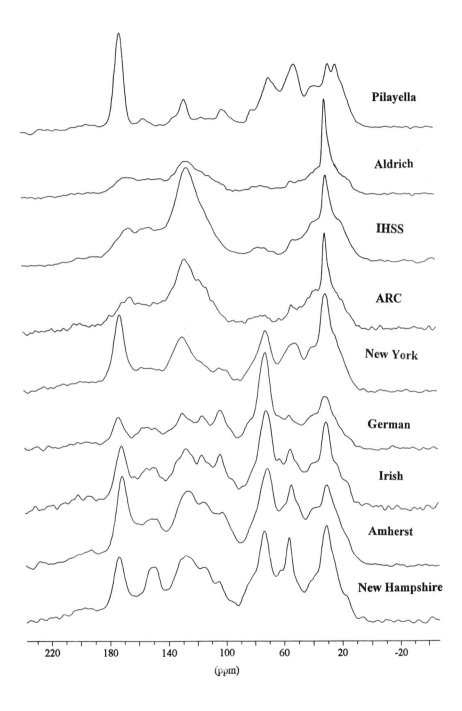

Figure 1 *CP-TOSS spectra of nine HAs*

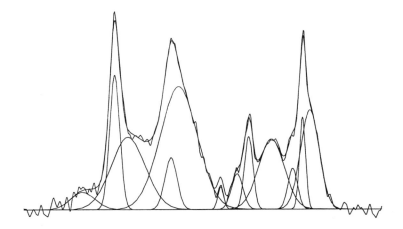

Figure 2 *Deconvolution of the New York HA DPMAS spectrum*

Because there were no bands within ranges 5 and 6, their percentages were zero. Using equations 2, 3 and 4, the %C, %H and %(N+O) were obtained: %C = 54.7, %H = 5.41 and %O = 39.9. Only the sideband correction was used for the spectra of ARC and IHSS HAs because of their complete relaxation.

Table 3 *Calculated data from New York HA NMR spectra*

Band position (ppm)	197	173	162	130	125	93	81	72	55	39	31	22
Deconvolution percentage (%)	4.50	13.9	14.3	20.2	9.62	0.85	2.86	3.76	11.8	2.53	3.60	12.0
Corrected percentage (%)	6.24	16.6	17.4	26.0	12.6	1.28	4.25	5.14	14.7	3.14	3.91	18.1
Normalized percentage (%)	4.82	12.8	13.5	20.1	9.75	1.00	3.29	3.98	11.3	2.43	3.03	14.0
Grouped percentage (%)	4.82	12.8	13.5	29.9[a]			8.27[b]		11.3	2.43	3.03	14.0
Corresponding ranges in Table 2	1	2	3	4			7		8	9	10	11

[a] 29.9 was the result of 20.1 plus 9.75 because they both belong to range 4 in Table 2; [b] 8.27 was the result of 1.00 plus 3.29 plus 3.98 because they all belong to range 7 in Table 2.

3.3 Comparison of Elemental Compositions from NMR Calculations and Chemical Analyses

The results of %C, %H and %(N + O) obtained from NMR calculation and chemical analysis are shown in Figure 3. The carbon and hydrogen results from the two methods agree well. The oxygen and nitrogen results of the two methods are also consistent. These comparisons indicate that DPMAS with CP/T_1-TOSS correction is generally a reliable technique. The deviations were largest for the HAs with long ^{13}C T_1 relaxation times (German, Irish and *Pilayella* in our case), which are not shown due to large uncertainties. For these HAs, the T_1 correction errors become significant.

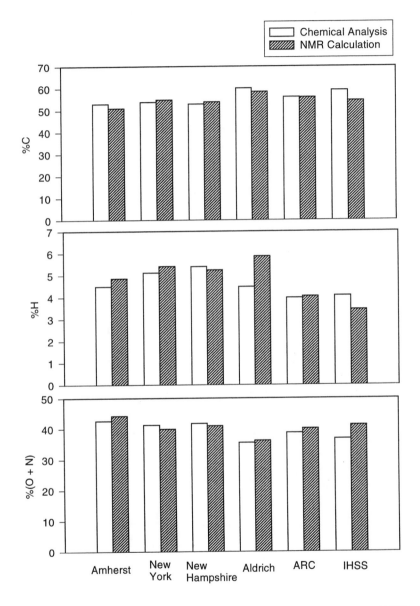

Figure 3 *Comparison of %C, %H, and %(O + N) between chemical analysis and NMR calculations*

3.4 Ratios of sp^3-C to sp^2-C

The ratios of sp^3-C to sp^2-C were calculated for different HA models (Table 4). Except for Schulten and Schnitzer's model,[31] all other models have fixed ratios. Fuchs's model[29] is based on the data for coal HAs and has a condensed ring system with attached COOH and OH groups. Hence, this model is inappropriate for soil HAs. Dragunov's model[32] suggests that the main structure of soil HAs should consist of aromatic rings of the di- or tri-hydroxyphenol-type linked primarily by -O-, -(CH$_2$)$_n$-, -NH- and -N-. This model also contains COOH and OH groups and quinone-type linkages. However, this model lacks aromatic COOH groups. Flaig's model[33] has a lot of phenolic OH and quinones but lacks COOH groups. Bergmann[29] proposed a model based on the data from activated sewage sludge which shows a high proportion of aliphatic structures with abundant COOH groups. HAs from sludge and aquatic environments have more aliphatic groups than those from agricultural soils. Schulten and Schnitzer's model[31] is consistent with the results from ^{13}C-NMR, analytical pyrolysis and oxidative degradation data. The structure consists of aromatic rings linked by long alkyl chains and many COOH and OH groups. The COOH and OH groups are attached to both aromatic rings and aliphatic side chains. The ratios of sp^3-C to sp^2-C of the HA models excluding Flaig's range from 0.5 to 0.84. The ratios of sp^3-C to sp^2-C of soil HAs ranged from 0.53 to 0.73 while those of commercial HAs range from 0.24 to 0.81 (Table 5). The wide range sp^3-C to sp^2-C ratios of HAs used in this study can not be represented by a single model in the literature.

Table 4 *Ratios of sp^3-C to sp^2-C in different HA models*[a]

The authors who proposed	sp^3-C/sp^2-C	Year
Fuchs	0.5	1931
Dragunov	0.68	1948
Flaig	0.1	1964
Bergmann	0.65	1985
Schulten and Schnitzer	0.68–0.84	1993
Jansen et al.[34]	0.56	1996

[a] We only list the models which can be used to calculate the ratios of sp^3-C to sp^2-C.

Table 5 *Ratios of sp^3-C to sp^2-C of HAs*[a]

HA	Amherst	New Hampshire	New York	Aldrich	ARC	IHSS
ratios	0.53	0.73	0.61	0.81	0.24	0.28

[a] Owing to the inconsistency between NMR calculation and chemical analyses, the ratios of German, Irish and *Pilayella* HAs are not quoted.

4 CONCLUSIONS

HA structures are closely related to their origins. The large differences in CP-TOSS spectra and the ratios of sp^3 to sp^2 carbons of different HAs imply that it would be difficult to seek a single, universal HA structure for all HAs. Commercial and plant HAs differ significantly from soil HAs. Comparison between the NMR calculations and chemical analyses indicates that DPMAS corrected by CP/T_1-TOSS is a good technique for quantitation of HAs with relatively short $T_1{}^C$ relaxation times. The effect of paramagnetic materials on detection of carbons by NMR is currently under investigation.

ACKNOWLEDGMENTS

We acknowledge the USDA Competitive Grant Program and a Faculty Research Grant from University of Massachusetts, Amherst for financial support.

References

1.   E. A. Ghabbour, A. H. Khairy, D. P. Cheney, V. Gross, G. Davies, T. R. Gilbert and X. Zhang, *J. Appl. Phycol.*, 1994, **6**, 459.
2.   A. Radwan, R. J. Willey, G. Davies, A. Fataftah, E. A. Ghabbour and S. A. Jansen, *J. Appl. Phycol.*, 1997, **8**, 545.
3.   G. Davies, A. Fataftah, A. Radwan, R. F. Raffauf, E. A. Ghabbour and S. A. Jansen, *Sci. Total Environ.*, 1997, **201**, 79.
4.   A. Radwan, G. Davies, A. Fataftah, E. A. Ghabbour, S. A. Jansen and R. J. Willey, *J. Appl. Phycol.*, 1997, **8**, 553.
5.   G. Davies, A. Fataftah, A. Cherkasskiy, E. A. Ghabbour, A. Radwan, S. A. Jansen, S. Kolla, M. D. Paciolla, L. T. Sein, Jr., W. Buermann, M. Balasubramanian, J. Budnick and B. Xing, *J. Chem. Soc., Dalton Trans.*, 1997, 4047-4060.
6.   D. F. Cameron and M.L. Sohn, *Sci. Total Environ.*, 1992, **113**, 121.
7.   B. Xing and J. Pignatello, *Environ. Sci. Technol.*, 1996, **30**, 2432.
8.   B. Xing and J. Pignatello, *Environ. Toxicol. Chem.*, 1996, **15**, 1282.
9.   B. Xing and J. Pignatello, *Environ. Sci. Technol.*, 1997, **31**, 792.
10.   B. Xing and J. Pignatello, *Environ. Sci. Technol.*, 1998, **32**, 614.
11.   C. M. Preston, *Soil Sci.*, 1996, **161**, 144.
12.   M. A. Wilson, 'NMR Techniques and Applications in Geochemistry and Soil Chemistry', Pergamon Press, Oxford, 1987.
13.   R. L. Wershaw and M. A. Mikita (eds.), 'NMR of Humic Acids and Coal', Lewis Publishers, Chelsea, MI, 1987.
14.   R. Fründ and H.-D. Lüdemann, *Sci. Total Environ.*, 1989, **81/82**, 157.
15.   P. Kinchesh, D. S. Powlson and E. W. Randall, *Eur. J. Soil Sci.*, 1995, **46**, 125.
16.   P. E. Pfeffer, W. V. Gerasimowicz and E. G. Piotrowsk, *Anal. Chem.*, 1984, **56**, 734.
17.   O. B. Peersen, X. Wu and S. O. Smith, *J. Magn. Reson.*, Ser. A, 1993, **106**, 127.
18.   G. Metz, X. Wu and S. O. Smith, *J. Magn. Reson.*, Ser. A, 1994, **110**, 219.
19.   G. Metz, M. Ziliox and S. O. Smith, *Solid State Nucl. Magn. Reson.*, 1996, **7**, 155.
20.   R. L. Cook, C. H. Langford, R. Yamdagni and C. M. Preston, *Anal. Chem.*, 1997, **68**, 3979.
21.   R. L. Cook and C. H. Langford, *Environ. Sci. Technol.*, 1998, **32**, 719.
22.   K. Schmidt-Rohr and H. W. Spiess, 'Multidimensional Solid-State NMR and Polymers', Academic Press, San Diego, 1994.
23.   E. O. Stejskal, J. Schaefer and J. S. Waugh, *J. Magn. Reson.*, 1977, **28**, 105.

24. D. E. Axelson, 'Solid State Nuclear Magnetic Resonance of Fossil Fuels', Multiscience Publications, Canadian Government Publishing Center, Supply and Services Canada, 1985.

25. A. Klute, 'Methods of Soil Analysis', Soil Science Society America, Madison, Wisconsin, 1986.

26. R. L. Wershaw, J. A. Leenheer, K. R. Kennedy and T. I. Noyes, *Soil Sci.*, 1996, **161**, 667.

27. J. A. Pierce, R. S. Jackson, K. W. V. Every and P. R. Grifiths, *Anal. Chem.*, 1990, **62**, 477.

28. E. Breitmaier and W. Voelter, 'Carbon-13 NMR Spectroscopy: High-Resolution Methods and Applications in Organic Chemistry and Biochemistry, VCH, New York, 1987.

29. F. J. Stevenson, 'Humus Chemistry', 2nd edn., Wiley, New York, 1994.

30. R. L. Malcolm, *Anal. Chim. Acta*, 1990, **232**, 19.

31. H.-R. Schulten and M. Schnitzer, *Naturwiss.*, 1993, **80**, 29.

32. M. Kononova, 'Soil Organic Matter', Pergamon, London, 1966.

33. W. Flaig, *Suomen Kem.*, 1960, **A33**, 229.

34. S. A. Jansen, M. Malaty, S. Nwabara, E. Johnson, E. Ghabbour, G. Davies and J. M. Varnum, *Mat. Sci. and Eng.*, 1996, **C4**, 175.

COMPARISON OF DESORPTION MASS SPECTROMETRY TECHNIQUES FOR THE CHARACTERIZATION OF FULVIC ACID

Teresa L. Brown, Frank J. Novotny and James A. Rice*

Department of Chemistry and Biochemistry, South Dakota State University, Brookings, SD 57007-0896

1 INTRODUCTION

Humic substances are heterogeneous mixtures of organic compounds formed as a result of the profound alteration of organic materials in the natural environment. These mixtures are believed to be so complex that each humic fraction (humin, humic acid, and fulvic acid) is itself a heterogeneous mixture that has defied all attempts to separate it into a quantitatively significant amount of any one, discrete compound. The heterogeneous nature of humic materials has made the identification of the components that comprise them an extremely difficult task. This difficulty with characterizing humic materials, for the most part, has arisen because analytical characterization techniques used to investigate their nature are based on the premise that the material being studied is a pure compound. For this reason, techniques commonly employed in structural determination such as infrared spectroscopy (IR) and ^{13}C nuclear magnetic resonance (^{13}C NMR) are limited in their application to humic substances. We now recognize "average" characteristics of each humic material based on the results of a large number of chemical and analytical characterization studies involving techniques such as elemental analysis, functional group analysis, chemical degradation, IR and ^{13}C NMR. For example, fulvic acid has been found to be the most polar and exhibits the greatest degree of aromatic character, the highest acidic functional group content (*e.g.* carboxylic acid functional groups), and the lowest average molecular weight (typically 0.8–1.2 kDa) of the three humic fractions.[1-3] While it is possible to identify the types and relative amounts of certain functionalities in a humic material with these techniques, it has not been possible to determine how these functionalities are actually joined together.

Mass spectrometry, primarily using pyrolysis (*i.e.* PYMS) to vaporize the sample, has been employed in the study of humic materials to try to identify its fundamental "building blocks". While mass spectrometry is capable of providing information about each component in a mixture if a high enough resolution can be attained, the information gained from PYMS is limited by the fact that pyrolysis produces mass spectra of the pyrolysate, not the sample itself. Though a pyrolysis mass spectrum is reproducible for a given set of instrumental conditions and can be used to "fingerprint" a sample, the possibility of gas phase reactions within the pyrolysate vapor make it very difficult to gain detailed information about the molecules which were pyrolysed, *especially* if the original sample was a mixture. If the advantages of mass spectrometry are to be fully exploited in the study of humic materials, a method of sample introduction needs to be identified that produces gaseous sample ions from the original molecules without significantly altering them. Traditional mass spectrometric techniques (*e.g.* electron impact or chemical ionization) have not been applicable to the study of humic substances because they are limited to the study of low molecular-weight, highly volatile compounds. However, recent developments in desorption ionization methods have led to the mass spectrometric analysis of thermally labile, high-mass and polar compounds. Desorption ionization techniques are "soft" ionization methods in that they result in little or no fragmentation of the components being analysed. They have been successfully applied to the mass spectrometric characterization of many different compound classes.[4–6]

The applicability of desorption ionization methods to the study of a variety of molecules suggests that they may also be applicable to the mass spectrometric study of humic materials. In fact, several desorption mass spectrometric characterizations of fulvic acid have been reported.[7–10] This paper compares the applicability of two desorption mass spectrometric techniques, Fast Atom bombardment (FAB) MS and laser desorption (LD) MS, to the study of fulvic acid.

2 MATERIALS AND METHODS

2.1 Fast Atom Bombardment Mass Spectrometry

Because fulvic acid has been subjected to intensive characterization and study,[11] is soluble in water, and is generally believed to be the lowest molecular-weight fraction of the humic materials, fulvic acid has been chosen for this study. Three fulvic acid samples were used in the evaluation of FAB MS. The samples were the extensively characterized Armadale fulvic acid,[12] a traditionally isolated soil fulvic acid,[13] and a groundwater sample isolated by the XAD-8 resin adsorption procedure.[14] All of the following FAB experiments were performed on a Kratos MS-25 mass spectrometer fitted with an IonTech saddle-field atom gun. A xenon Fast Atom beam was operated at a potential of 5 kV. Positive- and negative-ion spectra were recorded.

Initial FAB experiments focused on the determination of an appropriate matrix for the desorption of fulvic acid. Ten matrices of varying hydrophobicity were evaluated: glycerol, thioglycerol, the "magic bullet" [5:1 (w:w) dithiothreitol/dithioerythritol], polyethylene glycol (M_n 0.6 kDa), 3-nitrobenzyl alcohol, triethyl citrate, nitrophenyloctyl ether, diethanolamine, triethanolamine and nujol. The fulvic acid samples were individually dissolved in each matrix to give several matrix/analyte ratios. Background matrix spectra were acquired as well as the fulvic acid/matrix spectra.

A series of FAB experiments was then performed with a number of techniques to enhance ion formation and/or desorption. The following commonly employed techniques were evaluated for their utility in the study of fulvic acid: (1) adding acid or base to a sample/matrix mixture to form ions prior to desorption; (2) the addition of a cationic surfactant to the sample/matrix mixture to bring anionic counter-ions (*i.e.* the fulvic acid) to the matrix surface; (3) the removal of sodium through ultrafiltration[15] and complexation;[15-17] (4) increasing the sample:matrix ratio through continuous flow sample introduction;[18-20] and (5) methylation of the fulvic acid sample. These experiments are decribed in detail elsewhere.[13]

2.2 Laser Desorption Mass Spectrometry

The laser desorption MS experiments were conducted on four International Humic Substance Society (IHSS) reference fulvic acids; the Suwannee River, soil, Nordic aquatic, and peat fulvic acids. Laser-desorption Fourier-transform mass spectrometry (FTMS) experiments with a laser wavelength of 10.6 μm were performed on an Extrel FTMS 2000 mass spectrometer coupled to a pulsed CO_2 laser. Laser-desorption FTMS experiments at 1.06 μm and 355 nm were performed on an Extrel FTMS 2000 coupled to a pulsed Nd:YAG laser. The power densities at which the spectra were obtained at 10.6 μm, 1.06 μm, and 355 nm were ~10^7, 10^9, and 10^7–10^8 W/cm^2, respectively. Positive- and negative-ion spectra were recorded at all wavelengths for the four IHSS reference fulvic acids.

The fulvic acid samples were prepared by individually dissolving each sample in deionized, distilled water to give a concentration of ~5 mg/ml. Several drops of sample solution were placed on a stainless-steel probe-tip and air dried. Approximately ten laser events at different spots on the probe tip were averaged to increase the signal-to-noise ratio. All masses below 200 were ejected with chirp excitation.[21]

3 RESULTS

3.1 Fast Atom Bombardment Mass Spectrometry

Figure 1 presents the positive-ion glycerol matrix and the positive ion spectrum of the Armadale fulvic acid dissolved in a glycerol matrix. While there are several mass peaks that seem to have increased in intensity in Figure 1b, particularly between the mass range of 93–185 Da, there is no significant contribution of the Armadale fulvic acid to the fulvic acid/matrix mass spectrum in the molecular-weight range typically reported for fulvic acid. Comparison of the total ion current (TIC) for each matrix and the corresponding fulvic acid/matrix spectra suggested that the positive-ion fulvic acid/glycerol spectrum is the only spectrum with any measurable ion contribution from the fulvic acid.

Similar comparisons of matrix TIC and fulvic acid/matrix TIC were used to try to identify the contribution of each fulvic acid in the spectra acquired after the ion-formation enhancement techniques. No significant change in any of the fulvic acid FAB mass spectra was observed after the production of preformed ions, the addition of cationic surfactant to increase the concentration of fulvic acid at the matrix surface, the removal of sodium to reduce cationization (the process of cation attachment to a neutral molecule prior to mass analysis) that spreads the ion intensity out over several different species, decreasing the likelihood of observing any one of these ions, the use of continuous-flow FAB to increase the sample/matrix ratio, or methylation of the fulvic acid to increase its hydrophobicity, thereby increasing its concentration at the matrix surface. Fast Atom bombardment mass spectra with significant ion contributions from fulvic acid were not obtained under any of the experimental conditions employed.

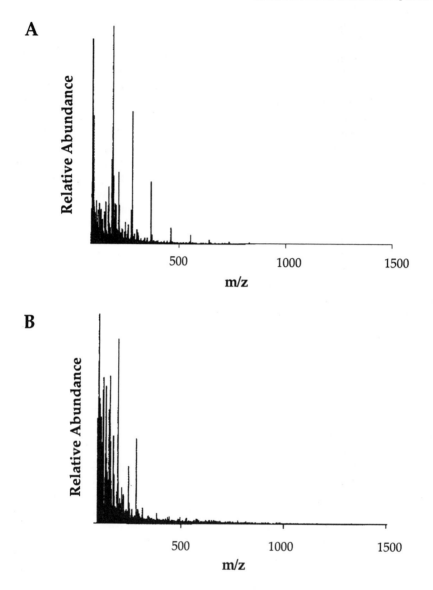

Figure 1 *The positive-ion FAB mass spectra of glycerol (A) and Armadale fulvic acid in glycerol (B)*

3.2 Laser Desorption Mass Spectrometry

Figure 2 shows the positive- and negative-ion LD MS spectra obtained for Suwannee River fulvic acid (SuFA) at a desorption wavelength of 10.6 µm. The spectra obtained for the other IHSS reference fulvic acids appear very similar to the spectra shown in Figure 2 though they do differ in the intensity of the masses observed. For each fulvic acid the positive-ion spectrum displays a higher molecular-weight distribution than the corresponding negative-ion spectrum. It has been suggested that positive-ion LD (10.6 µm) mass spectra might be representative of the fulvic acid's molecular-weight distribution.[22] Further studies, including the determination of ion charge and the extent of fragmentation due to thermal degradation, are necessary before this can be definitively ascertained.

Figures 3a and 3c show the positive-ion mass spectra obtained for SuFA at 1.06 µm and 355 nm. The positive-ion spectra at these wavelengths consist of a noncontinuous, low-mass ion distribution. Fragmentation of the SuFA sample is apparently occurring under these laser desorption conditions. Thus the production of a "representative" positive-ion spectrum of FA is dependent on the laser conditions employed.

The negative-ion mass spectra acquired at 1.06 µm and 355 nm are similar to the negative-ion spectra obtained at 10.6 µm (Figure 4); they do, however, display a lower mass distribution than the 10.6 µm negative-ion spectra.[22] Negative-ion fulvic acid laser desorption and ionization did not appear to exhibit a strong wavelength dependence.

4 DISCUSSION

4.1 Theory of Fast Atom Bombardment Desorption and Ionization

To put the results of the FAB characterization into context, it is necessary to understand the process of desorption and ionization associated with the FAB phenomena.

Fast Atom bombardment is a modification of secondary-ion mass spectrometry (SIMS).[23,24] The SIMS experiment, based on the collision of a high-velocity ion beam with a solid sample on a direct insertion probe, had been employed in the MS analysis of metal surfaces[25] and solid organic compounds deposited on metal surfaces.[26] Although SIMS resulted in the ionization of the solid species by sputtering the solid directly into the gas phase, the ion signals were often found to be weak and short-lived.[27] Barber *et al.*[4] significantly enhanced the efficiency of ion production and signal lifetimes through several relatively simple changes in the SIMS ionization method. Samples were prepared by depositing a solution or slurry of the analyte on a relatively non-volatile, viscous solvent support (the "matrix"). The resulting desorption/ionization process is depicted in Figure 5. Using this, Barber *et al.*[4] reported mass spectra of a variety of peptides, glycoside antibiotics, and organometallic compounds. Initially, the success of FAB over SIMS was credited to the use of a fast beam of atoms instead of ions. This belief was dispelled when Aberth *et al.*[27] showed that SIMS experiments performed with a liquid matrix could give results identical to those of FAB. For this reason FAB and liquid SIMS are rarely distinguished as separate ionization methods and the matrix and its interactions with the analyte have become the focus of understanding the FAB experiment.

Some studies have focused on the applicability of viscous, non-volatile compounds as FAB matrices (*e.g.* references 28–30), while comparatively few studies have examined the

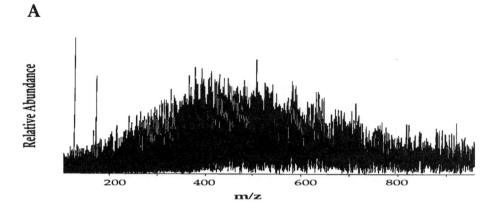

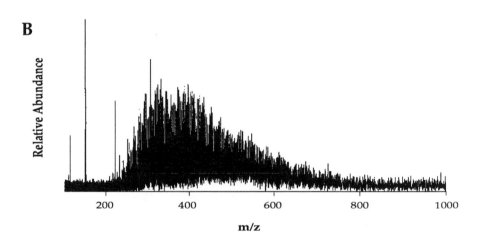

Figure 2 *The positive-ion (A) and negative-ion (B) 10.6 μm LD mass spectra of SuFA*

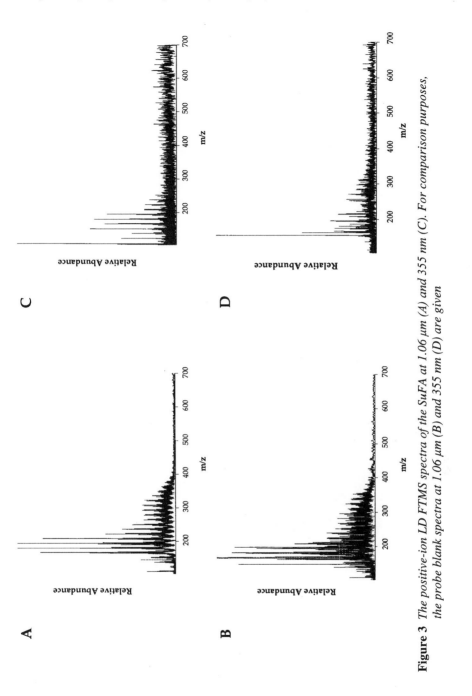

Figure 3 *The positive-ion LD FTMS spectra of the SuFA at 1.06 μm (A) and 355 nm (C). For comparison purposes, the probe blank spectra at 1.06 μm (B) and 355 nm (D) are given*

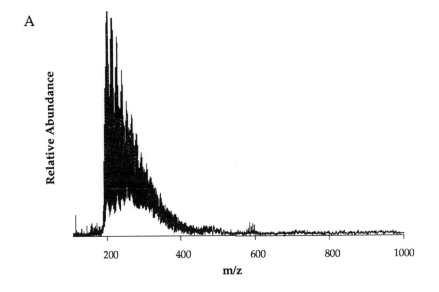

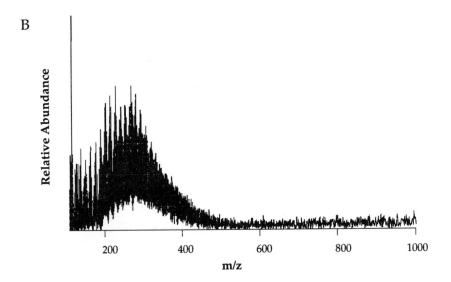

Figure 4 *The negative-ion LD mass spectra of SuFA at 1.06 μm (A) and 355 nm (B)*

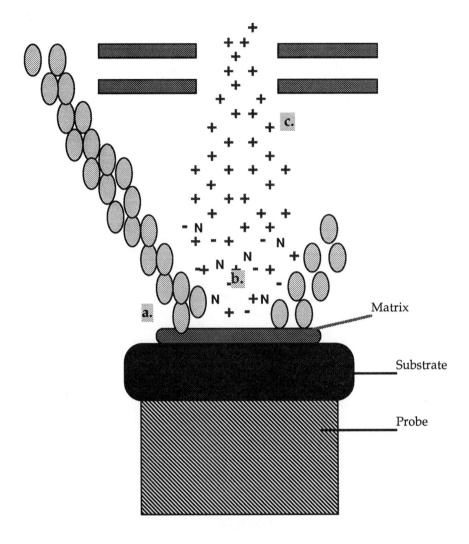

Matrix

Substrate

Probe

Figure 5 *The process of FAB desorption: (a) a high velocity atom beam collides with the matrix surface; (b) ions and neutrals are sputtered into the gas phase; (c) the ions to be detected are accelerated through a potential difference and focussed into the mass analyser*

actual nature of the sputtering process itself. Barber *et al.*[6] described the sputtering process as the result of the momentum transfer from the ion or atom beam to the liquid matrix upon collision. The energy transfer sets up a number of collision chains in the liquid surface, some of which lead to the ejection of ions or molecules into the gas phase. Ions which are preformed in the liquid matrix have been found to be sputtered most readily, followed by protonated or cationized species, and then odd-electron species.[31,32] The lack of significant molecular fragmentation during the sputtering process is probably a consequence of energy being dissipated in the matrix primarily as translational, rather than vibrational, energy.[33] This results in the ejection of ion-solvent clusters which can further dissipate energy through the loss of solvent molecules, rather than the breaking of intramolecular bonds.[34] However, Barber *et al.*[6] have stated that the majority of fragment ions observed in a FAB mass spectrum are the result of the unimolecular decomposition reactions of a molecular ion or species. Metastable-scanning experiments led them to conclude that the sputtered molecular species may have sufficient internal energy to undergo these decompositions in the gas phase.

While the sputtering of preformed ions may result in intense analyte signals, ionization of desorbed neutrals can also occur during the sputtering phenomena. The sputtering process creates a plasma of desorbed ions and neutrals. A number of chemi-ionization processes are thought to occur between the species of the plasma.[4] The result is the formation of $(M+H)^+$ and $(M-H)^-$ molecular ions through proton exchange and/or species such as $(M+Na)^+$ through cationization.

Because the sputtering phenomena results in the desorption of molecules or ions found at the matrix surface, the analyte ion-signal is enhanced when it forms a monolayer at the matrix surface. Monolayer coverage improves a mass spectrum by increasing the number of analyte ions sputtered while simultaneously reducing (or suppressing) the number of matrix ions sputtered. In the study of mixtures, one component with a significantly greater surface activity will dominate the matrix surface and thus the FAB mass spectrum until its concentration in the bulk matrix has been considerably reduced. This phenomenon is called the "suppression effect" and must be carefully considered in the FAB MS study of mixtures.[35] The choice of an appropriate matrix, as well as a number of other techniques to modify or control the surface chemistry of the matrix such as sample derivatization[36] or the addition of a surfactant of opposite charge,[36,37] can lead to an analyte-enriched surface layer.

The FAB matrix surface is dynamic. In order for sample signals to have prolonged lifetimes,[6] analyte molecules removed from the matrix surface through sputtering must be replaced. Analyte solubility in the bulk matrix is required for surface regeneration and is therefore one of the criteria used in the selection of an appropriate matrix compound for the FAB experiment. Other matrix requirements are that: (1) the matrix viscosity be low enough to permit diffusion of the analyte within it; (2) the matrix be relatively non-volatile so that it remains long enough in the high-vacuum region of the mass spectrometer to permit data acquisition; (3) the matrix spectrum should not mask the analyte spectrum, and; (4) the matrix should not interact (react) with the analyte in an undesirable way.[29,30]

4.2 Fulvic Acid Characterization by Fast Atom Bombardment

It is evident from this discussion of the theory and practice of the FAB experiment that its application to a mixture as complex as fulvic acid would require extensive evaluation of the experimental conditions employed. The results described earlier have shown that

FAB MS is not readily applicable to the characterization of fulvic acid. None of the commonly-employed liquid matrices that have been tested to date appear suitable for the FAB analysis of fulvic acid. All ion-enhancement procedures failed to significantly increase the ion contribution from fulvic acid in the acquired mass spectra. We believe this is due to the extremely heterogeneous nature of fulvic acid. It is highly probable that given the nature of fulvic acid,[38] no one component of fulvic acid is present at a high enough concentration to dominate the matrix surface in the FAB experiment, regardless of the enhancement procedure employed. Therefore, while a variety of fulvic acid components are probably found at the surface, none can be sputtered with high enough frequency to stand out against the background ion-signal produced by the FAB matrix.

FAB mass spectra of fulvic acid have been previously reported in the literature.[7-10] The published fulvic acid spectra of Thurman[7] and Dennett[9,10] are apparently produced by the computer-aided subtraction of the corresponding matrix spectrum from the fulvic acid/matrix spectrum. Background-spectrum subtraction procedures cannot be confidently applied to the study of unknown mixtures such as fulvic acid.[39,40] It is very likely that the resulting background-subtraction spectra of fulvic acid present in the literature are artifacts of the subtraction procedure and are not representative of the components of fulvic acid. Our conclusions are supported by Saleh *et al.*,[8] who were unable to produce a fulvic acid FAB mass spectrum with the use of a glycerol matrix.

4.3 Theory of Laser Desorption and Ionization

As with FAB, it is necessary to understand the desorption and ionization processes occurring in a LD MS experiment in order to evaluate its application to fulvic acid.

Laser desorption mass spectrometry was first applied to polar, non-volatile, bio-organic molecules by Posthumus *et al.*[5] The authors observed the production of cationized molecular ions for oligosaccharides, antibiotics, and chlorophyll after irradiation of the sample deposited on a metal surface with a microsecond laser pulse. Like FAB, laser desorption processes are poorly understood on a mechanistic level. The principal desorption mechanisms that have been described for the LD phenomena can be classified as thermal, shock-wave driven, or resonant desorption.[41-45] The ions produced in a desorption experiment may be related to one or more of these mechanisms; the primary mechanism of desorption for a given experiment is governed largely by power density and the ability of the sample to absorb the incident laser wavelength. Other important experimental parameters include the laser pulse duration and method of sample preparation.[46]

Thermal desorption is a temperature-dependent process thought to predominate at low power densities such as those typically observed with an infrared laser ($< 10^7$ W/cm^2).[41] In thermal desorption (Figure 6), the laser beam creates a thermal gradient in the substrate on which the sample has been deposited. When the substrate transfers sufficient heat energy to the sample so that the lattice binding energies of the sample can be overcome, desorption will occur. For thermally-labile compounds, the desorption yield depends on the competing phenomena of desorption and molecular decomposition. Under efficient desorption conditions, which are sample specific, the substrate temperature rises so quickly that desorption occurs before there is time for the sample species to decompose.

The thermal desorption process is further influenced by the ability of the analyte to absorb the incident laser wavelength as well as the ability of the substrate to conduct heat. These factors affect the thermal desorption process by controlling the efficiency with

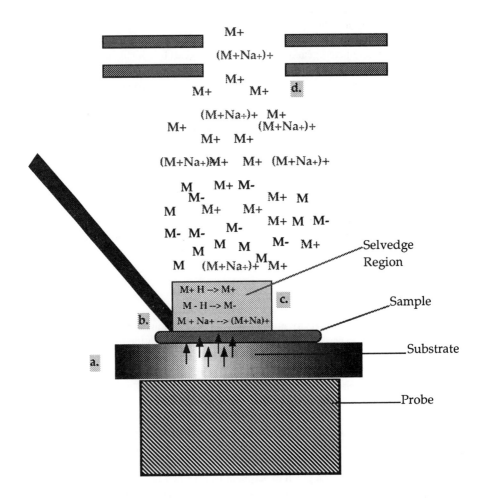

Figure 6 *The thermal desorption mechanism for a LD ion source: (a) the incident laser beam creates a thermal gradient in the substrate. The thermal gradient is represented by the color variation of the substrate; the white regions are 'hot' and the black regions are 'cool'; (b) the substrate transfers heat energy to the sample as demonstrated by the arrows; (c) when the lattice binding energies in the sample are overcome, ionic and neutral species are desorbed. Gas phase reactions occur in the selvedge region producing pseudomolecular ions; (d) the ions to be detected are focussed into the mass analyser*

which the incident light energy is transferred to the sample lattice. Sample species capable of absorbing the incident laser wavelength are found to have lower power density thresholds for the desorption of neutral molecules. Likewise, less incident power is required for desorption when the substrate employed has a high thermal conductivity.[41] Thus, energy transfer to the sample lattice is more efficient when the sample is capable of absorbing the incident light and the substrate is highly conducting.

Thermal desorption results in the production of a large number of neutral and ionic species. Following desorption, gas phase reactions in the plasma may result in the production of pseudomolecular ions through proton addition to the neutral molecules. The presence of alkali-salts in either the prepared sample or as a contaminant often results in cationization of neutral species. Inadvertent laser ablation of the metal substrate during desorption may also give rise to cationization through metal ion attachment.

In contrast to thermal desorption, shock-wave driven desorption is thought to be a temperature-independent process.[41] In this mechanism, ultrafast heating of the sample, not the substrate, produces an energy dissipating shock-wave that travels through and disturbs the binding potentials of the sample lattice (Figure 7). The vibrational energy imparted to the lattice results primarily in the desorption of intact molecules. Because of the rapid nature of this process, molecules do not remain in the lattice long enough to absorb energy sufficient for their fragmentation. Shock-wave driven desorption is often thought to be the dominant desorption mechanism under high power density conditions. This non-equilibrium process results in a short-lived plasma region which, along with the desorption of molecular clusters, leads to a decrease in the probability of gas phase proton attachment and/or cationization of desorbed neutrals.[41]

4.4 Fulvic Acid Characterization by Laser Desorption

From the previous discussion of desorption mechanisms, it is apparent that many factors influence the observed spectra of a laser desorption mass spectrometric experiment. Thus, it is not surprising that the utility of LD in the mass spectrometric determination of fulvic acid has been found to change with the desorption conditions employed (Figure 2). The observed wavelength dependence may be due to different mechanisms controlling desorption at the wavelengths and power densities used in this study.

Although further experimental evidence is required, it seems likely that thermal desorption mechanisms dominate at 10.6 μm, while other higher-energy mechanisms, such as shock-wave driven desorption mechanisms, are operating at 1.06 μm and 355 nm. The efficiency of desorption/ionization at 10.6 μm (943 cm^{-1}) may be enhanced by the ability of the fulvic acid samples to absorb this wavelength[47] which, under thermal desorption conditions, results in more efficient energy transfer to the analyte. The positive ions observed at 10.6 μm may be pseudomolecular ions produced by protonation or alkali-metal cationization to the desorbed neutral species (Na is ubiquitous because of the NaOH used in the isolation and extraction of most fulvic acid samples).

At 10.6 μm, the observed negative-ions may be $(M-H)^-$ ions resulting from deprotonation reactions occurring in the desorbed plasma of ions and neutrals. Previous studies have found that highly polar materials containing functional groups such as the carboxyl group, like fulvic acid, preferentially form $(M-H)^-$ ions during laser desorption.[48] As seen in Figure 2, the negative ion 10.6 μm spectrum is more dominated by lower mass ions than the positive ion spectrum for the same sample. This difference could be due to the existence of multiply-charged negative ions, or a real difference in the chemical

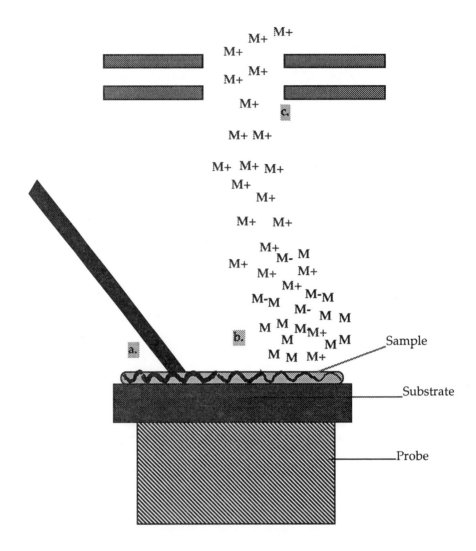

Figure 7 *The shock-wave driven desorption mechanism for a LD ion source: (a) the incident laser beam results in the ultrafast heating of the sample creating an energy-dissipating shock-wave which travels through the sample disturbing the binding potentials of the sample lattice. The energy of the shock-wave, represented by the thickness of the line, decreases as it travels away from the laser spot; (b) intact molecules are predominately desorbed. The selvedge region is short-lived, reducing the probability of gas phase protonization/cationization of the desorbed neutrals; (c) the ions to be detected are focussed into the mass analyser*

composition of the fulvic acid components that become positively charged, as opposed to negatively charged.

In contrast to the 10.6 μm analyses, the experimental power densities employed at 1.06 μm and 355 nm suggest that shock-wave-driven desorption may be the dominant mechanism at these desorption wavelengths. The positive-ion spectra observed at these wavelengths differ significantly from the 10.6 μm positive-ion spectra, probably for the following two reasons. First, the short lived selvedge region under such conditions decreases the probability of gas phase protonation or cationization which is necessary for FA positive-ion formation. Second, the power densities employed at these wavelengths may have exceeded the irradiance thresholds for these samples, which resulted in a large increase in the ratio of fragment ions to molecular ions in the resulting spectra.

4.5 Comparison of FAB and LD in the Mass Spectrometric Characterization of Fulvic Acid

Although FAB and LD are both categorized as desorption ionization techniques, the mechanisms resulting in desorption differ greatly between the two methods. In FAB MS, desorption occurs through the sputtering of ions and molecules from a viscous liquid surface. In contrast to the FAB experiment, laser desorption of the sample occurs directly off the substrate. Thus, the use of the matrix may be the factor which differentiates the applicability of these techniques to fulvic acid, not their desorption mechanisms. Owing to the extremely heterogeneous nature of fulvic acid it is likely that no one component of the mixture would be found in sufficient quantity to generate a signal high enough to be observed above the matrix background signal, regardless of the desorption mechanism operating. In the FAB experiment, fulvic acid components may very well be sputtered into the gas phase but they are subsequently "lost" in the intense matrix background signal. Because it does not require a matrix, the LD experiment has an inherently lower background signal than FAB. Consequently, detection of low intensity signals from the various fulvic acid components is much more likely to occur.

5 CONCLUSIONS

Although both FAB MS and LD MS have been extensively used in the mass spectrometric determination of non-volatile, high mass biomolecules, they apparently are not equally applicable to the MS analysis of fulvic acid. It has been shown that it is not possible to characterize fulvic acid using the commonly employed FAB matrices.[13] Furthermore, an exhaustive study of matrix modification and ion enhancement procedures has shown that it is not possible to enhance the production of fulvic acid ions such that they can be detected over the background of the FAB experiment.[13] This is believed to be a result of the extremely heterogeneous nature of fulvic acid. It is probable that no one component of fulvic acid is high enough in concentration,[38] or sufficiently surface active to dominate the matrix surface in the FAB experiment. Therefore, while a variety of fulvic acid components are probably found at the matrix surface, none can be sputtered with high enough frequency to overcome the matrix background inherent to a FAB analysis.

In contrast to FAB, the laser desorption ionization techniques discussed here have been shown to be unaffected by this background problem simply because they do not require a matrix for desorption to occur. Positive- and negative-ion fulvic acid spectra have been acquired at an infrared laser wavelength where thermal desorption mechanisms are thought to predominate. Further studies are needed to address the formation of multiply charged and fragment ions but it does appear that laser desorption techniques may provide a means for the mass spectrometric study of fulvic acid. The high resolution and ion-storage and -manipulation capabilities of Fourier transform mass spectrometry coupled with laser desorption ionization may prove to be a valuable tool in the characterization of these complex materials.

ACKNOWLEDGEMENT

This research was partially supported by the US Environmental Protection Agency through award no. 816962-01-0.

References

1. F. J. Stevenson, "Humus Chemistry", Wiley, New York, 1982.
2. F. J. Stevenson, in 'Humic Substances in Soil, Sediment and Water', G. R. Aiken, D. M. McKnight, R. L. Wershaw and P. MacCarthy, eds., Wiley, New York, 1985, p. 13.
3. M. H. B. Hayes, P. MacCarthy, R. L. Malcolm and R. S. Swift, in 'Humic Substances II: In Search of Structure', M. H. B. Hayes, P. MacCarthy, R. L. Malcolm and R. S. Swift, (eds.), Wiley, Chichester, 1989, p. 689.
4. M. Barber, R. S. Bordoli, R. D. Sedgwick and A. N. Tyler, *Nature*, 1982, **293**, 270.
5. M. A. Posthumus, P. G. Kistemaker, H. L. C. Meuzelaar and M. C. T. N. deBrauw, *Anal. Chem.*, 1978, **50**, 985.
6. M. Barber, R. S. Bordoli, G. J. Elliott, R. D. Sedgwick and A. N. Tyler, *Anal. Chem.*, 1982, **54**, 645A.
7. E. M. Thurman, 'Organic Geochemistry of Natural Waters', M. Nijhoff and Junk, Dordrecht, 1985.
8. F. Y. Saleh, D. Y. Chang and J. S. Frye, *Anal. Chem.*, 1983, **55**, 862.
9. K. E. Dennett, A. Amirtharaiah, T. F. Moran and J. P. Gould, in 'Proceedings of the Water Quality Technology Conference, Part 2', 1991, p. 1355.
10. K. E. Dennett, MS Thesis, Georgia Institute of Technology, 1990.
11. R. C. Averett, J. A. Leenheer, D. M. McKnight and K. A. Thorn (eds.), 'Humic Substances in the Suwannee River, Georgia: Interactions, Properties and Proposed Structures', USGS OFR 87-557, Washington, DC, 1989.
12. M. Schnitzer and S. U. Khan, 'Humus Chemistry', Marcel Dekker, New York, 1972.
13. F. J. Novotny, PhD Dissertation, South Dakota State University, 1993.
14. E. M. Thurman and R. L. Malcolm, *Environ. Sci. Technol.*, 1981, **15**, 463.
15. T. S. Chen, H. Yu and D. Barofsky, *Anal. Chem.*, 1992, **64**, 2014.
16. I. Fujii, R. Isobe and K. J. Kanematsu, *J. Chem. Soc., Chem. Commun.*, 1985, 405.
17. R. Orlando, *Anal. Chem.*, 1992, **64**, 332.
18. R. M. Caprioli and T. Fan, *Anal. Chem.*, 1986, **58**, 2949.
19. R. M. Caprioli, *Anal. Chem.*, 1990, **62**, 477.
20. C. E. Heine, J. F. Holland and J. T. Watson, *Anal. Chem.*, 1989, **61**, 2674.

21.   T. L. Brown, PhD Dissertation, South Dakota State University, 1998.
22.   J. A. Rice and D. Wiel, in 'Humic Substances in the Global Environment and Implications on Human Health', N. Senesi and T. M. Miano (eds.), Elsevier, Amsterdam, 1994, p. 355.
23.   M. Barber and J. C. Vickerman, *Surf. Defect. Prop. Solids*, 1976, **5**, 162.
24.   A. F. Dillion, R. S. Lehrle and J. C. Robb, *Adv. Mass Spectrom.*, 1968, **4**, 477.
25.   A. Benninghoven, D. Jaspers and W. Sichtermann, *Appl. Phys.*, 1976, **11**, 35.
26.   A. Benninghoven, *Proc. 9th Mater. Sym.*, NBS Spec. Publ., 1979, 627.
27.   W. Aberth, K. M. Straub and A. L. Burlingame, *Anal. Chem.*, 1982, **54**, 2029.
28.   E. DePauw, *Mass Spectrom. Rev.*, 1986, **5**, 191.
29.   J. L. Gower, *Biomed. Mass Spectrom.*, 1985, **12**, 191.
30.   A. Dell, *Adv. Carbohydr. Chem. Biochem.*, 1987, **45**, 19.
31.   C. Fenselau and J. Cotter, in 'IUPAC, Frontiers in Chemistry', J. K. Laider (ed.), Pergamon, New York, 1992.
32.   T. Keough and A. J. Destefano, *Anal. Chem.*, 1981, **53**, 25.
33.   R. A. Johnstone, I. A. S. Lewis and M. E. Rose, *Tetrahedron*, 1983, **9**, 1597.
34.   G. Puzo and J. C. Prome, *Org. Mass. Spectrom.*, 1984, **19**, 448.
35.   W. V. Ligon Jr., in 'Biological Mass Spectrometry', A. L. Burlingame and J. A. McCloskey (eds.), Elsevier, Amsterdam, 1985, p. 61.
36.   W. V. Ligon and S. B. Dorn, *Int. J. Mass. Spectrom. and Ion Proc.*, 1986, **78**, 99.
37.   W. V. Ligon and S. B. Dorn, *Int. J. Mass. Spectrom. and Ion Proc.*, 1984, **61**, 113.
38.   P. MacCarthy and J. A. Rice, in 'Scientists on Gaia', S. Schneider and P. J. Boston (eds.), MIT Press, Cambridge, MA, 1991, p. 339.
39.   B. L. Ackerman, J. T. Watson, J. F. Newton, J. B. Hook and W. E. Brasselton, *Biomed. Mass Spectrom.*, 1984, **11**, 502.
40.   B. L. Ackerman, J. T. Watson and J. F. Holland, *Anal. Chem.*, 1985, **57**, 2656.
41.   L. Van Vaeck, W. Van Roy, R. Gijbels and F. Adams, in 'Laser Ionization Mass Analysis', A. Vertes, R. Gijbels and F. Adams, (eds.), Wiley, New York, 1993, p. 177.
42.   B. Lindner and U. Seydel, *Anal. Chem.*, 1985, **57**, 895.
43.   F. Hillenkamp, M. Karas, U. Bahr and A. Ingendoh, in 'Ion Formation from Organic Solids', A. Hedin, B. U. R. Sundqvist and A. Benninghoven, (eds.),Wiley, Chichester, 1989, p. 111.
44.   R. J. Cotter, *Anal. Chem.*, 1984, **56**, 485A.
45.   F. Hillenkamp, in 'Ion Formation from Organic Solids', A. Benninghoven (ed.), Springer-Verlag, Berlin, 1983, p. 190.
46.   A. Vertes and R. Gijbels, in 'Laser Ionization Mass Analysis', A. Vertes, R. Gijbels and F. Adams, (eds.), Wiley, New York, 1993, p. 127.
47.   P. MacCarthy, and J. A. Rice, in 'Humic Substances in Soil, Sediment and Water', G. R. Aiken, D. M. McKnight, R. L. Wershaw and P. MacCarthy, (eds.), Wiley, New York, 1985, p. 527.
48.   D. M. Lubman (ed.), 'Lasers and Mass Spectrometry', Oxford University Press, New York, 1990.

THE RELATIVE IMPORTANCE OF MOLECULAR SIZE AND CHARGE DIFFERENCES IN CAPILLARY ELECTROPHORESIS OF HUMIC SUBSTANCES OF DIFFERENT ORIGIN

Maria De Nobili,[1] G. Bragato[2] and A. Mori[2]

[1] Dipartimento di Produzione Vegetale e Tecnologie Agrarie, University of Udine, via delle Scienze 208, 33100 Udine, Italy.
[2] Istituto Sperimentale per la Nutrizione delle Piante, Sezione di Gorizia, via Trieste 23, 34170 Gorizia, Italy.

1 INTRODUCTION

In conventional polyacrylamide gel electrophoresis (PAGE), molecular size differences can explain more than 95% of variations in the electrophoretic mobility of fractions of humic substances of reduced molecular weight polydispersity extracted from the same soil.[1] Capillary electrophoresis (CE) is, however, entirely peculiar in that specific factors such as the electro-osmotic flow (EOF) can affect not only the efficiency but even the mechanism of the separation itself. In particular, humic substances (HSs) that would naturally migrate towards the anode under the influence of the electric field are instead driven to the cathode by the EOF generated by the migration of cations near the negatively charged surface of the capillary.[2] However, the main factor influencing the capillary electrophoretic behaviour of HSs, even under conditions where the normal migration direction is reversed, is still the molecular size.[3,4] The enormous separation efficiency of CE might allow the detection of significant charge effects when comparing the mobilities of corresponding fractions of reduced molecular size polydispersity of HSs extracted from different peats and soils.

In the present work we examined the electrophoretic behaviour of humic substances in coated fused silica capillaries where the EOF had been suppressed by reaction of the free silanol groups on the inner capillary surface with a polyether coating.

2 MATERIALS AND METHODS

Fractions of reduced molecular weight polydispersity extracted from four different Histosols and from a spodic soil (Typic Haplorthod, Mount Sobretta, Italy) were prepared by ultrafiltration on Amicon YM membranes in the following apparent molecular weight ranges: 100–300, 50–100, 30–50, 10–30, 5–10 and 1–5 kDa. Sodium pyrophosphate (pH = 7.1) was used as eluent. Ultrafiltration was considered to be completed when the solution exiting the filter cell was colourless. Fractions were then exhaustively dialysed with distilled water to eliminate pyrophosphate, concentrated on the same membrane to about 1 mg ml^{-1} organic C and stored at -18 °C.

Capillary electrophoresis was performed in 100 µm i.d. µSil DB-WAX coated fused silica capillaries at a temperature of 30 °C under constant voltage (–14 kV). The detection wavelength was 400 nm. Samples were loaded by hydrodynamic injection (1 s) at the cathode. The running buffer was 50 mM tris-hydroxymethylaminomethane (Tris)-phosphate buffer solution at pH = 8.3. After each run the capillary was flushed with buffer for five minutes and eventually with buffer plus poly(ethylene glycol) (PEG) for another 5 minutes. The absolute mobilities of the fractions were calculated from migration times measured either in free solution or in the presence of different concentrations of PEG 4000 and PEG 20 000 at or above their entanglement threshold.

3 RESULTS AND DISCUSSION

The polyether coating of the capillaries effectively reduced the EOF so that HSs could be injected at the cathodic end of the capillary and detected at the anode. This is contrary to the behaviour observed in uncoated fused silica capillaries.[4] In the absence of a strong EOF, HSs therefore migrate to the anode, driven solely by the force of the electric field acting on their negative charge. It is therefore under conditions of suppressed EOF that it is possible to study the relative importance of charge and size differences.

By filling capillaries with a physical gel[2] produced by adding 2.5% PEG 4000 to the running buffer, strong linear correlations ($R^2 > 0.95$) were found between the absolute CE mobility of fractions of reduced molecular weight polydispersity extracted from a *Sphagnum* peat and from a Spodosol and the logarithm of their apparent molecular weight (M_w) as deduced from the 95% cut-off limits of ultrafiltration membranes (Figures 1 and 2).

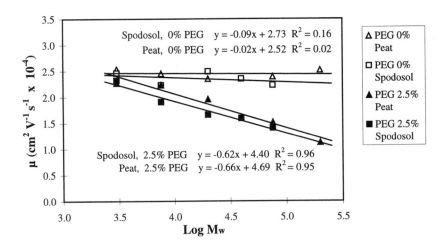

Figure 1 *Electrophoretic mobility at –14 kV constant voltage of HS fractions from a peat and a Spodosol versus the logarithm of their apparent molecular weight in 50 mM Tris-phosphate buffer solution (with or without 2.5% PEG 4000) in a 55 cm long (30 cm effective length) coated capillary.*

This behaviour was observed both on long (100 cm) and short (55 cm) polyether-coated capillaries. On the contrary, when run in free solution in a 55 cm long (30 cm effective length) capillary, all fractions displayed very close mobilities (Figure 1). These results show once again that the capillary electrophoretic behaviour of humic substances is governed in the first place by molecular size. The different fractions could, in fact, be separated only in the presence of a sieving medium. In free solutions (*i.e.* without PEG) separations are attributable to differences in charge/mass ratios or differences in the effective hydrodynamic size. The fact that the mobilities measured with the shorter capillaries were the same implies that HSs are an homologous series of molecules with very large hydration shells.

The use of a longer capillary actually enables detection of the effect of charge differences on the migration times of the same set of fractions run under similar experimental conditions. In fact, when run in free solution in a 100 cm long (75 cm effective length) capillary, the smaller size fractions appear to migrate faster than fractions of larger size even in the absence of a sieving medium (Figure 2).

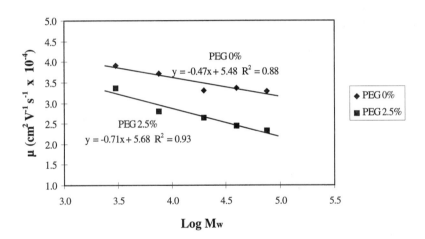

Figure 2 *Electrophoretic mobility at −14 kV constant voltage of HS fractions of the Spodosol versus the logarithm of their apparent molecular weight in 50 mM Tris-phosphate buffer solution (with or without 2.5% PEG 4000) in a 100 cm long (75 cm effective length) coated capillary.*

Increasing the length of the gel fibres in the physical gel, which can be achieved by using PEG 20 000 instead of PEG 4000, causes an apparent decrease in selectivity (Figure 3). The two regression lines are, however, not significantly different from one another, indicating a dependence of the available fractional volume on the gel concentration and molecular volume of HSs and not on the length of the gel fibres. The fact that the mobility of HSs is only marginally affected by the molecular size of PEG indicates that the separation is not affected by interactions of HSs with the polyether through hydrogen bonding.[5]

Three fractions with a mean apparent M_w of 7.5 kDa isolated from HSs extracted from other three *Sphagnum* peat samples of different geographical origin were also examined. All three fractions exhibited the same electrophoretic mobilities in 2.5% PEG when run on the 55 cm long capillary (not shown). This demonstrates again the predominance of molecular size differences on charge effects.

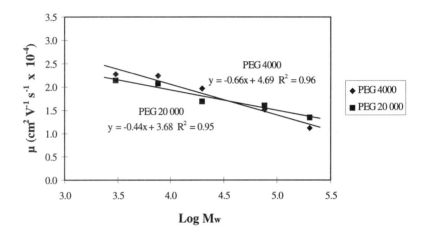

Figure 3 *Electrophoretic mobility at –14 kV constant voltage of HS fractions of a peat versus the logarithm of their apparent molecular weight in 50 mM Tris-phosphate buffer solution with 2.5% PEG 4000 or 2.5% PEG 20 000 in a 55 cm long (30 cm effective length) coated capillary.*

ACKNOWLEDGEMENTS

This research work was supported by a grant from the Ministry of Agricultural Policies, special project PANDA.

References

1.   M. De Nobili and F. Fornasier, *Europ. J. Soil Sci.*, 1996, **47**, 223.
2.   R. Weinberger, 'Practical Capillary Electrophoresis', Academic Press, Boston, 1993, Chapter 2, p. 17.
3.   M. De Nobili, F. Fornasier and G. Bragato, in 'The Role of Humic Substances in the Ecosystems and in Environmental Protection', J. Drozd, S. S. Gonet, N. Senesi and J. Weber (eds.), Polish Society of Humic Substances, Wroclaw, 1997, p.97.
4.   M. De Nobili, G. Bragato and A. Mori, *Acta Hydrochim. Hydrobiol.*, 1998, **26**, 1.
5.   Y. Esaka, Y. Yamaguchi, K. Kano and M. Goto, *Analytical Chem.*, 1994, **66**, 2441.

FLUORESCENCE DECAY OF HUMIC SUBSTANCES. A COMPARATIVE STUDY

Fritz H. Frimmel and Michael U. Kumke

Engler-Bunte-Institute, Division of Water Chemistry, University of Karlsruhe, Richard-Willstätter-Allee 5, 76131 Karlsruhe, Germany

1 INTRODUCTION

The fluorescence of aquatic humic substances (HSs) is commonly found in the wavelength region between 275 nm and 600 nm. Steady-state fluorescence techniques including emission, excitation and synchronous fluorescence spectroscopy mainly have been used for the investigation of HSs. These techniques have been successfully applied in the characterization of HSs of different origin and in the investigation of molecular interactions between HSs and xenobiotics. For example, steady-state fluorescence techniques were applied to characterize HSs of different origin, and aquatic and soil-derived HSs were distinguished.[1-3] Fluorescence anisotropy techniques were used to monitor association processes of HSs.[4] In order to get more specific information on fluorophores present in HSs, chemometric data evaluation approaches, e.g. rank analysis, were introduced.[5] Using fluorescence techniques, conditional binding constants for the interaction of metal ions and HSs were determined. Here, fluorescence quenching especially has proven to be a valuable tool.[6,7] The synchronous scan fluorescence technique was applied to investigate the competitive binding of different metal ions by HSs.[8] A combination of steady-state fluorescence techniques and chemometric data analysis (e.g. rank analysis) and evolving factor analysis has given a better understanding of the acid-base properties and the metal binding sites of HSs.[9-13] The fluorescence quenching approach was further used for the determination of interaction constants of polycyclic aromatic hydrocarbons (PAH) with HSs.[14-16] Here, the extrinsic fluorescence of the probe PAH was detected and HSs served as a quencher. Based on results obtained in fluorescence quenching experiments, models for the micro-organization of HSs were proposed.[17]

Recently, time-resolved fluorescence techniques have been applied for the investigation of the fluorescence decay of HSs and of the dynamics of interaction between HSs and environmental contaminants. The combination of steady-state and time-resolved fluorescence quenching experiments has proven that the observed interaction of PAH and HSs was due to a static ground state interaction.[15,16] Time-resolved laser-induced fluorescence spectroscopy was applied for characterization of the complexation of actinide and lanthanide ions with HSs. The method was used for the speciation of different ion/HSs-complexes with hydroxide and carbonate. The great sensitivity and selectivity allowed the distinction between individual carbonato complexes, *e.g.* $M(CO_3)^+$, $M(CO_3)_2^-$, and $M(CO_3)_3^{3-}$, respectively.[18–21] In these experiments the extrinsic fluorescence of the metal ions was used for the analysis.

Only a few results of investigation of the fluorescence decay of HSs themselves have been published. Time-resolved fluorescence anisotropy measurements of HSs were performed in order to monitor changes in size and shape of HSs in solution with variation of pH, ionic strength and HS concentration.[22]

Cook and Langford investigated the fluorescence of HSs quenched by metal ion complexation.[23] Highly complex fluorescence decay kinetics were found for all HSs investigated. In order to account for high kinetic complexity, three exponential decay terms were introduced. Although this is only a rough approximation, the approach was successful in describing conformational changes of the HSs. Based on the results, a three-component model for the binding of metal ions was suggested.[23]

Introduction of a limited number of exponential terms in the analysis assumes that the number of participating fluorescing sites is known. Considering the heterogeneity of HSs, this has to be considered very carefully. In a more sophisticated approach no pre-assumed number of decay terms is introduced in the analysis of the fluorescence decay of HSs. Here, the Maximum Entropy Method (MEM) and the Exponential Series Method (ESM) were used for the analysis of the HS fluorescence decay.[24,25] In both approaches to the fluorescence decay of the HSs investigated, a tri-modal fluorescence decay time distribution was found.

The scope of the present work was to evaluate further the capabilities of time-resolved fluorescence techniques and of ESM for the characterization of intra- and intermolecular interaction dynamics of HSs. In order to achieve the set objective, (a) characterization of fluorescing sites inside HSs of different origin is necessary and (b) a better understanding of the intramolecular processes involved in the fluorescence decay of HSs was needed.

2 EXPERIMENTAL

The HS samples investigated were of aquatic origin and the concentration of dissolved organic carbon (DOC) was 10 mg/L unless otherwise stated. In the investigation of original waters, the samples were filtered (0.45 µm). The isolation procedure for the fulvic acid (FA) samples was based on the XAD-method of Mantoura and Riley as described by Abbt-Braun *et al.*[26] The concentrations of the model compounds were in the range 10^{-4} – 10^{-5} M for the benzoic acid derivatives. The lignin sulfonic acid was used at a DOC concentration of 10 mg/L. All model compounds were of analytical purity and used as received. Table 1 summarizes basic experimental parameters of the samples investigated.

In the steady-state and time-resolved fluorescence experiments, a spectral bandwidth of 10 nm and 9 nm was applied, respectively, with a Perkin Elmer LS 5B luminescence spectrometer. A wavelength of 314 nm was chosen for the excitation of the samples in the steady-state and time-resolved experiments. The steady-state emission spectra were recorded between 330 nm and 550 nm, respectively.

The experimental set-up of the time-resolved fluorescence experiments is described elsewhere.[25] Briefly, a lifetime spectrometer (FL 900, Edinburgh Instruments) was set up in the single photon counting mode. A nitrogen filled flash lamp was used as the excitation

Table 1 *Model compounds investigated*

Name	λ_{ex}/nm	λ_{em}/nm	pH	Conc.
Salicylic acid	314	360 - 480	4 and 6	10^{-5}M
Benzoic acid	314	400 and 450	4 and 11	10^{-5}M
Methoxy benzoic acid	314	400 and 420	4 and 11	10^{-5}M
3,5-Dihydroxy benzoic acid	337	400 and 420	4 and 11	5×10^{-4}M
Lignin sulfonic acid	314	370 - 540	7	10 mg L^{-1}

light source (1 bar, 40 kHz). The experiments were performed with a final count number in the decay peak maximum of 5×10^4 counts per minute (CPM). In order to control the performance of the instrument, for example the stability of the excitation light pulse, the experimental data were collected in five cycles of 1×10^4 CPM. All fluorescence decay curves were deconvoluted with the time profile of the excitation light flash. The fluorescence measurements were performed with air-saturated samples.

2.1 Data Treatment

The fluorescence decay curves of the HS samples were analysed by two different approaches using a small number of exponential terms (i max. 4; discrete component approach, DCA) and a large number of exponential decay terms (i max. 100; distribution analysis).[25,27] The later approach was considered to be superior because no initial assumption about the number of fluorophores had to be made and the heterogeneity of the HS samples could best be taken into account. For the distribution analysis an exponential series method (ESM) was applied (Edinburgh Instruments, Level 2 software). In the ESM analysis no shift term was applied and a decay time range between 0.1 ns and 50 ns was introduced.

The overall mean fluorescence decay time τ_{mean} of the distribution was calculated from equation (1).

$$\tau_{mean} = \frac{\sum_i A_i \cdot \tau_i}{\sum_i A_i} \qquad \begin{array}{l} \text{with} \\ A_i = \text{relative fractional intensity} \\ \tau_i = i\text{th fluorescence decay time} \end{array} \qquad (1)$$

3 RESULTS AND DISCUSSION

3.1 General Aspects of the Fluorescence Decay of HSs

In Figure 1 the fluorescence decays of salicylic acid and a brown water HS sample are compared. The data evaluation of the fluorescence decay of salicylic acid yielded a fluorescence lifetime τ_f of 4.1 ns for all emission wavelengths investigated (380 nm < λ_{em} < 480 nm, step width 10 nm) using ESM and DCA, respectively. Contrary to the mono-exponential fluorescence decay of salicylic acid, which can be found for most solutions of a single fluorophore, the fluorescence decay of HSs was highly complex as shown in Figure 1. In order to represent the experimental data, at least 3 exponential decay terms

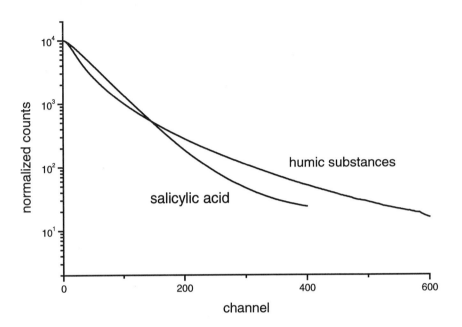

Figure 1 *Comparison of fluorescence decay curves of salicylic acid (λ_{ex}= 314 nm, λ_{em}= 425 nm, pH 6, fluorescence lifetime τ_f= 4.1 ns) and brown water HSs (λ_{ex}= 314 nm, λ_{em}= 420 nm, pH 6.5). The time calibration of the measurement was 0.095 ns/channel.*

were necessary (DCA). The disadvantage of analysing the fluorescence decay data of HSs with a small limited number of exponential decay terms is the introduction of an *a priori* model. Because of the heterogeneity of HSs, a self-modeling distribution analysis (ESM) of the fluorescence decay without a pre-assumed number of fluorescence decay terms was preferred in the data evaluation. Starting with a flat distribution of decay times, the ESM analysis optimized the amplitudes of each decay time, which were evenly spaced in the decay time range considered. In Figure 2 the results of the ESM analysis for a brown water fulvic acid (FA), lignin sulfonic acid and some benzoic acid derivatives are summarized.

The benzoic acid derivatives were chosen because related structures are considered as models for fluorophores present in HSs.[2] For an homogenous solution of a single compound, well resolved narrow peaks were found in the ESM. The mean fluorescence decay time of these peaks agreed with results obtained in a standard analysis fitting a single exponential decay (DCA). For lignin sulfonic acid and for HSs in general, less resolved broad fluorescence decay time distributions were found in the ESM analysis. Commonly for HSs, bi- or tri-modal fluorescence distributions are calculated with mean decay times centered around approximately 1 ns, 4 ns, and 10 ns, respectively.[25,28,29]

There are several reasons possibly causing the broad, poorly resolved decay time distribution found for HSs. They include the heterogeneity of the HSs and the limited capability of ESM to resolve closely located decay times with small fractional intensities

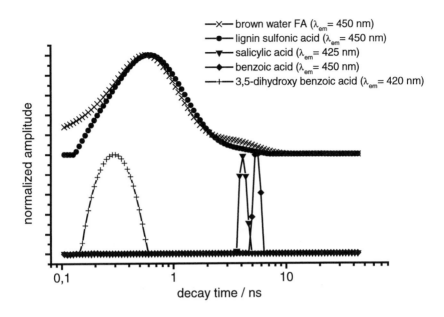

Figure 2 *Result of an ESM analysis for benzoic acid (pH 4), 3,5-dihydroxy benzoic acid (pH 4), salicylic acid (pH 4), lignin sulfonic acid (pH 7), and brown water FA (pH 7). For benzoic acid and its derivatives mean fluorescence decay times of 5.4 ns, 4.1 ns and 0.2 ns were calculated.*

(or small differences in their fractional intensities).[27] Owing to the heterogeneity of the HSs, a large number of different fluorophores in different molecular environments has to be considered. Furthermore, possible intra- and inter-molecular interaction processes, (*e.g.* energy transfer processes) have to be taken into account. For the model compounds that are assumed to be part of the 'fluorescing structures' of HSs, sharp narrow peaks were calculated. The decay times were located within the decay time distribution range found for HSs. However, for HSs no such sharp peaks were found. This makes it attractive to assume that the fluorophores present in HSs are interacting and non-isolated structures. Figure 2 also shows the fluorescence decay time distribution of lignin sulfonic acid. The distribution found agrees well in the first broad peak centered around 0.7 ns but no contributions at longer decay times were found. The fluorescence decay time distribution of HSs differed significantly. Brown water FAs are mainly built from plant material. This makes it tempting to relate part of the observed fluorescence decay of the brown water HSs to fluorescing structures originating from lignin.

3.2 Fluorescence Decay of HSs of Different Origin

In Figure 3 the fluorescence decays of HSs of different aquatic origin are compared with one another. The brown water and the soil seepage water are dominated by HSs derived from plant tissues. For these HSs the observed fluorescence decays are quite similar. However, different fluorescence decay curves were observed for HSs from the River Rhine and from a waste water effluent. For the latter two HSs, contributions of bacterial and animal tissues and of anthropogenic substances like detergents, optical brighteners and other classes of chemical compounds have to be considered as well.

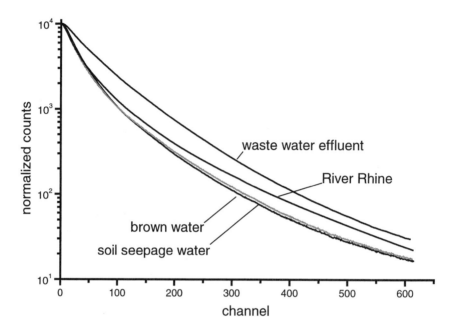

Figure 3 *Fluorescence decay curves of HSs of different origin (pH 7, λ_{ex}= 314 nm, λ_{em}= 450 nm, time calibration: 0.095 ns/channel).*

The fluorescence decay observed for HSs of the River Rhine at short times (channel < 100) resembles the fluorescence decay of brown water HSs and soil seepage water HSs. At longer times the fluorescence decay differs and approaches the decay curve observed for a waste water effluent. It is tempting to attribute the observed time-dependence of the River Rhine fluorescence to a combination of HS structures originating from (natural) plant tissues and anthropogenic compounds. The latter could be considered to influence the fluorescence properties of the waste water HSs more strongly.

To investigate intramolecular processes involved in the fluorescence decay of HSs, the influence of the emission wavelength was monitored. In the ESM distribution analysis an increase of the contribution of longer fluorescence decay times was found when the fluorescence emission was shifted to lower energy. To summarize the observed tendencies for different HSs, a mean fluorescence decay time τ_{mean} was calculated from equation (1).

The dependence of τ_{mean} on the emission wavelength is shown in Figure 4. Similar to the difference for the single decay curves monitored at λ_{em}= 450 nm (Figure 3), the observed emission wavelength dependence of τ_{mean} was related to the origin of the HSs. For all HS samples investigated, τ_{mean} increased with increasing emission wavelength. For the brown water and waste water FA the mean fluorescence decay time showed an increase up to approximately 470 nm. For higher emission wavelengths however, τ_{mean} decreased slightly again. For the ground water FA the increase of τ_{mean} was found for all emission wavelengths investigated.

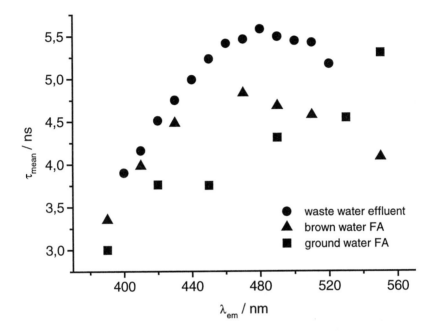

Figure 4 *Dependence of τ_{mean} on the emission wavelength λ_{em}. τ_{mean} was calculated according to equation (1) (λ_{ex}= 314 nm, pH 7).*

To prove the contribution of lignin-like structures to the fluorescence of HSs, the dependence of τ_{mean} on the emission wavelength was monitored over the range 370 nm < λ_{em} < 540 nm. Again, the decay time distribution found in an ESM analysis was shifted with increasing λ_{em} towards longer decay times. This emphasizes the role of plant-derived materials in the origin of the HS samples investigated.

Comparison with model compounds proved the capabilities of the ESM analysis and showed that the calculated broad decay time distribution pattern is caused by the properties of the HSs. Specifically, comparison of the fluorescence decay pattern of lignin sulfonic acid and HS samples showed a close relationship, indicating that for the samples investigated lignin-like structures were retained during the formation of the HSs. This indicates that lignin-related structures still determine the fluorescence of HSs derived from plant materials. It is tempting to attribute the observed complexity of the fluorescence decay of HSs to the heterogeneity of the HS structure as well as to intra- and intermolecular interaction processes. Attractive to assume are interaction processes like intra- and intermolecular energy transfer processes within the HS matrix. The observed dependence of the ESM decay time pattern on the emission wavelength could arise from such processes. Fluorophores with different energy levels are present in HSs. Fluorophores with high energy levels can fluoresce or transfer energy to a fluorophore in the molecular neighborhood that possesses suitable acceptor levels. To allow a successful energy transfer, several limiting conditions have to be satisfied: overlap of the donor-acceptor orbital, proximity of donor and acceptor, and so on. In energy transfer a decrease in the decay time of the energy donor would be observed. A longer decay time could be expected for the acceptor because it is 'excited' by the light flash and by the donor fluorophore which could be considered a kind of second 'light source'. As a consequence of the great heterogeneity of HSs and limiting experimental conditions, it impossible to monitor a kinetic build-up of an acceptor fluorophore.

Because of the heterogeneity of HSs it is tempting to assume a high density of energy levels of fluorophores. In HSs at short emission wavelengths (equal to high energy), a tendency of short decay times was found, whereas at longer emission wavelengths (*i.e.* lower energy) the mean fluorescence decay time increased and finally levelled off. This indicates that at very long emission wavelength ($\lambda_{em} > 480$ nm) energy transfer no longer occurs because of the low energy of the fluorophores.

More experiments are necessary to investigate the applicability of the suggested model summarized in Figure 5. Experiments with chemically modified HSs are in progress.

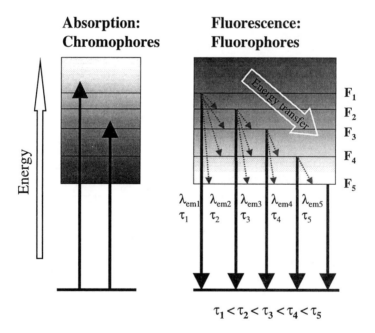

Figure 5 *Schematic representation of the proposed energy transfer interactions inside the HS matrix.*

On oxidation, the modified chemical structure of HSs should be obvious in changes of energy transfer processes. Complexation with metal ions and the effect of metals on the observed fluorescence decay of HSs being investigated will give further understanding of the dynamic processes of HSs.

ACKNOWLEDGEMENTS

The Deutsche Forschungsgemeinschaft (DFG) funded part of the work presented in the ROSIG research program. The authors wish to thank Dr. Gudrun Abbt-Braun for the supply of HS samples. They further appreciate the support of the staff of the Engler-Bunte-Institute, Division of Water Chemistry.

References

1. K. M. Spark and R. S. Swift, in 'Humic Substances in the Global Environment and Implications on Human Health', N. Senesi and T. M. Miano (eds.), Elesevier, Amsterdam, 1994, p. 153.
2. N. Senesi, T. M. Miano and M. R. Provenzano, in 'Humic Substances in the Aquatic and Terrestrial Environment', B. Allard, H. Boren and A. Grimwall (eds.), 'Lecture Notes in Earth Sciences', Springer Verlag, 1991, **33**, p. 63.
3. J. J. Mobed, S. L. Hemmingsen, J. L. Autry and L. B. McGown, *Environ. Sci. Technol.*, 1996, **30**, 3061.
4. R. R. Engebretson, T. Amos and R. von Wandruszka, *Environ. Sci. Technol.*, 1996, **30**, 990.
5. M. J. Pullin and S. E. Cabaniss, *Environ. Sci. Technol.*, 1995, **29**, 1460.
6. F. H. Frimmel and W. Hopp, *Fresenius' Z. Anal. Chem.*, 1986, **325**, 68.
7. C. F. Scheck, F. H. Frimmel and A. M. Braun, *Z. Naturforsch.*, 1992, **47b**, 399.
8. S. E. Cabaniss, *Environ. Sci. Technol.*, 1992, **26**, 1133.
9. J. C. G. E. da Silva, M. A. Ferreira and A. A. S. C. Machado, *Talanta*, 1994, **41**, 2095.
10. E. Casassas and I. M. R. Tauler, *Anal. Chim. Acta*, 1995, **310**, 473.
11. J. C. G. E. da Silva, A. A. S. C. Machado and C. S. P. C. O. Silva, *Anal. Chim. Acta*, 1996, **318**, 365.
12. J. C. G. E. da Silva, M. A. Ferreira, A. A. S. C. Machado and F. Rey, *Anal. Chim. Acta*, 1996, **333**, 71.
13. J. C. G. E. da Silva, M. A. Ferreira and A. A. S. C. Machado, *Analyst (Cambridge)*, 1997, **122**, 1299.
14 T. D. Gauthier, E. C. Shane, W. F. Guerin, W. R. Seitz and C. L. Grant, *Environ. Sci. Technol.*, 1986, **20**, 1162.
15. M. U. Kumke, H.-G. Löhmannsröben and T. Roch, *Analyst (Cambridge)*, 1994, **119**, 997.
16. U. Zimmermann, H.-G. Löhmannsröben and T. Skrivanek, in 'Remote Sensing of Vegetation and Water, and Standardization of Remote Sensing Methods, SPIE', G. Cecchi, T. Lamp, R. Reuter and K. Weber (eds.), 1997, **3107**, p. 239.
17. R. R. Engebretson and R. von Wandruszka, *Environ. Sci. Technol.*, 1994, **28**, 1934.
18. H. Wimmer, J. I. Kim and R. Klenze, *Radiochim. Acta*, 1992, **58/59**, 165.

19.   J. I. Kim, R. Klenze, H. Wimmer, W. Runde and W. Hauser, *J. Alloys Comp.*, 1994, **213/214**, 333.

20.   P. Panak, R. Klenze and J. I. Kim, *Radiochim. Acta*, 1996, **74**, 141.

21.   G. Bidoglio, I. Grenthe, P. Qi, P. Robouch and N. Omenetto, *Talanta*, 1991, **38**, 999.

22.   C. H. Lochmüller and S. S. Saavedra, *Anal. Chem.*, 1986, **58**, 1978.

23.   R. L. Cook and C. H. Langford, *Anal. Chem.*, 1995, **67**, 174.

24.   L. B. McGown, S. L. Hemmingson, J. M. Shaver and L. Geng, *Appl. Spectrosc.*, 1995, **49**, 60.

25.   M. U. Kumke, G. Abbt-Braun and F. H. Frimmel, *Acta Hydrochim. Hydrobiol.*, 1998, **26**, 73.

26.   G. Abbt-Braun, F. H. Frimmel and P. Lipp, *Z. Wasser-Abwasser-Forsch.*, 1991, **24**, 285.

27.   D. M. Gakamsky, A. A. Godin, E. P. Petrov and A. N. Rubinov, *Biophys. Chem.*, 1992, **44**, 47.

28.   M. U. Kumke, C. D. Tiseanu, G. Abbt-Braun and F. H. Frimmel, *J. Fluorescence*, 1998, submitted for publication.

29.   M. U. Kumke and F. H. Frimmel, in 'The Role of Humic Substances in the Ecosystems and in Environmental Protection', J. Drozd, S. S. Gonet, N. Senesi and J. Weber (eds.), Polish Society of Humic Substances, Wroclaw, Poland, 1997, p. 525.

EFFECT OF LIME ADDITIONS TO LAKE WATER ON NATURAL ORGANIC MATTER (NOM) IN LAKE TERJEVANN, SE NORWAY: FTIR AND FLUORESCENCE SPECTRAL CHANGES

James J. Alberts,[1] Dag O. Andersen[2] and Monika Takács[1]

[1] University of Georgia Marine Institute, Sapelo Island, GA 31327, USA
[2] Agder College, Tordenskjoldsgt. 65, Postuttak, 4604, Kristiansand, Norway

1 INTRODUCTION

Surface waters in many regions of the world are experiencing declining water quality due to acidification. This decline is due to accelerated inputs of acidic chemicals into poorly buffered systems. The lowered pH, which is often accompanied by increased heavy metal loading, results in declines in aquatic life to the point where the systems become "sterile" to many species of vertebrates and macroinvertebrates.[1-3]

In many areas, the poor buffering capacity of the surface waters results from the lack of watershed mineral bedrock or soils, which contain inorganic components (carbonates) that typically contribute to the buffering capacity of surface waters. The buffering in these systems is often controlled by natural organic matter derived from the vegetation of the watershed.

Abatement techniques to alleviate surface water acidification often involve the addition of carbonate minerals (liming) to increase the pH to a level that can sustain macrofauna. This technique is most commonly used in lakes, although liming of flowing waters is also practised. Extensive utilizations of liming techniques have been employed in Norway[4] and Sweden,[5] but the technique is also used in the United Kingdom,[6] Canada[7] and the USA.[7]

Waters receiving liming treatment often have high natural organic matter (NOM) loads, the majority of which are allochthonous in origin. This watershed derived material often comprises in excess of 90% of the total dissolved organic carbon (DOC), especially in polyhumic lakes.[8-10] Water residence time is a controlling factor for the retention of NOM in lakes[11,12] and humic matter is known to be removed from the water column by photochemical and biological degradation, as well as by flocculation.[13-17] However, the effect of liming on these systems is not well understood. Despite the fact that larger molecular size humic substances have been shown to precipitate with increasing salt concentrations,[18-20] limestone additions to acidic lakes have caused both reductions[21,22] and increases[23,24] in NOM concentrations.

We investigated the changes that occur to the NOM of Lake Terjevann in southeastern Norway as a result of liming. We have previously reported on the alteration of the size distribution of the NOM pool resulting from precipitation of organic matter during its residence in the lake basin.[25,26] In this report we investigate effects of lime additions on the infrared and fluorescence spectral characteristics of the NOM as it changes in the lake.

2 SITE CHARACTERISTICS

Lake Terjevann has been the subject of intensive study and a more complete description of the lake and its watershed is given elsewhere.[25,26] Briefly, it has a surface area of 0.09 km^2 and a catchment of 1.09 km^2. The latter is divided into two primary basins of significantly differing vegetation type (mixed deciduous/pine stands and a spruce forest established 30 years ago). The conductivity of the lake is relatively low (approx. 60–70 µS/cm) and dissolved organic carbon values range from 2.5 to 7.0 mgC/L. Estimated retention time of the lake is 0.72 years. The lake has received multiple liming treatments since 1980, with the most recent being in 1995. The effect of these treatments has been to raise the pH of the lake about 2 pH units, from 4.4 to 6.5.

3 METHODS

Three permanent sampling locations were established in the study area. One station at the mouth of each of the tributaries draining the two major catchment basins (Inlet 1, spruce uplands and Inlet 2, mixed deciduous/pine; respectively), and the third station (Outlet) located in the lake just before the weir that forms its outlet.

3.1 NOM Collection

Between 12th and 14th October 1996, NOM samples were collected by reverse osmosis (RO) using a PROS/2S portable RO unit. The NOM from 1.4, 1.7 and 1.8 m^3 (stations: Inlet 1, Inlet 2, Outlet, respectively) of water was concentrated[27] with estimated recoveries of 90, 87 and 85%, respectively. Samples were purified by filtration (0.45 µm) and freeze-dried.[26]

3.2 Fractionation of NOM by Ultrafiltration

Solid samples of each NOM isolate (100 mg) were diluted to 250 ml in deionized water and ultrafiltered over an XM50 ultrafilter (Amicon Corp.) which is reported to have a nominal molecular weight (NMW) cutoff of 50 kD as determined with globular protein standards. The filtration was conducted under N$_2$ pressure (8 psi). Sample volume was reduced to approximately 100 ml. One hundred ml of deionized water was then added to the retentate and the sample volume was again reduced under N$_2$ pressure to ~100 ml. This washing was repeated once again with a fresh 100 ml of deionized water. An aliquot of the retentate was reserved for DOC analysis and the remaining retentate solution was freeze-dried.

The ultrafiltrate from the XM50 ultrafiltration was ultrafiltered over a PM10 ultrafilter (Amicon Corp.), 10 kD NMW, in the same manner as above (N$_2$, 10–14 psi). Aliquots of the retentate and ultrafiltrate from this ultrafiltration were reserved for DOC analyses and the solutions were freeze-dried.

3.3 FTIR Spectroscopy

Solid potassium bromide was combined with the freeze-dried materials (2% w/w) and pressed into pellets for collecting the infrared spectra (Perkin-Elmer Paragon 1000). Peak assignments were made visually and with the Perkin-Elmer Spectroscopy software.

3.4 Ultraviolet–Visible Spectroscopy

Solid samples of the freeze-dried NOM were dissolved in 0.05N $NaHCO_3$[28] (200 ppm) and the absorbances at 465 and 665 nm were measured using a Hitachi Model 100-80A spectrophotometer in 1 cm cuvettes.

3.5 Fluorescence Spectroscopy

All fluorescence spectra were collected with a Hitachi Model 3010 spectrofluorometer in 1 cm cuvettes. Spectra were collected and data analysed using LabCalc software (Galactic Industries). Solid samples of NOM were dissolved in deionized water.

3.5.1 Excitation and Emission. Principal excitation and emission wavelengths were determined by pre-scanning the samples and then spectra were optimized for excitation and emission spectra. Excitation wavelengths were between 230 and 235 nm. Slit widths were 5 nm, recording speed 120 nm/min and instrument response factor 2.

3.5.2 Synchronous Scanning. Synchronously scanned spectra were collected by maintaining a constant difference between the excitation and emission wavelengths (Δ = 18 nm) and then scanning between 250 and 600 nm. Optimal resolution was obtained with spectral offset slit width of 3.

3.5.3 3-Dimensional Spectra. 3-dimensional spectra were collected for the NOM samples by collecting a full emission spectrum with an excitation wavelength of 220 nm and then repeating the emission spectrum scan but increasing the excitation wavelength by 6 nm. This collection of emission spectra was continued with subsequent 6 nm increases in excitation wavelength for a total of 29 scans. The emission wavelength range was 400–600 nm. Spectra collected in this manner were then combined to produce a 3-D representation of the fluorescence spectral behavior of the NOM mixtures.

3.6 Analyses

Ash contents of freeze-dried samples were determined gravimetrically after heating samples at 475°C for 5 hours. Dissolved organic carbon contents of the solutions used for fluorescence analyses were determined by high temperature combustion (Shimatzu Model 500 TOC Analyser). Carbon, hydrogen and nitrogen contents of the freeze-dried samples were determined on solid samples in duplicate (Perkin Elmer Model 2400 CHN Elemental Analyser).

4 RESULTS AND DISCUSSION

4.1 Elemental and Ultraviolet–Visible Evidence

The RO procedure, while being very efficient in isolating large quantities of dissolved organic matter from freshwater systems (approx. 90% recovery efficiency),[27] does suffer from the problem that certain divalent ions, particularly sulfate, are also concentrated by the process. This concentration leads to isolates with abnormally high ash contents as is the case with these samples (Table 1). While these ash contents may pose a problem in some analyses, many others are affected very little if at all. A recent symposium[29] reported the results of extensive analyses of nine Norwegian NOM samples collected by this technique and analysed by numerous international laboratories. The results of that work and similar preliminary studies[30,31] indicate that for the most part analyses that are conducted on re-dissolved samples are unaffected by the high ash contents.

Table 1 *Characteristics of Lake Terjevann NOM Samples and Solutions Used for Fluorescence Spectroscopy and Wavelengths of Maximum Excitation and Emission.*

Sample	pH	% Ash	C/N[a]	C mg/L	E_4/E_6	λ_{ex}	λ_{em}	Int.[b] mgC
Inlet 1								
Total	4.03	68.0	35.44	5.31	6.71	233	437	1.85
>50 kDa	4.83	53.9	46.87	12.8	ND[c]	233	439	0.64
<10 kDa	5.02	ND	24.69	8.77	ND	233	429	1.69
							350	
Inlet 2								
Total	4.16	68.2	31.34	4.92	13.5	236	431	2.57
>50 kDa	5.35	53.3	45.58	10.1	ND	238	438	1.11
<10 kDa	ND	ND	15.99	ND	ND	ND	ND	ND
Outlet								
Total	5.01	77.5	14.96	3.61	7.50	ND	ND	ND
>50 kDa	6.07	61.2	25.96	9.92	ND	233	429	1.17
<10 kDa	6.34	ND	15.99	8.54	ND	232	417	1.33
							349	

[a] atomic ratio; [b] relative fluorescence intensity/mg C in solution; [c] ND = not determined.

Ultraviolet–visible spectra of the samples showed the common monotonic decline of humic substances isolated from aquatic environments.[32] The absorbance ratios at 465 and 665nm (E_4/E_6) of the Lake Terjevann NOM isolates (Table 1) are characteristic of smaller size organic matter, usually believed to be fulvic acids,[33] although the lower value of the Inlet 1 NOM is indicative of the presence of some larger materials. There are no clear indications of loss of larger molecular size NOM as a result of in-lake precipitation processes, as was observed earlier using absorbance ratios at 254 and 410nm.[25]

The C/N atomic ratio for total NOM isolated from the Outlet corresponds to the values observed for the smallest size fractions from both Inlets, which is consistent with the hypothesis of concentration of the smaller fraction relative to the larger fractions as NOM traverses the lake by a process of in-lake precipitation of the larger NOM.[25,26] The fact that the values correspond and are much lower than the C/N ratios of the NOM in larger fractions indicates an increased nitrogen content relative to carbon of smaller sized NOM, which is also consistent with the hypothesis that allochthonous inputs of nitrogen to the lake NOM are less likely than concentration of the smaller fractions by in-lake precipitation processes.[26] This is particularly apparent in the value of the C/N ratio of the >50 kDa NMW fraction of the Outlet sample, which is almost the same as the <10 kDa NMW fraction of Inlet 1. In no case is there an indication of a significant amount of NOM in the Outlet sample with C/N ratio approaching that of the >50 kDa NMW fractions of both Inlets 1 and 2. The high C/N ratio means that the NOM in that fraction is relatively poor in nitrogen. Microorganisms have been shown to utilize humic substances as a nitrogen source rather than a carbon source.[34] These facts indicate that precipitation rather than biological activity is responsible for the changes in the total NOM pool at the outlet of the lake.

4.2 FTIR Evidence

Unfortunately, the FTIR spectral analyses were affected by the high salt contents of the RO isolates. Although there is some indication of carboxyl and amino structures in the spectra from all fractions of the NOM (Figure 1), the spectra are dominated by peaks consistent with hydrated inorganic sulfates, most probably sodium forms which are known to be concentrated by the isolation procedure.

4.3 Fluorescence Evidence

4.3.1 Emission Spectra. The emission spectra of the NOM samples all showed a broad featureless peak which centered around 435–440 nm (Table 1) and are typical of aquatic humic substances.[35] Fractionation of the NOM samples did not alter the spectra appreciably, although the <10 kD NMW fractions that were analysed did show a distinct shoulder at approximately 350 nm. The pH values of the sample solutions were very similar to those of the waters from which they were

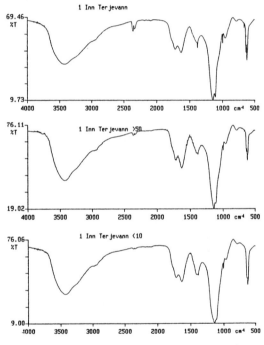

Figure 1 *FTIR spectra Inlet 1 NOM total and size fractions*

isolated. Despite the slightly higher pH value of the Outlet NOM sample, there was no indication of spectral alteration due to this pH shift as previously observed for humic substances,[35] nor of alteration of the NOM as a result of in-lake processes.

4.3.2 Synchronously Scanned Spectra. Unlike the emission spectra, synchronous scanning of the NOM isolates with a Δ = 6 nm provided significantly different spectra (Figure 2). The sample from Inlet 1 had a broader peak with absorbance at the longer wavelengths. The sample from Inlet 2 has a distinctly increased absorbance at shorter wavelengths while maintaining some of the characteristic shoulders observed in NOM from Inlet 1. The NOM from the Outlet has little evidence for peaks at the longer wavelengths and a slight blue shift of the major peak relative to samples from Inlets 1 and 2.

Absorbance at longer wavelengths is characteristic of larger, more highly conjugated systems. This is shown in the synchronous spectra of the NOM fractions from Inlet 1 (Figure 3). The > 50 kDa NMW fraction of the Inlet 1 NOM clearly shows the red shift of the spectra and enhanced importance of peaks from 400 nm upwards. Similarly, the spectrum of the < 10 kDa NMW fraction has much less red character and appears shifted to the blue wavelengths. Interestingly, the spectrum of the < 10 kDa NMW fraction from Inlet 1 is very similar to the spectrum of the total NOM sample from Inlet 2. This correspondence is in agreement with the molecular weight distribution study that showed 60–65% of the NOM fractions in the Inlets were in size fractions >10 kDa NMW, while only 30% of the Outlet NOM were in these fractions.[26]

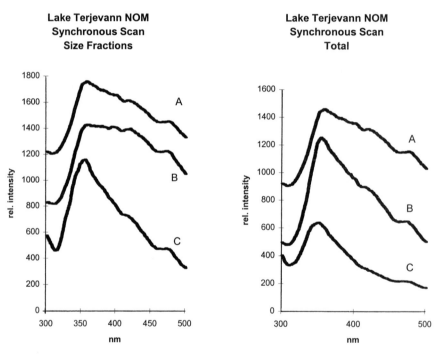

Figure 2 *Fluorescence spectra NOM*
A=Inlet1, B=Inlet2, C=Outlet

Figure 3 *Fluorescence spectra Inlet 1*
A=total, B=>50,000, C=<10,000

4.3.3 3-Dimensional Spectra. The 3-dimensional spectra of the NOM isolates from Lake Terjevann and their fractions were very similar, here represented by the total NOM isolate from Inlet 1 (Figure 4). The two characteristic peaks are common features of natural organic matter and humic substances isolated from a number of aquatic environments.[36-38] Peaks centered around λ_{ex} 340 nm, λ_{em} 440 nm are usually attributed to simple phenolic compounds, while the peaks centered around λ_{ex} 250 nm, λ_{em} 440 nm are typical of humic substances from many environments and may represent the fluorophores associated with aromatic structures.[39]

The positions of peak A for the total NOM isolates from all three sites (Table 2) are in reasonable agreement with those of the corresponding peak in a series of 11 isolates from 9 Norwegian surface waters.[40] However, while peak B for the Inlet 1 NOM sample is in the same location as in the isolates from the surface water series, both the Inlet 2 and Outlet NOM samples are shifted to lower excitation and emission wavelengths. A similar shift is noted for the Inlet 1 fractions as overall size decreases. As with the synchronously scanned spectra, the spectra of the <10 kDa NMW frac-tion from Inlet 1 is very similar to that of the total NOM isolate from the Outlet (Table 2), again corresponding to a loss of material by in-lake processes.

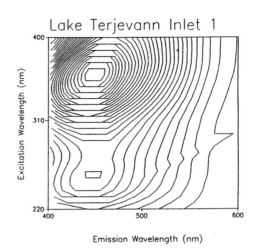

Figure 4 *Three-dimensional fluorescence spectra, Inlet 1 total NOM*

The peak ratios of the Inlet 1 NOM isolate also correspond to values found in the other Norwegian surface water isolates,[40] indicating that the simple phenolic peak is relatively more important than the more general absorbance peak A. However, at both Inlet 2 and the Outlet and in all size fractions, the relative importance of this peak becomes much more significant, and in fact is greater than the absorbance of peak B (Table 2).

Table 2 *Three-dimensional peak assignments and intensities for NOM samples and ultrafiltered fractions from L. Terjevann.*

Sample	Peak A λ_{ex}	λ_{em}	Int.[a]	Peak B λ_{ex}	λ_{em}	Int.	B/A[b]
Inlet 1							
Total	365	452	3.82	255	449	1.29	0.34
>50 kDa	347	446	0.64	238	441	0.69	1.08
<10 kDa	342	439	0.98	232	429	1.62	1.65
Inlet 2							
Total	345	438	1.57	231	436	2.28	1.45
>50 kDa	348	445	1.03	234	440	1.07	1.04
Outlet							
Total	333	432	1.50	233	425	2.84	1.89
>50 kDa	339	442	0.90	233	436	1.43	1.59
<10 kDa	329	425	0.62	233	420	1.67	2.69

[a] Int. = relative absorbance intensity normalized to mg C/sample; [b] B/A = ratio of Int. Peak B to Int. Peak A

This trend of decreasing peak A height has been noted in other aquatic fluorescence spectra and is indicative of "older" or more weathered organic matter. This interpretation would also support removal of NOM by in-lake processes, but in this case biological degradation cannot be excluded.

5 CONCLUSIONS

Natural organic matter (NOM) was isolated by reverse osmosis (RO) techniques from three stations in a lake in southeastern Norway. This lake has received lime treatments for over 15 years, resulting in an increase in pH of the outlet waters by almost 2 pH units to a value around 6.0. Elemental and multi-spectral analyses of the NOM isolates and size fractions of that material as defined by ultrafiltration indicate that in-lake processes, most likely precipitation, are altering the composition of the NOM by removing larger molecular size compounds. In addition, the NOM remaining in the lake is enriched in nitrogen relative to carbon. Thus, a potential result of liming of lakes with significant organic loadings would be the removal of larger organic matter, and possibly the pollutants associated with it. In addition, the increased nitrogen content of smaller molecules may allow heterotrophic activity to increase if these compounds are readily assimilated by microorganisms.

ACKNOWLEDGEMENTS

This work was supported by the Norwegian Directorate of Nature Management (DN). Grants from Andreas & K. Ludvig Endresens legate are appreciatively acknowledged. Janneke Aulie is gratefully thanked for co-operation in the field. J. J. Alberts wishes to thank the U.S.-Norway Fulbright Foundation for Educational Exchange for support of this work. This is Contribution No. 819 of the University of Georgia Marine Institute.

References

1.   R. F. Wright, *Water Quality Bull.*, 1983, **8**, 137.
2.   D. W. Schindler, *Science*, 1988, **239**, 149.
3.   A. H. M. Bresser and W. Salomons, 'Acidic Precipitation. International Overview and Assessment', Springer-Verlag, Berlin, 1990.
4.   A. Hindar and B. O. Rosseland, *Water, Air, Soil Pollut.*, 1988, **41**, 17.
5.   P. Nyberg and E. Thørneløf, *Water, Air, Soil Pollut.*, 1988, **41**, 3.
6.   G. Howells and T. R. K. Dalziel, 'Restoring Acid Waters: Loch Fleet', 1984-1990', Elsevier Applied Science, Amsterdam, 1992.
7.   H. Olem, 'Liming Acidic Surface Waters', Lewis Publishers, 1991.
8.   R. H. Hesslein, W. S. Broecker, P. D. Quay and D. W. Schindler, *Can. J. Fish. Aquat. Sci.*, 1980, **37**, 454.
9.   D. W. Schindler, S. E. Bayley, P. J. Curtis, B. R. Parker, M. P. Stainton and C. A. Kelly, *Hydrobiologia*, 1992, **229**, 1.
10.   M. Meili, *Hydrobiologia*, 1992, **229**, 23.
11.   D. R. Engstrom, *Can. J. Fish. Aquat. Sci.*, 1987, **44**, 1306.

12. E. T. Gjessing and J. E. Samdal, *J. Amer. Waterworks Assn.*, 1968, **60**, 451.
13. E. T. Gjessing and T. Gjerdahl, *Vatten*, 1970, **2**, 144.
14. D. J. Strome and M. C. Miller, *Verh. Internat. Verein. Limnol.*, 1978, **20**, 1248.
15. A. J. Stewart and R. G. Wetzel, *Arch. Hydrobiol.*, 1981, **92**, 265.
16. C. Steinberg and U. Muenster, in 'Humic Substances in Soil, Sediment, and Water', G. R. Aiken, D. M. McKnight, R. L. Wershaw and P. MacCarthy (eds.), Wiley, New York, 1985.
17. H. DeHaan, *Limnol. Oceanogr.*, 1993, **38**, 1072.
18. M. M. Kononova, 'Soil Organic Matter', 2nd edn., Pergamon Press, Oxford, 1966.
19. J. R. Ertel, J. J. Alberts and M. T. Price, 'Proceedings of the 1991 Georgia Water Resources Conference', K. J. Hatcher (ed.), University of Georgia Institute of Natural Resources, Athens, GA, 1991.
20. J. J. Alberts and C. Griffin, *Arch. Hydrobiol. Spec. Issues Advanc. Limnol.*, 1996, **47**, 401.
21. O. M. Brynildson, A. D. Hassler and J. A. Larson, *Wisconsin Conserv. Bull.*, 1952, **17**, 11.
22. C. T. Driscoll, J. R. White, G. C. Schafran and J. D. Rendall, *J. Environ. Engineer. Div. ASCE*, 1982, **108**, 1128.
23. R. F. Wright, *Can. J. Fish. Aquat. Sci.*, 1985, **42**, 1103.
24. O. Broberg, 'Liming of Lake Gårdsjøn', National Swedish Environmental Protection Board Report, **3426**, 1988.
25. D. O. Andersen, *Water, Air, Soil Pollut.*, in revision.
26. D. O. Andersen, J. J. Alberts and M. Takács, *Environ. Sci. Technol.*, in revision.
27. S. M. Serkiz and E. M. Perdue, *Wat. Res.*, 1990, **24**, 911.
28. Y. Chen, N. Senesi and M. Schnitzer, *Soil Sci. Soc. Am. J.*, 1977, **41**, 352.
29. Proceedings Workshop on NOM-Typing Multi-Method Characterization of Nine Norwegian NOM Isolates, Agder College, Kristiansand, Norway, June, 1998, to be published in *Environ. Internat.*
30. NOM-Typing Project, *NIVA Newsletter*, 1997, **1/97**.
31. E. T. Gjessing, J. J. Alberts, A. Bruchet, P. K. Egeberg, E. Lydersen, L. B. McGown, J. J. Mobed, U. Munster, J. Pemkowiak, E. M. Perdue, H. Ratnawerra, D. Rybacki, M. Takács and G. Abbt-Braun, *Wat. Res.*, in press.
32. Z. Filip and J. J. Alberts, *Sci. Total Environ.*, 1989, **83**, 273.
33. J. J. Alberts and Z. Filip, *Trends in Chem. Geol.*, 1994, **1**, 143.
34. Z. Filip and J. J. Alberts, *Sci. Total Environ.*, 1994, **144**, 121.
35. J. J. Alberts, S. Filip, M. T. Price, D. J. Williams and M. C. Williams, *Org. Geochem.*, 1988, **12**, 455.
36. P. G. Coble, C. A. Schultz and K. Mopper, *Mar. Chem.*, 1993, **41**, 173.
37. S. K. Hawes, MS Thesis, University of Florida, 1992.
38. J. J. Alberts and T. Miano, 1994-95 Annual Report, University of Georgia Marine Institute, 1995.
39. P. Blaser, Newsletter No. 1/97, Norwegian Institute for Water Research, Agder College, 1997.
40. P. Blaser, A, Heim and Jörg Luster, in Proceedings of the NOM-Typing Workshop, Agder College, Kristiansand, Norway, June 3-6, 1998.

A COMPUTATIONAL CHEMISTRY APPROACH TO STUDY THE INTERACTIONS OF HUMIC SUBSTANCES WITH MINERAL SURFACES

Leonid G. Akim,[1] George W. Bailey[2] and Sergey M. Shevchenko[3]

[1] National Research Council, c/o U.S. EPA
[2] U.S. Environmental Protection Agency, Athens, GA 30605-2700, USA
[3] University of British Columbia, Vancouver, BC V6T 1Z4, Canada

1 INTRODUCTION

There is a commonly held point of view that all environmental surfaces are either organic in nature or are mineral surfaces coated with organic, usually humic substances. If this is indeed so, does this coating mask these mineral surfaces such that their chemical functionalities are unavailable for reaction with organic and inorganic contaminants? One hypothesis is that, since humic substances have a very high affinity for the external surfaces of phyllosilicate and oxide minerals, they will be the predominant surface in equilibrium with solutes in solution or those in the vapor state. This will occur except in situations where organic carbon concentrations are very low. Humic substances in these circumstances may even coat the mineral surface with a monolayer of organic carbon. In the absence of organic carbon, oxides will coat external phyllosilicate surfaces and these surfaces will dominate in chemical reactions with solutes in solution. Generally it is the humic and fulvic acid surface chemical functionality that controls reaction with solutes dissolved in solution. Therefore, with regard to surface interactions it is organo-mineral aggregates that largely define the behavior and fate of chemical contaminants in the environment. We must develop a reliable computer simulation model of humic and fulvic acids, mineral surfaces and edges and organo-mineral aggregates to aid in predicting speciation, distribution, bioavailability, transport, and exposure concentrations of chemical contaminants in terrestrial and aquatic ecosystems.

In recent years the field of computational chemistry has made significant advances in the power, sophistication, and ease of software use. Parallel advances have been made in the computing power, lowered cost and enhanced availability of workstations to do conformational analyses and to study the energetics of chemical reactions at surfaces. Computational chemistry has been used to study the energetics of pure mineral surfaces and an attempt has been made to simulate the structures of model humic substances. There is an urgent need in both environmental soil chemistry and aquatic chemistry to simulate and predict the reactions of humic substances[1-4] with mineral surfaces,[5,6] as well as the reactivity of this composite surface with low molecular weight contaminants.

Interest in using computational methods to simulate humic substances and their interactions with minerals has been demonstrated by different research groups and a number of papers have been published on this subject. Different computer models of humic acids were proposed in the context of computer-assisted structure elucidation efforts[7] or to verify proposed earlier formulations.[4] Computational approaches have been used in conjunction with microscopic methods to propose secondary structures of humic substances.[8]

Recently, we have published several papers covering various aspects of modelling lignin–carbohydrate complexes, humic substances and non-bonded organo–mineral interactions.[1,9–11] In this paper we will review briefly our previously published methods and data, and integrate these separate efforts into a methodology to simulate and predict the chemical reactivity of humic substances, minerals and organo–mineral aggregates. The experimental details are given in our original papers.[1,9–11] Here we outline the most important results to give a holistic perspective of the developed technique, to show the methodological approaches and to briefly review previously obtained results. Furthermore, we show how the approach can be used to study the interaction of low molecular weight contaminants with organo–mineral surfaces.

2 COMPUTATIONAL METHODS

The calculations were performed using the SYBYL® software (Tripos Associates, Inc.) on two Silicon Graphics workstations: IRIS Indigo R4000 Elan and Indigo II R10000. These workstations are powerful enough to support molecular dynamic calculations of systems with up to 3900 atoms. Some systems with several water shells exceeded this size but the required computation time still stayed within a reasonable interval (*i.e.* about 3–4 days of continuous simulation).

The Tripos force field[12,13] has been used for molecular mechanical conformational calculations and for molecular dynamics simulations. The electrostatic interactions and hydrogen bonds were accounted for with atomic charges. These charges were calculated by employing empirical and semiempirical methods; the same approaches were used in a consistent way in each series of model structures (Gasteiger–Hückel[14,15] for small lignin models, Pullman[16,17] for polymer models and MNDO[18] for optimization and charge calculations on small molecules like water and organic pollutants). For muscovite the average atomic charges were ascribed to each type of atom in the mineral component based on a MNDO calculation of a smaller muscovite particle.[1]

Organo–mineral complexes or aggregates were compiled from optimized organic molecules, model muscovite structures and humic–muscovite structures using a "docking" procedure. Formed structures were fully optimized before any further operations were initiated. The hydration shells (the first hydration shells around a model muscovite structure contained 258 water molecules) were automatically generated through the droplet simulation procedure.[21]

Molecular dynamics calculations were performed using the simulated annealing algorithm.[19,20] In this technique a stochastic (Monte Carlo) search of conformational space at an initial high temperature is combined with an appropriate cooling schedule. Details of heating–cooling patterns varied for different models. The most typical consisted of rapidly heating the system to 700 K, holding for 1 ps at this temperature, and then cooling to 200 K in 1 ps. The procedure was repeated ten times. In cases where the procedures differed, the references to the original papers are provided with detailed description of all experimental methods.[1,9–11]

3 DISCUSSION

3.1 Construction of a Lignin-carbohydrate Complex

Lignin is an important precursor of humic substances. According to existing humification theories, a significant part of the aromatic structures in humic substances has originated from lignin, or, to some degree, from lignin-like aromatic fragments of suberin and from lignans.[22] The degree of preservation of lignin constituents in humic substances is still under discussion, but the amount persisting could be as much as 30%.[22,23] Owing to their high chemical reactivities, the lignin-derived structures play a key role in the chemical reactions of humic substances. Similarities and differences in structure and chemical reactivity between lignin and humic substances were recently reviewed as they relate to the existing humification hypothesis.[24] This accepted hypothesis provides us with the rationale to build models of humic acid from lignin and lignin–carbohydrate complex precursors.

Lignin is a natural cross-linked irregular polymer consisting of phenylpropane units connected by different types of carbon–carbon and ether bonds (Figure 1).[25] However, this polymer includes linear fragments where interunit linkages are represented by β-O-4

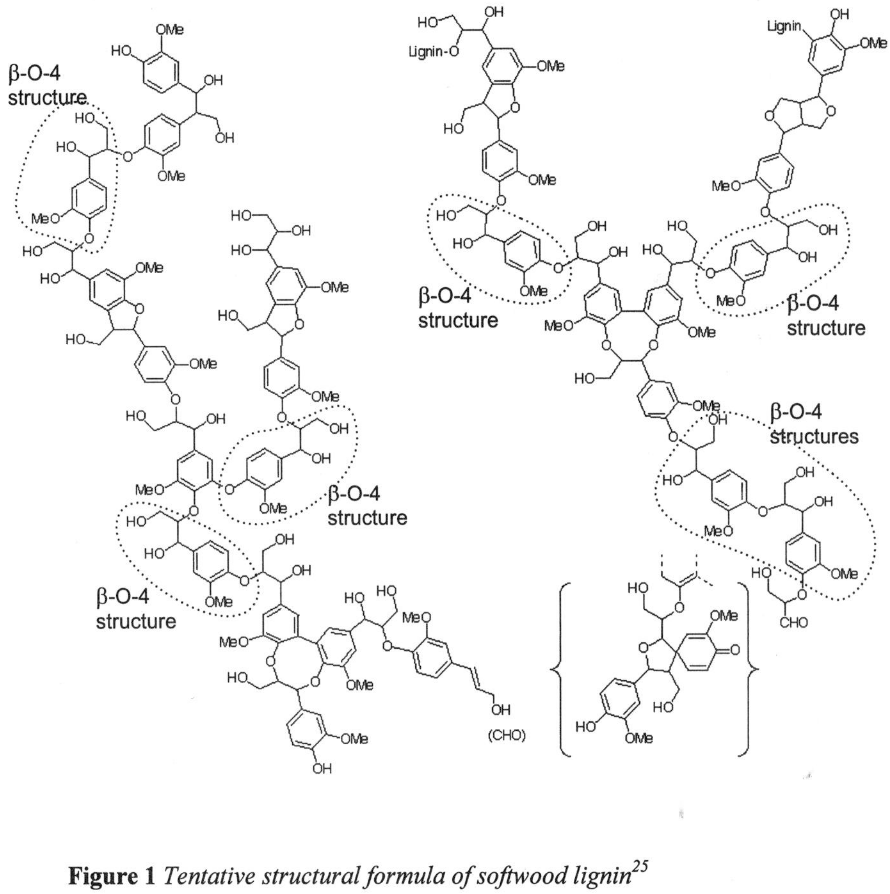

Figure 1 *Tentative structural formula of softwood lignin*[25]

etheric bonds between guaiacyl (softwood), and guaiacyl, syringyl and p-hydroxy-phenylpropane units (hardwood and non-wooden plants). This is the major type of interunit bonds in lignin (up to 60% of all such bonds in lignin).[25] It has been shown[26] that humic acids preserve a significant portion of β-O-4 bonds and that oxidized but not depolymerized lignin structures can play a major role in the reactivity of humic substances.

For simplification, we decided to build as lignin models regular linear oligomers constituted from guaiacyl units connected via β-O-4 bonds. The simplest structure that includes this linkage is a dimer I (Figure 2). This dimer may exist in *erythro* and *threo* diastereomeric forms. In our calculations[9] the *erythro* (SR-isomer) form has been chosen as representing the majority of these structures in lignin.[27] The *erythro*-configuration is also characterized by the lowest conformational energy according to Faulon *et al.*[28]

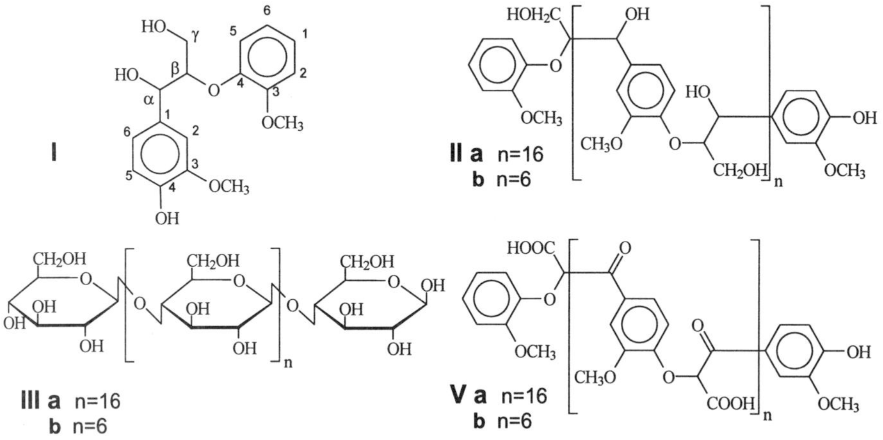

Figure 2 *Chemical structures of organic molecules studied; I - β-O-4 dimer, II - lignin chain, III - cellulose chain, V - oxidized lignin chain*

The dimer geometry was optimized using both the Tripos force field and the MM3 force field; the resulting global minimum conformers are shown in Figure 3. The geometry of **Ia** is reasonably close to the lowest energy geometry found in the calculations using the DREIDING force field and experimentally by crystallography.[28,29]

Conformer **Ib**, which corresponds to the global minimum when the MM3 force field was used, was of higher energy. The energy difference was small, though, and **Ib** was chosen as an alternative building unit.

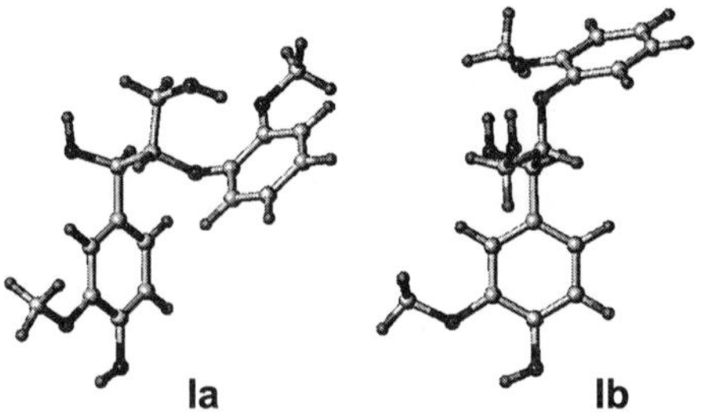

Figure 3 *Optimized conformations of β-O-4 dimer*

A detailed description of the computations is given in our previous publication.[9]

It should be noted that the absolute values of conformation or sorption energies obtained by molecular mechanics are not important. However, comparing these energies in a consistent series is helpful and can demonstrate meaningful patterns of behavior that the model follows. By comparing these data with those from laboratory experiments and with literature data, we can evaluate the validity of the model. If the model proves to be reliable, it can be used to study the chemical behavior and to predict properties of the system under consideration.

The optimized parameters of β-O-4 linkages obtained from the dimer structure have been used to construct the oligomeric chain. Conformer Ia produced an extended helix IIa, whereas conformer Ib yielded a broad helix IIb (Figure 4). These lignin chains correspond to the optimal regular structures. Preferential formation of helixes for lignin chains and polymers of similar nature has been already reported.[2,28] As can be seen from Figure 5, the hydroxylic groups exist on the exterior of the helix.

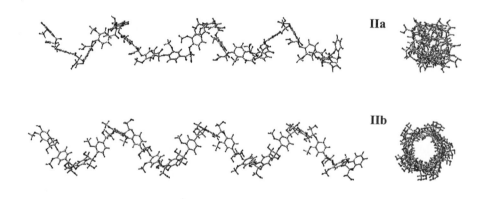

IIa

IIb

Figure 4 *Optimized lignin chain conformations: IIa - extended helix; IIb - broad helix*

Under natural conditions, thermal vibra-tions can result in significant conformational alterations of the molecule and an optimal structure may not be preserved. Molecular dynamics simulations have been used to model the thermal stability of the regular chains. It has been found that even under conditions of high-temperature dynamic simulations (20 ps at 1500 K) the conformer **IIb** preserved some helix-like segments.[9]

The broad helix has been chosen for construction of a model lignin–carbohydrate complex (LCC). A regular cellulose ribbon **III** has been used as a model of the carbo-hydrate part of the LCC. Although gluco-pyranoside structures make up only a fraction

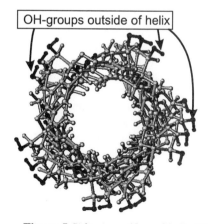

OH-groups outside of helix

Figure 5 *Side view of broad helix IIb*

of the actual LCC in plants, the cellulose oligomer was chosen as an initial model mostly because cellulose has previously been studied by computational methods.[9]

Several macroconformers involving different cases of mutual organization of lignin and carbohydrate chains have been studied. These cases represent possible ways of non-bonded attachment of these polymers in plant tissues. The energy of adhesion was significant in many cases due to the formation of hydrogen bonds and other types of non-bonding interaction. The complex IV (Figure 6) is an example of a stable LCC without covalent bonding between the polymeric components. In this complex the two chains are bound together topologically as well as by hydrogen-bond and non-bonding interactions. When such a highly stable complex is formed it is kinetically impossible and thermo-dynamically unfavorable to extract either component from the complex.

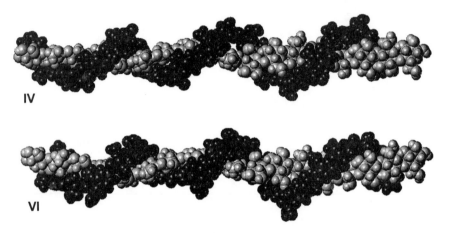

Figure 6 *Lignin–carbohydrate complex–cellulose chain inside wide lignin helix (**IV**) and oxidized lignin–carbohydrate complex (**VI**)*

To illustrate the possibility of formation of this inclusion complex we refer to plant morphology descriptors and to the lignification process.[24] The cell wall of a woody plant is represented schematically in Figure 7. Lignification occurs during the last stages of the wall formation when the cellulose microfibrils and hemicellulose chains are already present in an orderly structure. The formation of a lignin matrix goes on both between and on the carbohydrate chains. Under these circumstances it is likely that linear fragments of lignin macromolecules occasionally wrap around the carbohydrate chains. During this formation process conformational changes of polymers are hindered and limited to the mutual adjustment of the components. This nanoscale organization should strongly affect the macroscopic physico-mechanical properties of wood. Formation of topological LCC structures can also enhance the protective functions of lignin within plant tissues. The lignin structures, which undergo oxidation first due to the low ionization energies of aromatic units compared to carbohydrate units, more effectively preserve the structure of the carbohydrate chain with such a protective shell. This inclusion complex probably provides the best protection of both lignins and carbohydrates from depolymerization and their mutual stabilization during the humification process.

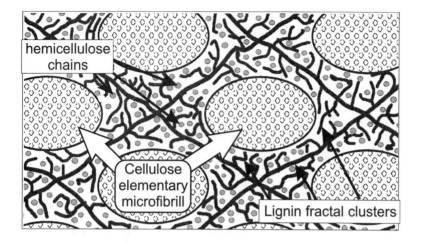

Figure 7 *Fragment of cell wall of a woody plant*

3.2 A Humic Acid Model Based on a Lignin–Carbohydrate Complex

In our approach the model humic polymers were derived from oxidation of the lignin chain. The α- and γ-hydroxyl groups in structures **IIb** and **IV** were transformed into carbonyl and carboxyl groups respectively (Figure 2). The conformations of the humic acid models **VI** and **V** are given in Figures 6 and 8 respectively.

In structure **VI**, representing the oxidized LCC, changes in spatial conformation due to oxidation were not significant. Reorganization of the polymer chains into the most favorable conformation for each chain conformation is hindered. We suggest that the initial

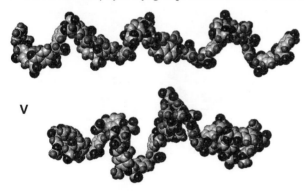

Figure 8 *Oxidized lignin chain (V) before (top) and after simulated annealing (bottom)*

spatial organization of a polymer involved in oxidative processes in the environment affects the eventual spatial organization of the product.

Under natural conditions, the pH of the soil strongly affects the properties of soil organic matter and organo–mineral aggregates. To study this phenomena a model polyanion **VII** was created starting from the oxidized LCC/humic acid (all of the COOH groups were replaced with COO⁻ groups). The neutral and ionized complexes were studied using molecular mechanics calculations and molecular dynamics (simulated annealing) in the gas phase and the hydrated state (using the first hydration shell generated around model structures according to ref. 11). The results are presented in Table 1 and Figures 8 and 9. Additional details of these experiments can be found in our earlier publications.[9,30] These

Table 1 *Calculated conformational energies U_{conf} and association energies $U_{assoc} = U_{conf}$ (LCC) – U_{conf} (Lignin) – U_{conf} (carbohydrate), kcal mol^{-1}*

Model	U_{conf}	U_{assoc}
IIa	160.5	-
IIb	168.1	-
V	82.7	-
IV	283.3	−91.6
VI	211.0	−78.5

results clearly indicate that neither oxidation nor ionization drastically affect the molecular shape of lignin-derived humic acids (Figures 6 and 9). However, molecular dynamics simulations suggest that some significant irreversible changes can develop over time (Figure 8) due to the low rotation barriers and the high flexibility of polymers. The oxidized LCC and its ionized counterpart demonstrate higher stability under a simulated annealing procedure (Figure 9). This can be explained by the occurrence of interpolymeric interactions inside the aggregate. During thermal annealing the aggregate shape and spatial conformation change significantly, but the topological entanglement of the components is preserved. The degree of change is expected to depend on properties of the polymers involved: the longer the polymer chains, the fewer the changes. A simple explanation of the effect is that the "unwinding" process starts from the ends of the polymer chains: the longer the chain length, the longer it takes to disengage the polymer components.

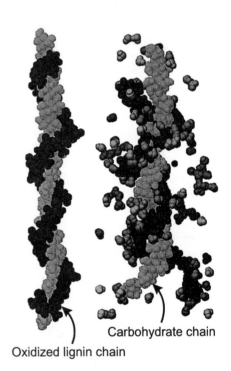

Carbohydrate chain

Oxidized lignin chain

Figure 9 *Oxidized ionized LCC **VII** before (left) and after simulated annealing (right, shown with one hydration shell)*

3.3 Modeling of Muscovite and Organo–Mineral Aggregation

To study organo–mineral interactions the model humic acid (oxidized lignin and oxidized LCC) was docked onto a mineral surface. As a mineral model, the crystallographic structure of the $2M_1$ polymorph of muscovite $KAl_2(Si_3Al)O_{10}(OH)_2$ was used.[1] This mineral was chosen because muscovite is the typical structural model of phyllosilicate-type soil and sediment constituents,[31,32] whose surfaces are normally coated by dissolved or colloidal-size organic matter. The model consisted of one layer of muscovite $(K_{41}Al_{72}Si_{170}O_{538}H_{139})$ and contained 28 whole ditrigonal cavities on each basal surface (Figure 10). This size was a reasonable compromise, providing a sufficient basal surface area of the mineral layer and keeping computational time at a reasonable level.

The partial atomic charges were calculated using a semi-empirical quantum chemical method (MNDO) on a small muscovite structure segment (one complete hexagonal unit on each 001 surface).[1]

Based on this calculation, average atomic charges were ascribed to each type of atom in

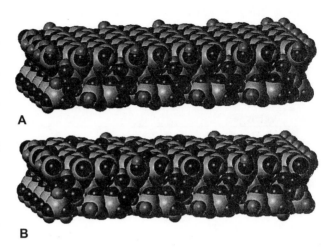

Figure 10 *Model muscovite: A- with lower basal surface completely packed with K⁺ and upper one without K⁺; B- with both equally packed with cations*

the mineral component: Al +1.10, Si +1.80, O –0.85, H +0.22. Potassium cations (K +1) were used to compensate the negative charge on the mineral.Two models were used: (**A**) one with both basal surfaces equally packed with cations, and (**B**) one basal surface fully packed with potassium ions and the other basal surface without cations (Figure 10).

The Tripos force field model lacks some parameters necessary to calculate mineral models correctly. For this reason during all optimization procedures and molecular dynamics simulations the mineral framework was considered rigid (frozen). In this case atom positions corresponded to crystallographic geometrical data. At the same time the interactions between the oxygen (basal) surface and organic molecules were calculated correctly. The movements of potassium ions and water molecules were unrestrained.

3.4 Interactions Between Humic Model and Muscovite

Owing to the size of the mineral model, the simulation experiments were limited to 8 monomeric units for both the oxidized lignin chain and the carbohydrate chain. The polymer chain was docked onto the basal mineral surface, geometry-optimized and the simulated annealing procedure was applied (10 cycles). The resulting aggregate was compared with the initial state. Conformational and sorption energies were calculated and the resulting values are given in Table 2. Different combinations of possible interactions between organic polymers and mica have been studied, including the various positioning of the polymer chains on the mica surface, different counter-ion arrangements and a comparison of the behavior of neutral and ionized humic acids.[1]

If the organic complex is neutral, the carbohydrates, with a higher polarity, have a higher affinity for the mineral surface than do the more flexible helical aromatic polymers. Calculated sorption energies for the carbohydrate chain are two times larger than those for the lignin chain.[1] Thus, the carbohydrate component assures sorption of the polymer complex on the mineral surface.

Table 2 *Conformational energies of organic oligomers U_{conf} and association energies*
$U_{assoc} = U(aggregate) - U(organic) - U(muscovite)$, kcal mol^{-1}

Aggregate	Before simulated annealing		After simulated annealing	
	U_{conf}	U_{assoc}	U_{conf}	U_{assoc}
VI/B	4.6	−80.3	79.3	−258.0
VII/A	9.0	55.5	0.0	−45.8
VII/B	0.0	−232.3	83.3	−567.6

When a higher pH is modelled and the oxidized lignin component is negatively charged,[11,30] then electrostatic forces prevail. In the absence of charge-compensating metal cations on the mineral surface, the lignin-derived anion drifted away from the surface due to electrostatic repulsion. Such a system exemplifies a known process of desorption upon changing the pH to higher values.

In soil systems, humic and fulvic acids exist as loose ion pairs. When cations are present, the ionized lignin-derived oligomer develops strong attractive organo–mineral interactions through the formation of cation bridges (Figure 11). This mechanism was first postulated by Peterson.[33] During dynamic processes the oligomers reorganize in such a way as to orient ionized carboxyl groups toward the mineral surface, allowing the counter-ions to form a bridge. This factor significantly influences the geometry of the oxidized lignin part of humic acid and through it the geometry of the carbohydrate segment as well. The binding effect of cations should be even greater for divalent cations like Ca^{2+} and Mg^{2+}.

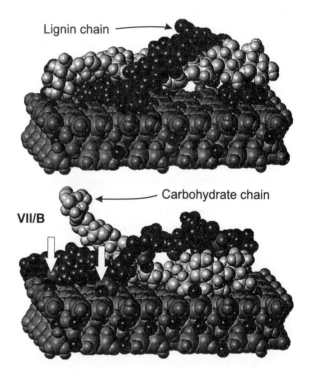

Figure 11 *Oxidized ionized LCC (**VII**, 8 units) on mica basal surface (**B**) before (top) and after simulated annealing (bottom). Cation bridges formed are arrowed*

3.5 Applications of Organo–Mineral Model and Current Perspectives of Development

Organo–mineral aggregates in soil play key roles in the sorption of organic and inorganic pollutants. Modelling of the sorption processes is possible with the humic acid-mineral models described above and the same computational approaches that have been explored here.

The preliminary computational experiments with small exogenic organic molecules were described in an earlier publication.[10] The mineral surface and the organo–mineral surface (neutral oxidized lignin–carbohydrate on muscovite) were used to evaluate the sorption properties of low molecular weight compounds. Three compounds were chosen as representative models: benzene as a non-polar organic molecule, sodium benzoate as an organic salt and atrazine (2-chloro-4-ethylamino-6-isopropylamino-s-triazine) as both a sample pesticide and polar organic molecule. As expected, the sodium benzoate had the

highest affinity for the basal muscovite surface through formation of ionic bridges. For small organic molecules on the organo–mineral aggregate (Figure 12), the results show that the organic coating does not decrease sorption energies except for ionized benzoate molecules, which develop weaker electrostatic interactions with organic matter compared with the mineral.[10]

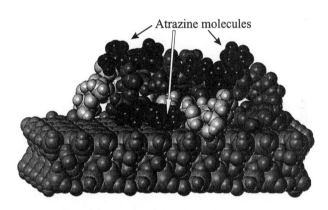

Atrazine molecules

Figure 12   *Atrazine molecules sorbed on organo–mineral aggregate*[25]

We are continuing this research by including other classes of pesticides, toxic organics and hazardous waste organic contaminants. We will simulate their behavior on hydrated mineral surfaces and on hydrated organo–mineral surfaces. Another promising direction for this work is introducing more mineral models and simulating their interactions with water, humic substances and with low molecular weight anthropogenic molecules. Kaolinite (a nonexpandable phyllosilicate) and goethite (an iron oxide) are current candidate surfaces for study. Any mineral can be studied as long as the appropriate crystallographic data are available, including unit cell dimensions, space group and atomic coordinates. This will allow us to assess processes in soils with different types of mineralogy and possibly the formation of different types of organo–mineral aggregates.

4 CONCLUSIONS

Computational chemistry provides the capability to simulate chemical reactivity, behavior and energetics of humic substances and their interactions with mineral surfaces. Similarly, computational chemistry permits the study of the interactions between organo–mineral aggregates and low-molecular weight chemical contaminants.

A helical structure is preferred for lignin/humic chains and the broad helix has sufficient internal space to accommodate the linear carbohydrate structure inside. This inclusion complex demonstrates a high stability over time and probably provides the best protection of the humic substances from depolymerization.

Flexible linear polymers were found to undergo drastic conformational changes when approaching the mineral surface to gain in sorption energy. A gain in aggregation (sorption) energy always outweighs a loss in the conformational energy. Molecular dynamics simulations suggest high stability of the organic polymer coatings on mineral surfaces.

The computation chemistry approach has great applicability in simulating and predicting the chemical reactivity of organic contaminants, which provides the basis for predicting the speciation, bioavailability, transport and transformation of organic chemical contaminants in terrestrial and aquatic ecosystems. The potential use of this approach is only limited by the availability of crystallographic data. This methodology can be used to increase our knowledge of soil genesis, the humification process, the leachability of low molecular weight organics in soils and to evaluate the pollution potential of a newly synthesized organic prior to its introduction into the environment.

References

1.   S. M. Shevchenko and G. W. Bailey. *Supramol. Science*, 1998, **5**, in press.
2.   S. A. Jansen, M. Malaty, S. Nwabara, E. Johnson, E. Ghabbour, G. Davies and J. M. Varnum, *Materials Sci. Eng. C*, 1996, **4**, 175.
3.   J.-L. Faulon, G. A. Carlson and P. G. Hatcher, *Org. Geochem.*, 1994, **21**, 1169.
4.   H.-R. Schulten, *Int. J. Environ. Anal. Chem.*, 1996, **64**, 147.
5.   B. J. Teppen, K. Rasmussen, P. M. Bertsch, D. M. Miller and L. Schafer, *J. Phys. Chem.*, 1997, **101B**, 1579.
6.   J. Greathouse and G. Sposito, *J. Phys. Chem.*, 1998, **102B**, 2406
7.   P. G. Hatcher, J. -L. Faulon, D. A. Clifford and J. P. Mathews, in 'Humic Substances in the Global Environment and Implications on Human Health', N. Senesi and T. M. Miano (eds.), Elsevier, Amsterdam, 1994, p. 133.
8.   G. W. Bailey, S. M. Shevchenko, Y. S. Yu and H. Kamermans, *Soil Sci. Soc. Am. J.*, 1997, **61**, 92.
9.   S. M. Shevchenko and G. W. Bailey, *J. Mol. Struct. (Theochem)*, 1996, **364**, 197.
10.   S. M. Shevchenko and G. W. Bailey, *J. Mol. Struct. (Theochem)*, 1998, **422**, 259.
11.   S. M. Shevchenko, L. G. Akim and G. W. Bailey, *J. Mol. Struct. (Theochem)*, submitted for publication.
12.   M. Clark, R. D. Cramer III and N. Van Opdenbosch, *J. Comput. Chem.* 1989, **10**, 982.
13.   J. G. Vinter, A. Davis and M. R. Saunders, *J. Computer-Aided Mol. Design*, 1987, **1**, 31.
14.   J. Gasteiger and M. Marsili, *Tetrahedron*, 1980, **36**, 3219.
15.   W. P. Purcel and J. A. Singer, *J. Chem. Eng. Data*, 1967, **12**, 235.
16.   H. Berthod and A. Pullman, *J. Chem. Phys.*, 1965, **62**, 942.
17.   H. Berthod, C. Giessner-Prettre and A. Pullman, *Theor. Chim. Acta*, 1967, **8**, 212.
18.   J. J. S. Stewart, *J. Computer-Aided Mol. Design*, 1990, **4**, 1.
19.   S. Kirkpatrick, C.D. Gelatt and M.P. Vecchi, *Science*, 1983, **220**, 671.
20.   H. J. C. Berendsen, J. P. M. Postma, W. F. van Gunsteren, A. Dinola and J. R. Haak, *J. Chem. Phys.*, 1984, **81**, 3684.

21. M. Blanco, *J. Computat. Chem.* 1991, **12**, 237.
22. I. Kögel, R. Hempfling, W. Zech, P. G. Hatcher and H.-R. Schulten, *Soil Science*, 1988, **146**, 124.
23. R. Hempfling, W. Ziegler, W. Zech and H.-R. Schulten, *Z. Pflanzenernaehr. Bodenkd.*, 1987, **150**, 179.
24. S. M. Shevchenko and G. W. Bailey, *Critical Rev. Environ. Sci. Technol.*, 1996, **26**, 95.
25. G. Brunow, in 'Lignin and Lignin Biosynthesis', N. G. Lewis and S. Sarkanen (eds.), American Chemical Society, Washington, DC, p. 131.
26. L. G. Akim, P. Shmitt-Kopplin and G. W Bailey, *Organic Geochemistry*, 1998 (in press).
27. B. Saake, D. S. Argyropoulos, O. Beinhoff and O. Faix, *Phytochemistry*, 1996, **43**, 499.
28. J.-L. Faulon and P. G. Hatcher, *Energy & Fuel*, 1994, **8**, 402.
29. R. Stromberg and K. Lundquist, *Nordic Pulp Paper J.*, 1994, **9**, 37.
30. S. M. Shevchenko, L. G. Akim and G. W. Bailey, *Proc. 9th Intern. Symp. Wood Pulping Chem.*, 1997, **2**, 100/1.
31. V. C. Farmer, in 'The Chemistry of Soil Constituents', D.J. Greenland and M.H.B. Hayes (eds.), Wiley, New York, 1978, p. 405.
32. A. C. Schindler, *Rev. Mineral.*, 1990, **23**, 281.
33. J. B. Peterson, *Soil Sci. Soc. Amer. Proc.* 1947, **12**, 29.

DETERMINATION OF TRACE METALS BOUND TO SOIL HUMIC ACID SPECIES BY SIZE EXCLUSION CHROMATOGRAPHY AND INDUCTIVELY COUPLED PLASMA MASS SPECTROMETRY

Peter Ruiz-Haas,[1] Dula Amarasiriwardena[1] and Baoshan Xing[2]

[1] School of Natural Science, Hampshire College, Amherst, MA 01002, USA
[2] Department of Plant and Soil Sciences, University of Massachusetts, Amherst, MA 01003, USA

1 INTRODUCTION

Humic acids (HAs) are well known for their metal ion binding properties, which account for their role in soil nutrient regulation.[1] The study of HA-metal binding is of particular relevance when determining the speciation status of trace metals in contaminated waters[2] and soils.

The complexation of HAs with metals can be either beneficial or deleterious in its effect on metals in soils and waters. The speciation of metals can be affected by the type of complex as well as oxidation–reduction reactions in solution. The soil solution is the medium from which plant roots take up metal ions. Metal ions can be sorbed on organic and inorganic components in soil and sorbed ions can be released (desorbed) into the soil solution. In addition, these ions can complex, precipitate, and be sorbed onto minerals and organic matter. Desorption and/or dissolution make elements mobile and available for plant uptake. Adsorption, complexation and precipitation reduce the availability and mobility of metals in soil.

Size exclusion chromatography (SEC) is a useful method for separation of molecular species present in a particular sample. Inductively coupled plasmas (ICPs) have developed into a dominant atom excitation and ionization source for elemental analysis.[3] ICP mass spectrometry (ICPMS) and ICP atomic emission spectrometry (ICPAES) are ideally suited detectors for elemental analysis because of their sensitive multielement detection capabilities. During the last decade, ICPMS has become a powerful technique for elemental and isotope analysis. It offers multielement capabilities, a large dynamic range and very low detection limits.[3,4]

Inductively coupled plasma mass spectrometry is now a well-established chromatographic detector for SEC,[5–9] enabling a wide variety of trace metal speciation studies, particularly in biological applications.[8] Commonly, ICPMS or ICPAES is connected by a length of inert tubing to the outlet of the SEC column or the UV–visible detector of many liquid chromatography systems. Care must be taken in column selection and method because the SEC must be able to separate the metal species unaltered to study fully the trace metal speciation phenomena. Furthermore, the rate at which the sample is aspirated into the ICP nebulizer must match the SEC flow rate. In addition, timing of both mass spectral and UV–visible chromatograms needs to be synchronized to account for the void volume caused by the tubing connection.[7] Column interactions also decrease the effectiveness of ICPMS detection at ultra-trace (ppt) levels.[7]

Speciation studies with trace metal-bound HAs are limited. Zernichow and Lund[10] used graphite furnace atomic absorption spectroscopy (GFAAS) to detect Al species. They collected SEC eluted fractions at short intervals and determined their Al content by GFAAS. They discovered that Al was bound to a wide molecular size range of humic acids in water and the presence of inorganic polymeric forms of aluminum. They were not able to determine molecular weights of the organic matter due to the unavailability of suitable molecular weight standards for the SEC column calibration. Rottman and Heumann[11,12] used an on-line SEC–isotope-dilution–ICPMS method to detect Cu and Mo interactions with dissolved organic matter (DOM) in natural waters. They found that Mo species are bound to high molecular weight fractions of DOM.[12] Cu interacted with low molecular weight fractions. The sum of the speciated fractions agreed with the total Mo and Cu levels, which indicates the reliability of the method.

Extensive applications of molecular fraction separations were accomplished using high performance-SEC (HPSEC) with aquatic organic matter.[13–15] Yet few studies have explored the on-line SEC analysis of soil HAs. Furthermore, to our knowledge, trace metal analysis by ICPMS of separated soil HA species has not been reported.

The purpose of this study was to develop an analytical methodology for interfacing HPSEC and ICPMS to determine qualitatively trace metal binding profiles in HA separated by SEC. The method derived is illustrated in Figure 1. On-line interfacing of HPSEC and ICPMS should achieve fractionation of HA by molecular size. On-line detection of trace metal bound HA species should be obtained by ICPMS detection.

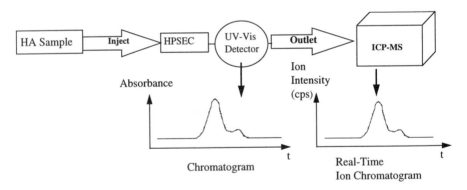

Figure 1 *Principle of interfacing between HPSEC and ICPMS as a detection system for HPSEC to determine trace metal-bound-humic acid profiles.*

2 MATERIALS AND METHODS

2.1 Sample Selection and Extraction

Humic acids were extracted from seven samples: (a) two soil samples: [from Pittsfield, Massachusetts (MASHA) and Chelsea, Michigan (MISHA), both from the surface horizon]; (b) three compost samples [household wastes: one with 15 days of maturation (15DCHA), one with 6 weeks of maturation (6WKCHA), and a mature compost sample (MatCHA). The less mature compost samples were obtained from a composting reactor; mature compost sample was obtained from a local composting facility]; (c) Leonardite (LHA) and Pahokee Peat (PHA), both obtained from the International Humic Substances Society (IHSS).

Samples were air dried for several days and subsequently ground in a water-cooled sample grinder (Analysenmühle A10, Ika Labortechnik, Staufen, Germany) for approximately 30 s. The grinding blade and inner compartment were of tungsten construction. Humic acids were extracted with the method described by Chen and Pawluk.[16] Distilled deionized water (18 MΩ cm^{-1}) was used throughout the extraction process and in the preparation of all solutions for this study. Polypropylene (PP) flasks and centrifuge vials were used. All glassware was leached in 50% (v/v) HNO_3 to eliminate additional trace metal contamination during the extraction. The total ash content of each HA was determined by heating HA samples in oven-dried, acid-leached, pre-ashed porcelain crucibles for 6 h at 730 °C (temperature ramp: 5 °C/min) in a muffle furnace (Thermolyne 6000 Series).

2.2 Spectroscopic Characterization of Humic Acids

2.2.1 UV–Visible Spectrophotometric Analysis. HA samples (0.2–0.4 mg) were dissolved in 10 ml of 0.05 M $NaHCO_3$ solution,[1] transferred to a 1 cm path length quartz cuvette and scanned from 190 to 820 nm wavelengths using a diode-array spectrophotometer (Hewlett Packard Model 8452A). The E_4 (absorbance at 465 nm) was obtained by averaging the absorbance at 464 and 466 nm; E_6 (absorbance at 665 nm) by averaging absorbance at 664 and 666 nm. Three ($n = 3$) replicate HA samples were measured for each HA type.

2.2.2 Diffuse Reflectance Fourier Transform Infrared (DRIFT-IR) Analysis. DRIFT-IR analysis was performed in an infrared spectrophotometer (Midac Series M 2010) with a DRIFT accessory (Spectros Instruments). All HA samples were dried in a dessicator. Each humic acid sample was mixed with oven-dried spectroscopic grade KBr (3–4% w/w) and finely ground in an agate mortar and pestle. At the time of analysis, the sample was carefully inserted into the sample holder and smoothed with a glass microscope cover slide. The analysis compartment was purged for *ca.* 5 min before each analysis with N_2 to eliminate interference from CO_2 and moisture. A small jar (20 ml) containing anhydrous $Mg(ClO_4)_2$ was placed inside the sample compartment to further reduce atmospheric moisture.

Samples were scanned 50 times at a resolution of 16 cm^{-1}. The blank consisted of finely powdered KBr stored under the same environmental conditions as the HA–KBr mixtures. Absorption spectra were converted to Kubelka–Munk functions using the Grams/32 software package (Galactic Corporation).

2.3 Elemental Analysis

2.3.1 High Pressure/Temperature (HPA) Nitric Acid Digestion. Approximately 95–100 mg of dried HA samples were digested in 3.0 ml of sub-boiled nitric acid (Optima grade; Fisher) at a pressure of 1700 psi and temperature of 230 °C using a high pressure/temperature asher (HPA) (Anton Paar). All glassware and digestion vessels were acid-leached (50% HNO_3 v/v) for > 24 hours, rinsed several times in distilled deionized water (18 $M\Omega$ cm^{-1}), and oven-dried before use. All manipulations were done in a class 100 clean room. Four replicates of HA sample and one blank digestion per load were carried out to evaluate reagent and digestion blank levels. The operational principles of HPA are described previously.[17,18] Details of the high pressure/temperature nitric acid digestion used in this study will be published elsewhere.

2.3.2 Semi Quantitative Analysis (SQA). Rapid, semi-quantitative analysis[19] of all metal elements was performed to obtain an overview of elemental species present in HA samples. A Perkin Elmer/Sciex Elan 6000 ICPMS system equipped with Elan NT software and the Total Quant II application was used for this analysis. The instrument response factors for elements were updated with the new response factors obtained by running an external standard containing 20 ng/ml ^{7}Li, ^{24}Mg, ^{51}V, ^{58}Cu, ^{114}Cd, ^{139}La and ^{208}Pb in 2% v/v sub-boiled HNO_3. The elements selected for the external standard cover the mass range *m/z* 7–239 of all elements of interest. Samples were transported into the ICPMS nebulizer by a variable speed peristaltic pump (Rabbit, Rainin Instruments). An internal reference standard containing 20 ng/ml ^{9}Be, ^{45}Sc, ^{115}In and ^{210}Bi used to correct for instrumental drift and matrix interferences was simultaneously transported with blank, sample and external standard solution streams through a "T" junction and a mixing coil.[20] The ICPMS detector was programmed to scan in peak hop mode, with a dwell time of 50 ms per amu, 10 sweeps per reading and 1 reading/replicate with 1 replicate.

2.3.3 Quantitative Analysis. Humic acid samples were quantitatively analysed for ^{208}Pb, ^{75}As, ^{64}Zn and ^{63}Cu. Linear calibration functions were obtained using 4 point multielement standards (0.1, 1, 10 and 100 ng/ml in 2% v/v HNO_3) and a 20 ng/ml ^{9}Be, ^{45}Sc, ^{115}In and ^{210}Bi internal standard in 2% v/v HNO_3 to correct for instrument drifts. These multielement standards were prepared by appropriately diluting and mixing of 1000 µg/ml stock standards (Spex). The ICPMS detector operating parameters were set as follows: mass scan in peak hop mode, with a dwell time of 50 ms per amu, 20 sweeps per reading, and 1 reading/replicate with six replicate sample readings. Sample solutions were diluted by a factor of 10 prior to analysis to obtain readings within the calibration range.

2.4 Size Exclusion Chromatography

Equipment consisted of a Beckman Gold® high performance liquid chromatography (HPLC) system with a model 168NM diode array detector, Autosampler (Model 507e), Solvent Module (Model 126NM) with PEEK™ (polyetheretherketone) plumbing and pumps. The chromatographic column used was a Superose 12/30 HR size exclusion chromatography column (Pharmacia Biotech). Humic acid samples (0.02–0.04 mg) were dissolved in 10 ml 0.05 M $NaHCO_3$. The column was equilibrated with at least 3 column volumes (*ca.* 75 mL) of eluent prior to use. Eluent was distilled deionized water (18 $M\Omega$ cm^{-1}) with a flow rate of 0.45 ml min^{-1} and the upper pressure limit was set at 2.0 MPa (410 psi). Each chromatogram was programmed to run for 45 min. All solutions were vacuum filtered through a 0.22 µm filter (Millipore).

The diode-array detector was programmed to scan the entire spectrum from 190–350 nm and display and analyse chromatograms at 254 and 280 nm. A 50 µl aliquot was injected into the column through a 100 µl sample loop using the autosampler.

The SEC column was calibrated using globular proteins (Sigma): bovine serum albumin (BSA) (M_W 67 kDa), α-chymotrypsinogen (M_W 25 kDa), α-lactalbumin (M_W 15.5

kDa) and cytochrome-c (M_W 12.4 kDa). The eluent used for column calibration was a 50:50 v/v mixture of 0.1 M sodium acetate/acetic acid buffer.

2.5 HPSEC/ICPMS Coupling

The core of this project was interfacing of HPSEC and ICPMS to enable on-line trace metal analysis of SEC eluted molecular fractions. The ICPMS and HPSEC were interfaced by a PEEK transfer line and a 4 port-switching valve (Upchurch Scientific). Tubing length was short (*ca.* 10 cm) to minimize peak broadening and void volume. The switching valve was necessary to ensure a constant flow of solute into the ICPMS nebulizer and to facilitate switching between on-line HPSEC/ICPMS eluent and to introduce the blank (2% v/v sub-boiled nitric acid) or elemental standards for calibration purposes. A read-delay was programmed into the ICPMS system to account for the void volume delay between HPSEC detector and ICPMS detector. This was accomplished by connecting a serial port trigger read cable from the HPSEC autosampler to the ICPMS central processing unit (CPU). Our setup was experimentally optimized with a read delay of 2 s between HPSEC autosampler contact closure and ICPMS data acquisition. The mass spectrometer software was set up to scan for ^{208}Pb, ^{75}As, ^{64}Zn, ^{63}Cu and ^{55}Mn to produce real-time ion chromatograms. These ion chromatograms revealed the trace metal binding profiles of the HAs fractionated by SEC (Figure 2).

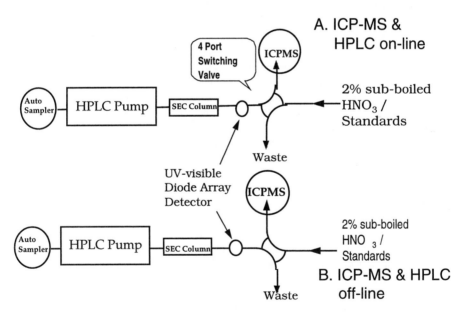

Figure 2 *Diagram of HPSEC/ICPMS interface with a 4-port switching valve*

The mass spectrometer dual detector was programmed at 30 sweeps/reading, 450 readings/replicate and a dwell time of 25 ms per amu in peak hop mode. ^{57}Fe and ^{27}Al were determined for select samples to establish the method's viability for determining other trace elements in HAs. Instrument operation, analysis, as well as data acquisition (ion chromatograph recording) was handled by Perkin-Elmer Elan NT software. Ion chromatograms were imported by individual masses into Perkin-Elmer Turbochrom™ software for further analysis. Ion chromatograms and mass spectral data were analysed using Grams/32™ software with the Perkin-Elmer Views™ option.

3 RESULTS AND DISCUSSION

3.1 Soil and Compost Humic Acid Characterization

The ash contents of the soil and compost humic acid samples ranged from 0.7 to 4.2% w/w. They are within the range of values reported in the literature.[1,21] All humic acid samples were characterized by UV–visible spectroscopy and DRIFT-IR to evaluate the quality of our extracted humic acids.

    3.1.1 UV–Visible Spectroscopic Analysis. The results from E_4/E_6 ratios fall between 5.8 and 2.8, consistent with ratios reported in the literature,[1,21] but it was difficult to discern the degree of humification of different humic acids with these values.

    3.1.2 DRIFT-IR Spectroscopic Analysis. DRIFT-IR spectral data of our extracted HA samples enable verification with those values reported in the literature[22,23] to confirm that we indeed extracted HAs and to assess the structural composition of our HAs. They exhibit typical HA spectra as reported in the literature. Comparison of different HA spectra is the most popular application of infrared analysis in the study of HAs, as subtle differences in functional makeup might be detected. It also is possible to calculate functional group ratios based on peak intensities and thus to elucidate HA makeup further. For instance, an investigator may choose to compare the intensity of carboxylic peaks against the intensity of aliphatic or aromatic peaks to determine HA activity in soil.[22]

    DRIFT-IR spectra of HA samples are presented in Figures 3a,b. Significant peaks found and their assignments are summarized in Table 1. Essentially, the spectra obtained can be classified by attributes, for example HAs with differing intensity and/or number of characteristic major functional groups. A noteworthy feature in our compost humic acid spectra is the increase in number of aromatic peaks and decrease in aliphatic groups with HAs of increasing maturity. The 15 day immature compost sample (15DCHA) (Figure 3a) exhibits strong peaks at 2925 and 2855 cm^{-1} that are indicative of aliphatic CH_2 asymmetric stretching bonds. The two broad peaks observed in a 15 day immature compost HA at 2925 and 2855 cm^{-1} are typical of humic substances that are not very mature and are common in composts.[24] There is a likelihood that these peaks are also caused by extraneous material such as lignins, as the humic acid (15DCHA) DRIFT-IR spectra resemble that of raw compost.[25] Much of the extracted 15DCHA had some plant residue that was not eliminated in the extraction process. The intensity of these peaks decreased with increasing maturation of the HA sample. A peak at 1054 cm^{-1} is assigned to aliphatic alcohol groups. Furthermore, this HA lacks peaks in the 1600 cm^{-1} region (C=C aromatic bond) that can be interpreted as a lower degree of aromaticity in the sample.

    The spectrum of 6WKCHA compost humic acid shows a peak at 1600 cm^{-1} that indicates a higher degree of aromaticity than the previous sample and has no clearly defined amide peak at 1530 cm^{-1}. Amide peaks at 1530 cm^{-1} decrease with increasing maturity of the compost HA sample, indicating the degradation of -NH_2 groups associated with the proteinaceous material in compost (see Figure 3a). The soil sample from Massachusetts (MASHA) also displays this peak (see Figure 3b), suggesting that this HA

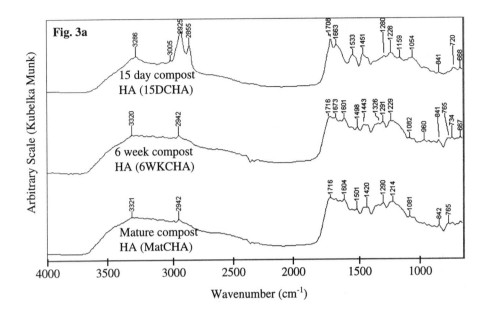

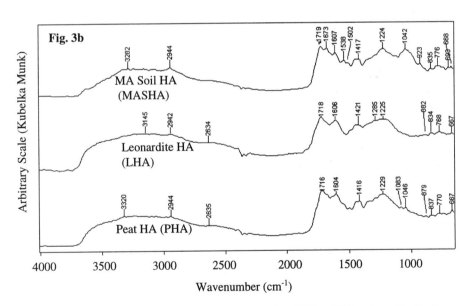

Figure 3 *DRIFT-IR spectra of compost (Fig. 3a) and soil (Fig. 3b) humic acids. Peak markings on both figures indicate wavenumbers associated with each peak. The vertical axes of spectra are represented in Kubelka–Munk units.*

Table 1 *Diffuse Reflectance Fourier Transform Infrared (DRIFT) spectra peak assignments for HA. Assignments based on Wander and Traina,[22] Stevenson,[1] and Baes and Bloom.[23]*

Wavenumber (cm⁻¹)	Bond type	Humic acid samples						
		LHA	PHA	MASHA	MISHA	15DCHA	6WKCHA	MatCHA
3279–3340	Phenol OH, amide N–H	✓	✓	✓	✓	✓	✓	✓
2962–2950	CH₂ symmetric stretch	✓	✓	✓	✓	✓	✓	✓
2924–2930	CH₂ asymmetric stretch	✓		✓		✓	✓	
2850	CH₂ symmetric stretch					✓		
2500	CO–OH bond	√(2600)	√(2600)		√(2600)			√(2600)
1850	C=O stretch							
1735–1713	C=O ketonic, COOH	✓	✓	✓	✓			
1650	C=O, C=O–H bonded, amide H	✓		✓	✓	✓	✓	✓
1630–1608	C=C aromatic	✓	✓	✓	✓	✓	✓	✓
1550	aromatic ring, amide							
1535–1520	C=C aromatic ring, amide		✓	✓		✓		
1509	aromatic ring, amide		✓	✓	✓			
1457	CH₃ asymmetric stretch, CH bend							
1420	aromatic ring stretching	✓	✓	✓	✓	✓	✓	✓
1400	COO salt, COOH	✓	✓	✓	✓		✓	✓
1379–1327	COO-. CH₃, symmetric stretch	✓	✓	✓	✓	✓	✓	✓
1260–1240	CO, COOH, COC, phenol OH	✓	✓	√(1224)	✓	✓	√(1291 & 1229)	√(1221)
1190–1127	aliphatic, alcoholic OH	✓	✓		✓	✓	✓	✓
1080–1050	CO aliphatic alcohol	✓	✓	✓	✓	✓	✓	✓
1030	aliphatic COC, aromatic ether, SiO	✓	✓	✓	✓	✓	✓	✓
918–912	OH, COOH, Al–OH		✓	✓	✓	✓	√(960 & 922)	✓
850–830	CH aromatic bend, Al–O–Si	✓	✓	✓	✓	✓	✓	✓
797	Fe-O-Si							
779	CH aromatic out-of-plane bend	✓	✓	✓	✓	✓	✓	✓
750	unknown mineral peak	✓	✓	✓	✓	✓	✓	✓
694	unknown mineral peak	✓	✓	✓	✓	✓	✓	✓

is not a very mature type as well. Other relevant findings include strong peaks at 1040 cm^{-1} for the MASHA humic acid sample, which represents aliphatic ether (C-O-C) and alcohol groups. Obviously, peat and leonardite humic acids (PHA and LHA) appear to be the most humified samples as they display peaks that are assigned to aromatic groups (*e.g.* 1600 cm^{-1}, 1420 cm^{-1}) and a lack of aliphatic and amide groups 1535 and 1520 cm^{-1} (see Table 1 and Figure 3b).

All HAs display to a varying degree carboxylic acid groups (1735–1713 cm^{-1}) that are important in forming complexes with metal ions. Compost HAs have fewer peaks associated with these functional groups. For instance, absorption in the 1240–1260 cm^{-1} range is weaker in the compost HA samples when compared with the mature soil HA. Also, all HAs exhibit peaks with varying intensity at 1720 cm^{-1}, which indicates the presence of ketonic (C=O) and carboxylic acid (-COOH) functional groups. Immature compost humic acid (15DCHA) and soil HAs have prominent peaks in this area, groups that are important in forming metal–HA complexes. Most oxygen-containing functional groups are thought to be involved in HA reactions with metals, as well as sorption mechanisms controlled by physical characteristics of the HA, like coiling of the molecule in solution.[26,27]

More detailed information on the reactivity of HAs is provided by calculation of O/R ratios, which are the intensities of oxygen-containing functional groups *vs.* aliphatic and aromatic (referred to as recalcitrant) groups.[22] Humic acids of increased maturity show higher O/R ratios, as they display a greater degree of oxygen-containing functional groups (Table 2). There is a distinct increase in the R_1 ratio (Table 2) as HAs become increasingly mature, owing to the decrease in peak intensities of aliphatic groups. The R_2 ratio also

Table 2 *Ratios of selected peak heights from DRIFT-IR Spectra of humic acids (HA) O/R ratios constructed after Wander and Traina[22] and Baes and Bloom[23]*

	O/R ratios[a]	
HA sample	R_1 $$\frac{1727 + 1650 + 1160 + 1160 + 1127 + 1050}{2950 + 2924 + 2850 + 1530 + 1509 + 1457 + 1420 + 779}$$	R_2 $$\frac{1727}{1457 + 1420 + 779}$$
15 day compost (15DCHA)	0.66	1.65
6 week compost (6WKCHA)	0.72	0.60
Mature compost (MatCHA)	0.66	0.65
Massachusetts Soil (MASHA)	1.34	0.80
Michigan Soil (MISHA)	1.86	1.45
Peat (PHA)	1.64	1.50
Leornadite (LHA)	1.27	0.82

[a] O/R (oxygen/recalcitrant) Ratios = R

shows a similar trend except for the immature compost sample (15DCHA). This finding is particularly relevant to the study of trace metal binding capacities, since metal ions bind with oxygen-containing sites like -COOH as well as other groups having lone pair electrons (*i.e.* R-NH$_2$, R-S-R) and/or charges in HA molecules.[27,28] It is reasonable to expect increased binding capabilities of matured HAs with trace metal ions as opposed to HAs with a high degree of aliphaticity. The increased functionality of these also reflects the HA's ability to hold water and regulate nutrient cycling in soil.[22]

3.2 Elemental Analysis

Ash represents the mineral fraction and total trace metals present in the humic acid samples that are not dissolved during the extraction process and thus remain as impurities in the HA. These mineral residues are responsible for some of the DRIFT-IR peaks observed at 750 and 694 cm^{-1}. High temperature/pressure nitric acid digestion of HA samples is satisfactory. The digested samples were clear and colorless except for white siliceous residue in some humic acid samples, which was not attacked by nitric acid. A detailed description of this digestion approach will be presented in a forthcoming publication.

3.2.1 Semi-quantitative Analysis (SQA). SQA allows rapid pre-screening of samples to determine metal elements present with an accuracy of ± 20%.[19] In addition to the trace metals quantitatively determined, semi-quantitative analysis results of nitric acid-digested HA samples are presented in Table 3. The presence of Ba, Co, Cr, Ce, Mo, Ni, Sr, Ti, and V in humic acid samples was demonstrated by SQA in addition to those elements detected by quantitative analysis.

3.2.2 Quantitative Analysis. Total trace metals analysis provided a means of corroborating observations associated with HPSEC–ICPMS speciation studies. Results of trace metals analyses for Pb, Cu, Zn are shown in Table 4. The data reveal elevated concentrations of metals (Pb and Cu) in the mature compost sample (MatCHA) as well as elevated amounts of Cu in the soil humic acid samples (MASHA and MISHA). Lead levels recorded for MatCHA are the highest of all samples analysed. Higher metal concentrations in MatCHA probably indicate a higher degree of Pb and Cu binding in these humic acid samples. This is reflected in the increased degree of aromaticity of this HA with respect to the less mature compost HA samples, as discerned from UV–vis and DRIFT-IR analysis above. Copper and Zn appear to bind to the HAs in the soils that MISHA and MASHA were extracted from.

3.3 Speciation of Humic Acid Bound Trace Metals by HPSEC–ICPMS

3.3.1 High Performance Size Exclusion Chromatography (HPSEC). Size exclusion chromatograms were obtained for soil and compost humic acids. They exhibited a simple, monomodal peak with a few shoulders and one or two small peaks before or after the main peak (Figure 4). The size exclusion chromatogram patterns recorded for our HAs are similar to those reported for aquatic humic substances by Chin and co-workers[15] and Peuravouori and Pihlaja.[13] Size exclusion chromatogram column calibration with proteins resulted in a linear relationship between the retention volume (V_R) and the logarithm of the molecular weight. Molecular weight (M_W) estimates of sampled HAs, as determined by column calibration, are shown in Table 5. $M_W = 10^{(50.371-V_R/7.0355)}$, $r^2 = 0.993$ for the calibration, where M_W = molecular weight.

Table 3 *Element concentrations of digested HA samples determined by semi quantitative ICPMS analysis*

Element	15DCHA	6WKCHA	MatCHA	MASHA	MISHA	PHA	LHA
			Concentration ($\mu g/g$)				
Na	n.d.[a]	2.1	48	222	10	7503	28
Mg	23	12	126	527	34	385	230
Ca	322	153	38	163	382	6197	196
Ti	28	25	40	298	21	29	36
V	0.5	2.4	4.7	14	3.4	4.2	0.5
Cr	5.4	10	15	39	9.6	7.7	6.0
Mn	7.1	8.9	5.5	59	1.5	2.0	0.7
Fe	198	448	1259	5182	383	1663	64
Co	5.6	20	8.7	2.5	2.2	2.4	2.3
Ni	1.1	13	12	n.d.	11	n.d.	1.7
Sr	1.4	0.7	2.1	2.6	133	293	0.4
Zr	2.4	1.2	2.4	4.2	1.7	2.4	0.8
Mo	3.0	5.8	5.7	1.5	8.8	2.9	10
I	4.4	0.5	4.4	74.0	2.6	4.4	4.5
Ba	1.2	3.3	11	103	9.9	32	12
Ce	1.0	1.7	10	488	4.2	1.3	0.8
W	5.0	5.5	13	1.4	0.4	n.d.	0.3
Hg	0.1	0.2	0.5	2.2	0.2	n.d.	0.3
Pb	1.6	27	436	9.9	4.7	2.7	2.4

[a] n.d. not detected

Table 4 *Quantitative analysis of total trace element concentrations in HA materials by ICPMS after high pressure/temperature (HPA) digestion.[a]*

HA Sample	Cu	Zn	As	Pb	n[b]
		Element concentration ($\mu g/g$)			
Massachusetts Soil (MASHA)	1054 ± 44	22 ± 1	3.5 ± 0.8	12 ± 2	3
Michigan Soil (MISHA)	564 ± 77	6.6 ± 1.7	4.6 ± 0.8	4.4 ± 0.2	4
Peat (PHA)	6.6 ± 0.2	2.9 ± 0.6	4.1 ± 0.1[c]	2.5 ± 0.1	4
Leonardite (LHA)	16 ± 2	0.6 ± 0.1	0.6 ± 0.1[d]	2.7 ± 0.4	4
15 day compost (15DCHA)	28 ± 1	11 ± 1	0.15 ± 0.01	2.0 ± 1.4	4
6 week compost (6WKCHA)	79 ± 8	53 ± 5	0.7 ± 0.1	28 ± 3	5
Mature compost (MatCHA)	516 ± 31	25 ± 3	3.2 ± 0.2	800 ± 43	4

[a] $\pm$ one standard deviation; [b] n = digestion replicates ; [c] $n = 2$; [d] $n = 3$

Determination of molecular weight of HA fractions by HPSEC is still a much debated topic, since different HA matrices behave differently in SEC columns. HA uncoil in different ways in the mobile phase, depending upon ionic strength and pH,[13,28] which complicates comparison of results when using different eluents. There appears to be an increasing consensus that globular proteins do not resemble the molecular configuration of HAs in solution. Several authors, including Chin *et al.*,[15] Knuutinen *et al.*,[29] and Zernichow and Lund[10] argue that there is a better correlation between the HA and random-coil polymers such as polystyrene-sulfonates (PSS). Swift[30] commented on other investigators'[31] successful use of polysaccharides for SEC column calibration. As a result, our data only show apparent molecular weights, which we classified into M_1, M_2 and M_3 fractions according to the relative retention time displayed by the HPSEC chromatogram (Figures 4 and 5). However, M_W obtained for the M_2 range from 985–1700 kDa and are somewhat higher than M_W reported elsewhere. For the purposes of this research project, this discrepancy does not overshadow the purpose of SEC HA analysis to separate HAs into different molecular weight fractions (regardless of actual weight) to determine trace metal binding profiles on different HA species.

3.3.2 HPSEC–ICPMS Interfacing. We have demonstrated the ability to separate different molecular weight fractions of soil and compost HA by SEC and on-line detection of humic acid bound trace metals by ICPMS. The instrument communications setup between two instruments and ICPMS timed read–delay allowed for both instruments to work in concert, thus displaying ion and UV–visible chromatograms whose peaks are detected at nearly identical times and have similar curve areas.

3.3.4 Humic Acid Bound Trace Metal Profiles. The ability to determine qualitatively the trace metal bound soil HA fractions are demonstrated in Figures 6 and 7. Total trace metal analysis of the HA samples shows measurable concentrations of the trace metals (see Table 4) whose HA binding patterns were analysed chromatographically. This confirms that the metal ion signals recorded by SEC-ICPMS interfacing are not caused by artifacts. Results from speciation studies of HA bound trace metals in peat and immature humic acids (15DCHA) show the following characteristics.

(1) Zn and Cu are strongly bound to all HA fractions (see Table 5). In particular, M_2 is the fraction (intermediate molecular weight) that consistently elutes at similar times and demonstrates HA bound trace metals. It is noteworthy to examine binding behavior of these elements with heavy (M_1) and light (M_3) humic acid molecular weight fractions, because comparatively strong ion intensities are recorded at the M_1 and M_2 peaks, as is the case for peat (PHA) and immature compost humic acids (Figures 5 and 6).

(2) Of the HA samples analysed, Pb exclusively binds to the intermediate molecular weight fraction (M_2 – the fraction represented by the largest chromatographic peak) (Figures 5 and 6).

(3) As and Mn appear not to be bound to any HA molecular weight fraction, no discernible peaks resolve in the elemental ion chromatograms for these elements, and their signal/noise ratios do not allow for valid statistical analysis of the vaguely resolved peaks. However, low levels of As and Mn are recorded by total trace metal analysis of HA samples (Table 5). Signal/noise ratios for Zn, Cu and Pb (see Figures 5 and 6) provide evidence of binding of these elements to HA fractions.

(4) Ion peak intensities are strongest for Cu, Zn and Pb, which confirms our data with the complexing constants expressed for divalent cations in the Irving–Williams series.[32] Nonetheless, a quantification of elemental species still needs to be made to further refine this hypothesis.

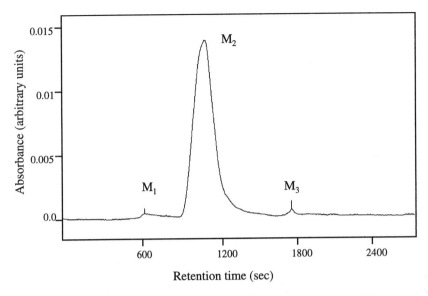

Figure 4 *Typical HPSEC chromatogram of an HA sample detected at 254 nm. Eluent flow rate was set at 0.45 mL/min, P_{max}= 2.0 MPa, eluent used 18MΩ cm^{-1} water. M_1, M_2, M_3 are molecular weight fractions.*

Table 5 *Elements bound to each humic acid molecular weight fractions (M_1, M_2, M_3)*

| | | Humic acid molecular weight fraction | |
HA sample	M_1	M_2	M_3
Massachusetts Soil (MASHA)	none	Cu, Pb, Zn	Cu, Zn
Michigan Soil (MISHA)	none	Cu, Pb, Fe, Al, Cd	Cu
Peat (PHA)	Cu, Zn	Cu, Pb, Al, Cd	Cu, Zn
15 day compost (15DCHA)	Cu, Zn	Cu, Pb, Zn	Cu, Zn
6 week compost (6WKCHA)	none	Cu, Pb, Zn	Zn
Mature compost (MatCHA)	Cu	Cu, Pb, Zn, Fe, Al	Cu, Zn

(5) Further experimental runs of this method with other elements (*e.g.* Al, Fe and Cd) show the potential application to analysis of other metal species bound to HAs. The software allows a maximum of twelve elements to be recorded per chromatographic run.

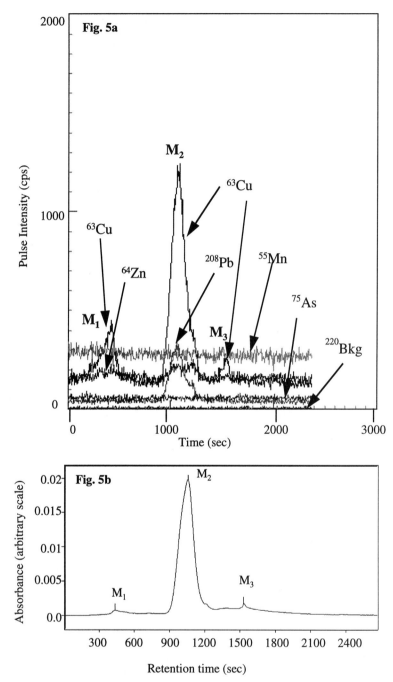

Figure 5 *ICPMS ion chromatogram (Fig. 5a) and size exclusion chromatogram (Fig. 5b) of peat humic acid sample (PHA). M_1, M_2, and M_3 are separated molecular fractions. See text for details.*

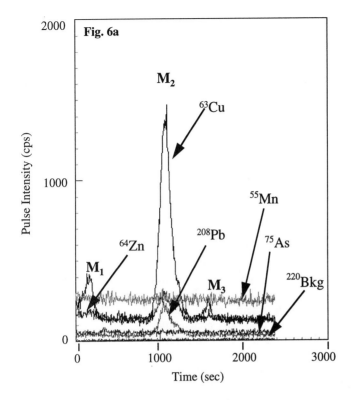

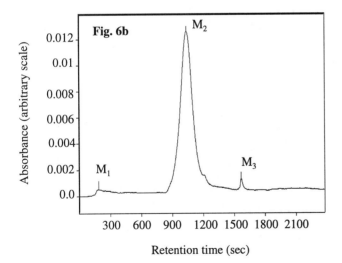

Figure 6 *ICPMS ion chromatogram (Fig. 6a) and size exclusion chromatogram   (Fig. 6b) of 15 day compost humic acid sample (15DCHA). M₁, M₂, and M₃ are separated molecular fractions. See text for details.*

4 CONCLUSIONS

Humic acids from soil, compost, peat and Leonardite were successfully extracted with base and characterized by UV–visible and DRIFT-IR spectroscopy. The results show typical characteristics of soil humic acids. In addition, comparative analysis of the HAs shows that compost HA can become increasingly humified with increased compost maturity.

Total trace metal analysis of HAs is a complex task owing to the need of complete dissolution of HAs prior to spectrochemical analysis. Digestion of HAs that contain more than 50% C in matrices is satisfactory. This work demonstrates the ability of ICPMS to quantitatively determine Pb, As, Cu and Zn by after high pressure/temperature digestion. We obtained clear, colorless final solutions that are ideal for spectrochemical analysis after digesting 0.1 g of HA at 1700 psi (120 bar) with 3 ml of nitric acid. Semiquantitative analysis of all trace metal elements by ICPMS reveals the presence of Mn, Al, Fe, Mo, Co, Cr, Ti, Sr and V in humic acid samples. Quantitative total trace metal analysis by ICPMS was performed for Zn, Cu, As and Pb. The metals Cu and Pb were present in higher concentrations.

The SEC and ICPMS setup developed in this study has the capability to determine trace metal bound HA species. On-line determination of HA bound trace metals by HPSEC–ICPMS was successful. We determined Zn, Cu, Pb and Al profiles bound to different HA fractions. All metals studied bind preferentially, if not exclusively, to the intermediate molecular weight HA fraction (M_2). Cu, Zn and Al bind to all molecular fractions, whereas Pb appears to be exclusively bound to the M_2 fraction.

ICPMS is a versatile tool for elemental and speciation analysis of a wide variety of HA samples and soil organic matter fractions. Furthermore, results of this investigation agreed with the Irving–Williams binding series. We were able to determine qualitative binding affinities of different elements by evaluating the peak intensities in ion chromatograms. Metal speciation information in humic acids is useful for better understanding of mobilization and transport mechanisms of trace metals for environmental remediation efforts and soil nutrition studies.

ACKNOWLEDGMENTS

Dula Amarasiriwardena would like to thank the National Science Foundation (Grant: BIR 9512370), a Pittsburgh Memorial College Grant, the Howard Hughes Medical Institute (HHMI), the George Alden Trust, the Kresge Foundation, and the Keck Foundation for instrument and financial support for this project. The undergraduate research grant awarded by the HHMI to Peter Ruiz-Haas is gratefully acknowledged. Prof. Ramon Barnes very kindly provided us access to the high pressure/temperature asher and gave valuable comments on the manuscript.

References

1. F. J. Stevenson, 'Humus Chemistry: Genesis, Composition and Reactions', 2nd edn., Wiley, New York, 1994.
2. T. M. Florence, *Talanta*, 1982, **29**, 354.

3.   R. S. Houk, *Anal. Chem.,* 1986, **58**, 97A.

4.   D. Beauchemin, *Spectroscopy (Eugene)*, 1992, **7**, 12.

5.   K. Sutton, R. M. C. Sutton and J. A. Caruso, *J . Chromatogr. A.*, 1997, **789**, 85.

6.   P. Uden, *J. Chromatogr. A*, 1995, **703**, 393.

7.   A. Seubert, *Fresenius' J. Anal. Chem.*, 1994, **350**, 210.

8.   B. Gercken and R. M. Barnes, *Anal. Chem.,* 1991, **63**, 283.

9.   S. C. K. Shum and R. S. Houk, *Anal. Chem.*, 1993, **65** , 2972.

10.   L. Zernichow and W. Lund, *Anal. Chim Acta* , 1995, **300**, 167.

11.   L. Rottmann and K. G. Heumann, *Anal. Chem.*, 1994, **66**, 3709.

12.   L. Rottmann and K. G. Heumann, *Fresenius' J. Anal. Chem.*, 1994, **350**, 221.

13.   J. Peuravouori and K. Pihlaja, *Anal. Chim. Acta*, 1997, **337**, 133.

14.   Y. P. Chin and P.M. Gschwend, *Geochim. Cosmochim. Acta,* 1991, **55**, 1309.

15.   Y. P. Chin, G. Aiken and E. O'Loughlin, *Environ. Sci. Technol.,* 1994, **28**, 1853.

16.   Z. Chen and S. Pawluk, *Geoderma*, 1995, **65**, 173.

17.   G. Knapp and A. Grillo, *Am. Lab.,* 1986, **3**, 76.

18.   R. T. White, Jr., *J. Assoc. Off. Anal. Chem.,* 1989, **72**, 387.

19.   D. Amarasiriwardena, S. F. Durrant, A. Lásztity, A. Krushevska, M. D. Argentine and R. M. Barnes, *Microchem. J.*, 1997, **56**, 352.

20   M. Viczian, A. Lásztity and R. M. Barnes, *J. Anal. At. Spectrom.,* 1990, **5**, 293.

21.   D. S. Orlov, 'Humic Acids of Soils', Amerind Publishing Co., New Delhi, 1985, p. 126.

22.   M. M. Wander and S. J. Traina, *Soil Sci. Soc. Am. J.,* 1996, **60**, 1087.

23.   A. U. Baes and P. R. Bloom, *Soil Sci. Soc. Am. J.,* 1989, **53**, 695.

24.   J. Niemeyer, Y. Chen and J. M. Bollag, *Soil Sci. Soc. Am. J.,* 1992, **56**, 135.

25.   S. Deiana, C. Gessa, B. Manunza, R. Rausa and R. Seeber, *Soil Sci.,* 1990, **150**, 419.

26.   F. J. Stevenson, 'Cycles of Soil: Carbon, Nitrogen, Phosphorus, Sulfur, Micronutrients,' Wiley, New York, 1986.

27.   N. Senesi, in 'Biogeochemistry of Trace Metals', D. Ariadno (ed.), Lewis Publishers, Boca Raton, FL., 1992.

28.   H. K. J. Powell and E. Fenton, *Anal. Chim. Acta,* 1996, **334**, 27.

29.   J. Knuutinen, L. Virkki, P. Mannila, P. Mikkelson, J. Paasivirta and S. Herve, *Wat. Res.*, 1988, **22**, 985.

30.   R. S. Swift, Organic Matter Characterization in 'Methods of Soil Analysis: Part 3-Chemical Methods,' Soil Science Society of America Book Series No. 5, Madison, 1996, p. 1011.

31.   R. S. Cameron, R. S. Swift, B. K. Thornton and A. M. Posner, *J. Soil Sci.,* 1972, **23**, 342.

32.   M. B. McBride, 'Environmental Chemistry of Soils', Oxford University Press, New York, 1994, p. 144.

FORMATION AND VOLTAMMETRIC CHARACTERIZATION OF IRON–HUMATE COMPLEXES OF DIFFERENT MOLECULAR WEIGHT

L. Leita,[1] M. De Nobili,[2] L. Catalano[2] and A. Mori[1]

[1] Istituto Sperimentale per la Nutrizione delle Piante, Ministero Italiano per le Politiche Agricole, 34170 Gorizia, Italy
[2] Dipartimento di Produzione Vegetale e Tecnologie Agrarie, Universita', 33100 Udine, Italy

1 INTRODUCTION

Metal complexation by fulvic and humic acids is a subject of great nutritional and environmental importance.[1,2] In fact, one of the most significant properties of these ubiquitous natural compounds is their ability to interact with metal ions to form complexes with different solubility as well as chemical and biochemical stability.

Humic acids play important roles in soil-forming processes (*e.g.* podzolization) and in the availability of essential or toxic elements to plants since they affect immobilization and release equilibria of many cations in soil.

In particular, the higher solubility of humic substances with a molecular weight lower than 5 kDa makes them of considerable biological importance because they are involved in many processes that take place in the soil–plant system and contribute to the mobilization of heavy metals in the environment.[3–5]

The physico-chemical characterization of metal–humic complexes is not simple since humic substances have to be considered as macromolecular ligands characterized by polyfunctionality, polyelectrolyte properties and conformational factors that play a key role in metal ion binding.[6–12]

Monovalent cations are held primarily by simple charge interaction resulting in the formation of salts with carboxyl groups (R-COO$^-$ M$^+$), while multivalent cations are able to form coordinative bonds with organic molecules.[13]

The use of spectral techniques like FTIR, ESR, NMR has improved our knowledge of metal–humic complexes.[14]

Many investigations have been carried out by voltammetric methods that are well suited for the determination of total concentration and speciation of trace elements because of their high sensitivity and the direct relations between voltammetric signals and physico-chemical properties of the electroactive species.[9,15–17]

In this work we studied the complexation of ferro- and ferri-cyanide by different molecular size humic substance fractions.

2 MATERIALS AND METHODS

2.1 Extraction and Fractionation of Humic Substances

Humic substances (HSs) were extracted from a commercial sample of sphagnum moss peat (Lithuania) with 0.5 M NaOH under nitrogen flow; the extract was filtered on a 0.45 μm Whatman filter and treated with a cation exchange resin (Amberlite IR 120 H$^+$, Serva, Germany) in order to remove the excess of sodium and to lower the pH to 7.[18]

Fractionation was conducted by multistage ultrafiltration on Amicon YM membranes of the following M_W cut-offs: 1, 10, 30, 50, 100, 300 kDa. Ultrafiltration was carried out with 0.1 M $Na_4P_2O_7$ adjusted to pH 7.1 with concentrated H_3PO_4. The ultrafiltration of each fraction was stopped only when the outflowing liquid was colourless. Fractions were dialysed with distilled water to eliminate pyrophosphate until the conductance of the outflowing liquid was less than 30 μS. After concentration on the same membrane to about 1/10 of the original volume, fractions were stored at –18 °C.

2.2 Iron–Humate Formation and Characterization

A three electrode polarographic analyser (EG&G 264) connected to a 303 EG&G polarographic cell was employed.

Cyclic voltammetric measurements were performed at a glassy carbon stationary electrode (GCSE) as the working electrode, a Pt wire as the counter electrode and Ag/AgCl as the reference electrode. The test aqueous solutions were 0.5 mM $K_3[Fe(CN)_6]$ containing 0.1 M $NaClO_4$ as the supporting electrolyte, to which increasing aliquots of organic carbon of HS fractions were progressively added.

Electrode cleaning was performed after each set of additions by polishing with alumina and performing an electrochemical cleaning (scanning excursion from +1.4 V to –0.4 V). The reliability of the working electrode was periodically checked by recording voltammograms of standard solutions of ferricyanide.

UV–visible spectra of $[Fe(CN)_6]^{3-}$ solutions and of $[Fe(CN)_6]^{3-}$–HS solutions were recorded on a Varian Cary 1E spectrophotometer with distilled water or solutions containing the appropriate amount of HS as references.

The relative stability constants of $[Fe(CN)_6]^{3-}$–HS complexes were calculated according to the Yu and Ji[19] equation 1, where ΔE_p is the potential shift caused by the

$$\Delta Ep = \frac{0.059}{z} n \log L + \frac{0.059}{z} \log K - \frac{0.059}{z} \log \frac{i_M}{i_{ML}} \qquad (1)$$

ligand addition, i_M is the peak current in the absence of the added ligand HS, i_{ML} is the peak current in the presence of ligand, n is the average number of ligands in the complex, L is the ligand concentration expressed in mol l^{-1} of organic carbon present in the dissolved HS, z is the number of electrons involved in the electrochemical process (z = 1 in our case) and K is the relative stability constant.

By plotting the quantity y = ΔE_p + 0.059 log (i_M/i_{ML}) relative to the cathodic and anodic processes *vs.* x = log L, the stability constant K can be easily calculated from the intercept of the straight line extrapolated to log L = 0. The regression relative to HS fraction >300 kDa showed clear bimodal behaviour (Figure 1) and two relative stability constants have been calculated.

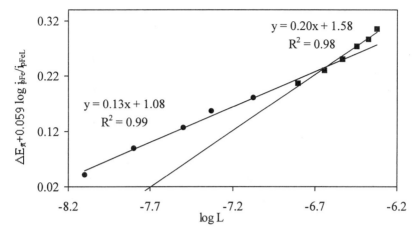

Figure 1 *Plot of ΔEp+0.059 log ipFe/ipFeL vs. logarithm of ligand concentration for determination of relative stability constant of [Fe(CN)₆]⁴⁻–HS (> 300 kDa)*

3 RESULTS AND DISCUSSION

Cyclic voltammograms recorded on ferricyanide solutions displayed a well defined cathodic peak at +0.160 V (Fe^{III}–Fe^{II} reduction) associated with an anodic peak at +0.230 V (Fe^{II}–Fe^{III} oxidation). Amperometric titrations of potassium ferricyanide with increasing aliquots of HS fractions showed a significant depression of both oxidation and reduction peak currents together with a cathodic shift of the Fe^{III} reduction peak and an anodic shift of the associated Fe^{II} oxidation peak even after the first addition of each HS fraction (Figure 2). The trends of ΔE_p (E_{pa}–E_{pc}) values upon addition of HS varied with the apparent mean molecular weight of the fractions (Figure 2). The potential shift of the Fe redox couple increased sharply on addition of the HS fraction of lower molecular size, whereas no significant changes were observed after supplemental additions. Results of the amperometric titration with the HS fraction of nominal molecular weight >300 kDa showed a progressive increase of the potential shift, suggesting different dynamics of $[Fe(CN)_6]^{3-}$–humate formation at larger HS concentrations.

The depression of both i_{pc} and i_{pa} were concomitant with a shift of peak potentials. The dynamics of complexation given by the relative ratio of ipFe/ipFe-HS were different. In particular, Figure 3 shows the trend for $[Fe(CN)_6]^{3-}$ complexation with HS fractions of apparent molecular weight of 1–10 kDa and > 300 kDa. Results obtained reflected the trend of peak potential shifts and revealed a more pronounced complexing capacity of HS fraction of lower apparent molecular weight.

UV–visible spectra were recorded in parallel with voltammetric measurements in order to get more information about the $[Fe(CN)_6]^{3-}$ interaction with HSs. The spectrum of the ferricyanide standard solution differed from that of $[Fe(CN)_6]^{3-}$–HS complexes. The spectra are nearly coincident in the region 250–500 nm, but they differ markedly at wavelengths lower than 250 nm. The ferricyanide complex displays a sharp band at 210 nm which was progressively depressed. As shown in Figure 4, the absorbance values recorded at 210 nm decreased more sharply after addition of HSs of lower molecular weight, confirming the larger complexing capacity of these HS fractions.

Natural ligands such as humic and fulvic acids have to be considered as non-uniform species with peculiar features of polyfunctionality, polyelectrolytic nature and conformational properties favouring the formation of very stable metal complexes and also for configurational reasons.[20-22]

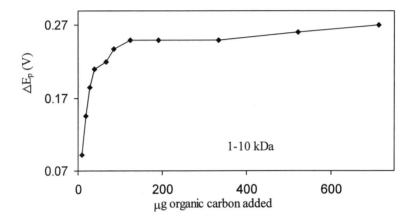

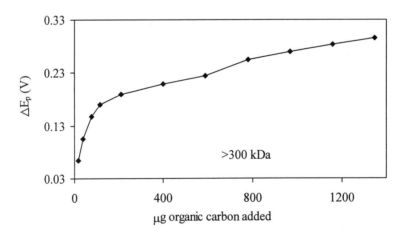

Figure 2 *Cathodic peak potential shift of [Fe(CN)$_6$]$^{3-}$ after stepwise additions of HS fractions: 1–10 kDa (above) and > 300 kDa (below)*

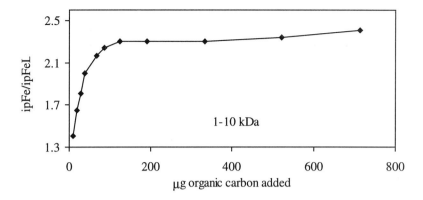

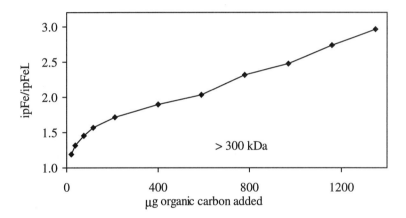

Figure 3 *Relative ratios of cathodic current peaks of $[Fe(CN)_6]^{3-}$ and $[Fe(CN)_6]^{3-}$–HS after stepwise additions of HS fractions: 1–10 kDa (above) and > 300 kDa (below)*

The calculation of the relative stability constants showed that $[Fe(CN)_6]^{3-}$–HS and $[Fe(CN)_6]^{4-}$–HS complexes are much more stable than ferro- and ferri-cyanide complexes even considering the contribution of kinetic control of ΔE_p.[23] Relative stability constants varied with the molecular weight of the fractions and showed a maximum for the 50–100 kDa fraction. As expected, stability constants for Fe(II) were lower than for Fe(III). The number of ligand groups (n) averaged over the whole set of fractions was 3.0 for $[Fe(CN)_6]^{3-}$ and 2.7 for $[Fe(CN)_6]^{4-}$ (Table 1). We emphasize that the n values found must be considered with care because of the complex nature of humate ligands.

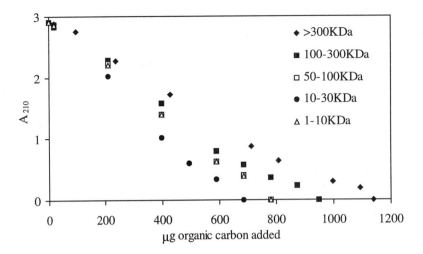

Figure 4 *Complexation capacity of HS fractions determined by stepwise addition of HS fractions to a 0.5 mM ferricyanide solution*

Table 1 *Values of relative stability constants of $[Fe(CN)_6]^{3-}$–HS and $[Fe(CN)_6]^{4-}$–HS and calculation of average numbers of ligands[a]*

	M_w (kDa)	1–10	10–30	50–100	100–300	> 300	
Fe(III)–HS	log K_{rel}	25.6	23.7	29.9	18.7	18.2	27.8
	n	3.6	3.1	3.6	2.2	2.1	3.5
Fe(II)–HS	log K_{rel}	15.7	18.3	24.9	22.8	18.4	26.8
	n	2.2	2.4	3.1	2.7	2.2	3.4

[a] $K_3[Fe(CN)_6]$ log K stability = 31; $K_4[Fe(CN)_6]$ log K stability = 24.

References

1.   N. Senesi, *Anal. Chim. Acta*, 1990, **232**, 51.
2.   N. Senesi, *Anal. Chim. Acta*, 1990, **232**, 77.
3.   R. F. Boyer, H.M. Clark and S. Sanchez, *J. Plant. Nutr.*, 1989, **12**, 581.
4.   R. L. Chaney and P. F. Bell, *J. Plant Nutr.*, 1987, **10**, 963.
5.   Y. Chen and F. J. Stevenson, 'The Role of Organic Matter in Modern Agriculture', Martinus Nijhoff Publishers, 1986, p. 73.
6.   G. S. P. Ritchie, A. M. Posner and I. M. Ritchie, *J. Soil Sci.*, 1982, **33**, 671.
7.   S. Deiana, C. Gessa, B. Manunza, R. Rausa and V. Solinas, *Europ. J. Soil Sci.*, 1995, **46**, 103.
8.   G. Tyler, *Soil Biochem.*, 1981, **5**, 371.
9.   J. Buffle and F. L. Greter, *J. Electroanal. Chem.*, 1979, **101**, 231.
10.   J. E. Gregor and H. K. J. Powell, *Anal. Chim. Acta*, 1988, **211**, 141.

11. J. E. Gregor, H. K. J. Powell and R. M. Town, *J. Soil Sci.*, 1989, **40**, 661.

12. M. Plavsic, B. Cosovic and S. Miletic, *Anal. Chim. Acta*, 1991, **255**,15.

13. F. L. Greter, J. Buffle and W. Haerdi, *J. Electroanal. Chem.*, 1979, **101**, 211.

14. F. J. Stevenson, 'Micronutrients in Agriculture', Soil Science Society of America, Madison, WI, 1991, **6**, p. 145.

15. N. Senesi, 'Biogeochemistry of Trace Metals', Lewis Publishers, Boca Raton, FL, 1992, **16**, p. 429.

16. Z. Navratilova and P. Kula, *Anal. Chim. Acta*, 1993, **273**, 305.

17. A. M. Mota, A. Rato, C. Brazia and M. L. Simoes Goncalves, *Environ. Sci. Technol.*, 1996, **30**, 1970.

18. M. De Nobili, L. Leita and P. Sequi, *Sci. Total Environ.*, 1987, **62**, 85.

19. T. R. Yu and G. L. Ji, 'Electrochemical Methods in Soil and Water Research', Pergamon Press, Oxford, 1993, **13**, p. 405.

20. J. P. Pinheiro, A. M. Mota and M. L. S. Goncalves, *Anal. Chim. Acta*, 1994, **284**, 525.

21. M. Filella, J. Buffle and H. P. van Leeuwen, *Anal. Chim. Acta*, 1990, **232**, 209.

22. Z. Navratilova and P. Kula, *Anal. Chim. Acta*, 1993, **273**, 305.

23. J. Bierrum, G. Schwarzenbach and L. G. Sillen, 'Stability Constants of Metal-ion Complexes, with Solubility Products of Inorganic Substances', Chemical Society, London, 1958, p. 34.

NONLINEARITY AND COMPETITIVE SORPTION OF HYDROPHOBIC ORGANIC COMPOUNDS IN HUMIC SUBSTANCES

Baoshan Xing

Department of Plant and Soil Sciences, University of Massachusetts, Amherst, MA 01003

1 INTRODUCTION

Humic substances (HSs) are a class of relatively high molecular weight, brown-to-black colored substances formed by secondary reactions.[1] They are ubiquitous in water, soils and sediments. Humic substances play a significant role in formation of soil aggregates, control of soil acidity, cycling of nutrients, soil moisture retention, detoxification of hazardous compounds, the fate and binding of solutes, sustainable agriculture and environmental quality. Thus, their properties and quantity directly or indirectly affect our environment and human health. This paper focuses on the aspect of sorption of organic compounds by HSs.

Hydrophobic organic compounds (HOCs) have a high affinity for HSs in soils and sediments. Mineral surfaces are generally unimportant for sorption of HOCs unless the soil organic carbon fraction is $< 0.01\%$.[2] This is particularly true in aqueous systems due to the strong dipole interactions of water molecules with these mineral surfaces.[3,4] Most soils contain organic carbon higher than 0.01% and are somewhat moist-to-saturated. Therefore, it is important to study and understand interactions between HSs and HOCs in order to predict the fate and behaviors of HOCs in soil and thus develop effective remediation technologies.

Partitioning (dissolution) has been considered to be the prevailing mechanism for HOC sorption in HSs at least since the late 1970s. This mechanism has explained the positive correlation of sorption coefficients with soil organic carbon content or n-octanol-water distribution coefficients (*i.e.* linear free energy relationships).[3] Other experimental results in support of partitioning are linear uptake isotherms, lack of competition between solutes and sorption enthalpies more positive than corresponding condensation enthalpies.

Sorption enthalpies, however, cannot be used as direct evidence to differentiate sorption mechanisms (*e.g.* partitioning and surface adsorption). Some surface *ad*sorption can have more positive enthalpy than the enthalpy of condensation. For example, adsorption of 19 nonpolar HOCs by three minerals (calcium carbonate, hematite, and corundum) had lower heats of sorption than their corresponding heats of condensation.[5] Similarly, some partitioning can also have substantially negative enthalpy, which has been traditionally used as the criterion for surface adsorption. Specific interactions between solutes and solvents (*e.g.* hydrogen bonds) likely produce negative solution enthalpies, *e.g.* phenol in triethylamine and quinoline and substituted pyridines in methanol.[6,7] Hence the assumption that positive solution enthalpy is a feature or criterion of partitioning is oversimplified. A more comprehensive review of this subject is given elsewhere.[8]

The driving force for partitioning is the hydrophobic effect, which results from reduction of free energy when a HOC molecule is transferred out of water. But the hydrophobic effect plays an important role in all physisorption processes from aqueous solution, whether site-specific or not, *e.g.* sorption of hydrocarbons by metal oxides[5] and cation exchange of organic ions on clays.[9] This is because the hydrophobic effect originates from solution-phase rather than sorbent-phase interactions. Therefore, the experimental demonstration of linear free energy relationships does not constitute proof that partitioning occurs exclusively. More details on this topic were given in a review article.[10]

Recent research indicates nonlinear isotherms when a wide range of solute concentration is used.[11-14] Competitive sorption between solutes also has been observed.[15,16] These observations suggest an adsorption-like process in addition to partitioning. This paper reports further experimental evidence for isotherm nonlinearity and competitive sorption of HOCs in humic substances.

2 EXPERIMENTAL

Well-humified Pahokee peat was purchased from the International Humic Substance Society. Humic acid (HA) and humin of the peat were extracted with the procedure outlined by Chen and Pawluk.[17] Selected properties are given in Table 1. The peat was also treated with HF/HCl to reduce the ash content and the resulting peat (trt peat) had a composition of 53% C, 4.0% H, 3.6% N and 1.2% ash (as compared with 6.9% ash of the untreated peat). Another HA (AHA) was extracted from a mineral soil (Mollisol) from Alberta, Canada. Its composition was 53% C, 3.96% H, 3.54% N, 38.6% O and 0.6% ash. A mineral soil was a Cheshire loam from Connecticut. The dry soil composition was 56% sand, 36% silt, 8% clay and 1.4% organic carbon. Organic chemicals were toluene, chlorinated benzenes, naphthalene, 2,4-dichlorophenol, atrazine [2-chloro-4-ethylamino-6-isopropylamino-*s*-triazine], prometon [2,4-bis(isopropylamino)-6-methoxy-*s*-triazine], and metolachlor [2-chloro-*N*-(2-ethyl-6-methylphenyl)-*N*-(2-methoxy-1-methylethyl) acet-amide].

Table 1 *Selected properties of peat and its humic fractions*

	%C	%H	%N	%Ash	Aromatic-C^a	Alkyl-C^a	Carboxyl-C^a
Peat	45	4.7	3.1	6.9	17	32	7
HA	52	3.2	3.3	5.6	25	21	12
Humin	44	4.1	3.1	25	17	32	7

[a] % of total detected carbon.

The elemental contents (C, H, N) were determined with a Carlo Erba CHNS-O EA 1108 elemental analyser. The ash content was determined by heating the samples at 740 °C for 4 hr.

The solid-state ^{13}C NMR spectra were obtained using cross-polarization (CP) and magic angle spinning (MAS) techniques. The NMR spectrometer was a Bruker AM 300 instrument operating with CP (1 ms contact time) and MAS (5 kHz spinning rate) and a HP WP 73A probe, at 75 MHz frequency, with a pulse width of 5.50 µs and 70 ms acquisition time. The rotor was a 4 mm/18 zirconia rotor with a Kel-F cap. Within the 0–220 ppm chemical shift range, C atoms were assigned to alkyl C (0–50 ppm), O-alkyl C (50–107), aromatic C (107–165), carboxyl C (165–190), and carbonyl C (190–220). Aromatic, alkyl and carboxyl carbon percentages are reported in Table 1.

Sorption experiments were conducted with a batch equilibration technique in 8-mL screw-cap vials (minimal headspace) with Teflon-lined septa. The solution was 0.01 M $CaCl_2$ containing 200 mg/L $HgCl_2$ as a biocide to minimize microbial activity. The solution to solid ratio was adjusted to achieve 30–80% uptake of a test compound and the initial concentration ranges were over at least 2.5 orders of magnitude. Two blanks without sorbent were run for each initial concentration. The sorbent and test compound suspensions were shaken in hematology mixers giving rocking-rotating motions. After appropriate mixing, the vials were centrifuged at 900 x *g* for 20 min and supernatants were sampled for liquid scintillation counting of ^{14}C-labelled compounds or for liquid chromatography analysis of unlabelled compounds. Hexane extraction was used for gas chromatography analysis of 1,3-dichlorobenzene and trichloroethylene. Because of minimal sorption by vials and no biodegradation, sorbed concentrations were calculated by the mass difference. Isotherms were constructed using the Freundlich equation $S = KC^N$, where S and C are the sorbed amount and equilibrium solute concentration, respectively, K is a constant and N is the exponent. The parameters K and N were determined by linear regression of log-transformed data.

Competitive sorption was determined by two methods. First, the concentration of the principal solute was maintained while varying the competitor (co-solute) concentrations. In this way, competitive sorption was determined by measuring the single concentration point solid-solution distribution ratio (S/C, K_d) of the principal solute. The S/C ratio should remain constant with varying co-solute concentrations if partitioning is the only sorption mechanism. Second, the concentration of co-solute was kept the same while several concentrations of the principal solutes were used. In the second method, the competitive effect was determined by measuring the change of isotherm parameters, K and N or the change of the S/C ratio of the principal solute over a concentration range. Again, for partitioning, no change should be observed. All experiments were conducted using the batch equilibration technique. Further details are given elsewhere.[15]

3 RESULTS AND DISCUSSION

Freundlich isotherm parameters are shown in Table 2 and examples of log-transformed Freundlich isotherms are displayed in Figures 1–3. As one can see, all isotherms were nonlinear (*i.e.* $N < 1$) regardless of polar or apolar solutes. Even sorption of apolar HOCs in nearly 100% organic sorbents was also nonlinear, *e.g.* toluene in peat, naphthalene and chlorinated benzenes in peat humic acid and 1,3-dichlorobenzene in the HF/HCl treated peat. These apolar compounds are incapable of exhibiting specific sorption (*e.g.* H-bonding). Thus, mineral surfaces and specific sorption can be discounted for the nonlinearity. The high quality of experimental data (see the r^2 values, standard deviations

Table 2 *Isotherm parameters from single-solute experiments*

Solute/sorbent	K, $(\mu g/g)(\mu g/mL)^{-N}$	N	r^2
Atrazine/soil	2.15	0.923±0.004	1.00
Atrazine/peat	74.1	0.916±0.007	0.999
Prometon/soil	2.60	0.930±0.035	0.985
Prometon/peat	123	0.799±0.018	0.995
Metolachlor/soil	1.62	0.852±0.005	0.999
Metolachlor/peat	76.4	0.890±0.006	0.999
2,4-Dichlorophenol/soil	7.86	0.806±0.006	0.999
2,4-Dichlorophenol/peat	319	0.777±0.005	0.999
Atrazine/AHA	186	0.886±0.011	0.998
Prometon/AHA	123	0.858±0.024	0.992
Toluene/peat	74.3	0.846±0.012	0.998
1,2-Dichlorobenzene/peat	332	0.833±0.007	0.999
1,2-Dichlorobenzene/HA	140	0.896±0.007	0.999
1,2-Dichlorobenzene/humin	422	0.780±0.006	0.999
1,3-Dichlorobenzene/peat	340	0.852±0.008	0.998
1,3-Dichlorobenzene/HA	158	0.928±0.008	0.998
1,3-Dichlorobenzene/humin	403	0.766±0.007	0.999
1,3-Dichlorobenzene/trt peat[a]	402	0.856±0.006	0.999
1,3-Dichlorobenzene/trt peat[b]	468	0.818±0.004	0.999
Naphthalene/peat	398	0.790±0.008	0.998
Naphthalene/HA	230	0.903±0.008	0.999
Naphthalene/humin	517	0.751±0.006	0.999

[a] isotherm at 1 day contact time; [b] isotherm at 30 days contact time.

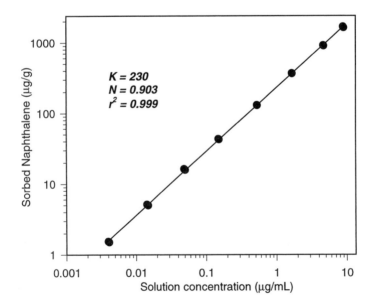

Figure 1 *Sorption isotherm of naphthalene in peat humic acid*

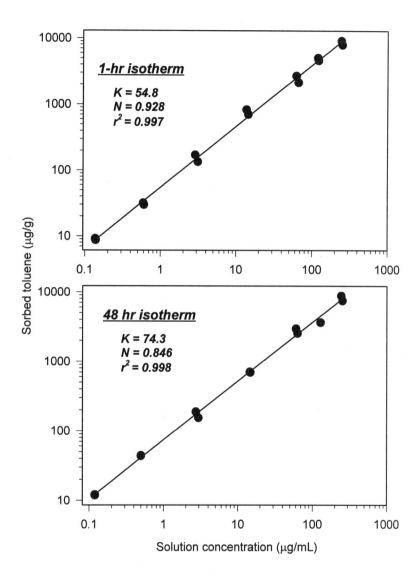

Figure 2 *Sorption isotherms of toluene in peat at two different contact times*

of N, and regression in isotherm plots) suggests little or no experimental artifact. Furthermore, nonlinear isotherms are not an artifact of insufficient contact time because isotherms typically become more nonlinear with increasing time (Table 2 and Figure 2). Non-linear isotherms indicate a distribution of sorption site energy in HSs or natural organic matter.

Previously, we conducted the experiments on sorption of HOCs in rubbery and glassy polymers.[11,15] Under dilute aqueous conditions, isotherms of both polar and apolar HOCs were linear in rubbery polymers (indicative of partitioning) but nonlinear in glassy polymers indicative of a hole-filling mechanism. This latter term has been used to describe sorption of small to medium size molecules in glassy polymers.[18] The hole sites are believed to be unrelaxed free volume in the form of nanometer-size voids within the glassy polymers. Penetrant molecules filling the voids undergo adsorption-like interactions with the polymer strands making up the void walls. By analogy, we proposed the dual-mode model for sorption of HOCs in HSs; that is, HSs consist of expanded (swollen), rubbery and condensed domains.[13,15] Partitioning occurs in both domains and hole-filling (adsorption-like) only in the latter domains. The hole-filling process leads to nonlinear sorption in HSs. This model can explain numerous nonlinear isotherms observed in the literature and the biphasic nature of diffusion rates in organic matter.[19] Also, the dual-mode model is consistent with the proposed chemical structure and configuration of HSs, a water-swollen, three-dimensional assembly of polydisperse, complex, heterogeneous macromolecules.[20] This assembly has a Gaussian distribution of molecular mass with the greatest mass density at the center and decreasing to zero at the outer edges. The hole sites would be more concentrated in the dense center, and the outer edges would be more like rubbery polymers for partitioning. A detailed description of the dual-mode model is provided elsewhere.[10,21]

High surface area carbonaceous materials (HSACM) have recently been proposed to account for isotherm nonlinearity.[22] HSACM, if present, could lead to some nonlinearity. But sorption of 1,3-dichlorobenzene in polyvinyl chloride containing no HSACM still showed nonlinearity.[11] In addition, the humin from the peat contained much less soot

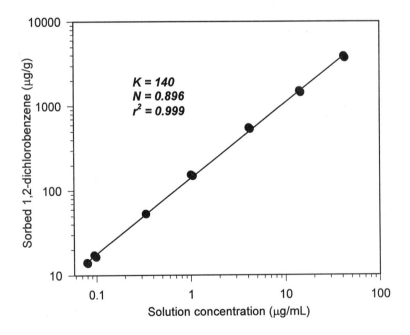

Figure 3 *Sorption isotherm of 1,2-dichlorobenzene in peat humic acid*

carbon (carbonaceous residue from incomplete combustion) than peat, but showed greater nonlinearity and competitive sorption.[10] This argues against HSACM or soot particles playing a significant role in these behaviors.

Inability to interrupt or break strong polar contacts within HSs molecules could lead to physically limited sorption domains. These domains have been suggested to explain isotherm nonlinearity of apolar compounds.[23] Strong polar contacts within HSs are not in conflict with the dual-mode model. These contacts may well be the cause for relatively condensed domains or dense molecular mass within HSs through inter- and intra-molecular interactions and cross-linking.

Nonlinearity increased with increasing contact time: the N value of toluene in peat decreased from 0.928 at two hr to 0.846 at 48 hr (Figure 2) and of 1,3-dichlorobenzene in the treated peat from 0.856 at one day to 0.818 at thirty days (Table 2). These results are in line with the dual mode model, indicating that hole sites are relatively concentrated within the HS matrix. For short contact times, penetrant molecules primarily react with exterior expanded, less cross-linked regions of HSs and isotherms are more linear due to more partitioning in these regions. With increasing contact time, the molecules diffuse into the HS matrix and have access to sorption (hole) sites in condensed regions (center) and isotherms became increasingly more nonlinear because of more "site-specific" sorption or *ad*sorption. Similar results were reported for phenanthrene sorption in soils by Weber and Huang.[24] Weber and co-workers independently proposed a "dual reactive domain model".[25,26] The condensed ("crystalline") domains caused nonlinearity, slow sorption and hysteresis while partitioning and fast sorption took place in "amorphous" domains.

Molecular diffusion may also have contributed to the increasing nonlinearity with time. For solids with nonlinear sorption, the effective diffusion coefficient is concentration dependent.[27] Thus, diffusion at higher solute concentrations is faster than at lower concentrations. At a given time before equilibrium, sorption from higher concentration solutions would be relatively closer to equilibrium as compared to the sorption from less concentrated solutions. As a result, the N value of sorption isotherms can decrease with time before reaching equilibrium. However, it is difficult to determine the exact contribution from diffusion on temporal changes of the N value because of the heterogeneity and complexity of HSs.

Competitive sorption plots are shown in Figures 4 and 5. Peat and its humic fractions and soil all demonstrated competitive sorption. The competitive effects declined exponentially with increasing co-solute concentrations (Figure 4). This indicates that competition between the two chlorinated benzenes was not due to an artifact from co-solute effects on the properties of the sorbent and aqueous solution. If that were the case, competition would be weakest (smallest reduction of K_d) at lower concentrations of co-solute. Figure 4 also shows that the competitive effect increased in the order humic acid < peat < humin. This was the same order of their nonlinearity (Table 2). These results are consistent with the notion that the $Na_4P_2O_7$ extraction would preferentially remove loosely-bound fractions first (fulvic and humic acids) and leave behind the strongly-bound, highly cross-linked, dense fractions (humin). Moreover, these results indicate that the number of postulated hole sites increases from humic acid to peat to humin, and that removing fulvic and humic acids exposes more hole sites. We also observed that adding HSs precursor molecules (*e.g.* vanillic acid) could occupy and block adsorption-like sites of HSs.[13]

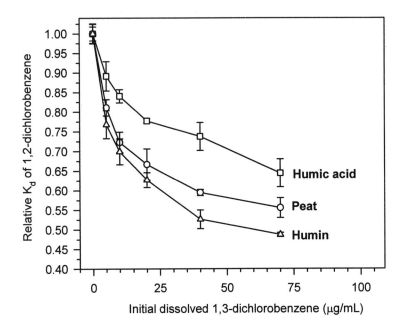

Figure 4 *Competitive effect between two chlorinated benzenes in HSs indicated by decrease of the relative distribution ratio (S/C, K_d) of 1,2-chlorobenzene with increasing initial concentrations of 1,3-chlorobenzene (constant initial concentration of 1,2-chlorobenzene = 2 µg/mL)*

Prometon not only linearized the atrazine isotherm but also reduced the sorption coefficient (K) through competition (Figure 5A). Prometon molecules might compete for and block adsorption sites in the condensed domains of HSs and not affect partitioning domains. Thus, the atrazine isotherm was more linear in the presence of prometon. Figure 5B displays the effect of prometon on the *S/C* ratio. Without prometon, the *S/C* ratio of atrazine decreased with increasing solution concentrations due to distribution of sorption site energy. With prometon, the *S/C* ratio of atrazine remained practically constant, indicative of partitioning as a prevailing sorption mechanism. This again is because prometon molecules blocked the adsorption sites but not the partitioning domains. These competitive results are in agreement with the proposed dual-mode sorption model.

There are other data supporting the dual-mode sorption model for HSs. The presence of nanoporosity in humic acids was reported using CO_2 to measure surface area.[21,28] The glass transition temperature was observed for a humic acid in both dry and moist conditions.[29] X-Ray diffraction revealed a 0.35 nm peak of humic acids which was believed to be condensed aromatic carbons.[30,31] Nonlinearity decreased with increasing temperature and co-solvents.[21] Furthermore, molecular modelling indicates that small molecules (*e.g.* atrazine) can be trapped in the voids or cavities of humic substances.[32,33]

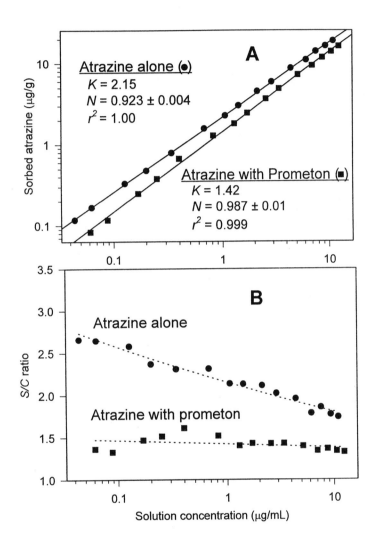

Figure 5 *Competitive sorption between atrazine and prometon in soil (constant initial concentration of prometon = 130 µg/mL). A = effect of prometon on isotherm nonlinearity of atrazine; B = effect of prometon on distribution ratio (S/C, K_d) of atrazine. The dotted lines are the Freundlich fits*

4 CONCLUSIONS

Humic substances are assemblies of polydisperse, heterogeneous macromolecules. This heterogeneous nature dictates sorption behaviors of hydrophobic organic compounds and results in nonlinear isotherms and competitive effects. These sorption behaviors can be explained by the dual-mode sorption model, in which both partitioning and hole (site)-filling mechanisms take place and competition occurs only in the condensed domains of HSs. The condensed domain may cause slow kinetics, hysteresis, reduced bioavailability with ageing, and reduced exposure risk due to slow diffusion and energetic sites.

ACKNOWLEDGMENTS

I thank Dr. Joseph J. Pignatello for his collaboration and discussion, Dr. Ellen R. Graber for her discussion on diffusion, Mr. Zhaowei Jin and Ms. Tina Arounsack for technical assistance and the U.S. Department of Agriculture National Research Initiative Grant Program for financial support.

References

1.   F. J. Stevenson, in 'Humus Chemistry: Genesis, Composition, Reactions', 2nd edn. Wiley, New York, 1994.
2.   B. T. Mader, K. Goss and S. J. Eisenreich, *Environ. Sci. Technol.*, 1997, **31**, 1079.
3.   C. T. Chiou, in 'Reactions and Movement of Organic Chemicals in Soils', B. L. Sawhney and K. Brown (eds.), Spec. Publ. 22, Soil Science Society of America, Madison, WI, 1989, p.1.
4.   B. Xing, W. B. McGill, M. J. Dudas, Y. Maham and L. G. Hepler, *Environ. Sci. Technol.*, 1994, **28**, 466.
5.   K.-U. Goss and S. J. Eisenreich, *Environ. Sci. Technol.*, 1996, **30**, 2135.
6.   E. M. Arnett, B. Chawla, L. Bell, M. Taagepera, W. J. Hehre and R. W. Taft, *J. Am. Chem. Soc.*, 1977, **99**, 5729.
7.   E. M. Arnett, L. Joris, E. Mitchell, T. S. S. R. Murty, T. M. Gorrie and P. v. R. Schleyer, *J. Am. Chem. Soc.*, 1970, **92**, 2365.
8.   M. D. Borisover and E. R. Graber, *J. Environ. Qual.*, 1998, **27**, 312.
9.   R. P. Schwarzenbach, P. M. Gschwend and D. M. Imboden, 'Environmental Organic Chemistry', Wiley, New York, 1993.
10.   J. J. Pignatello, in 'Kinetics and Mechanisms of Reactions at the Mineral–Water Interface', D. L. Sparks and T. J. Grundl (eds.), ACS Symposium Series, Washington, DC, in press.
11.   B. Xing and J. J. Pignatello, *Environ. Toxicol. Chem.*, 1996, **15**, 1282.
12.   W. J. Weber, Jr., P. M. McGinley and L. E. Katz, *Environ. Sci. Technol.*, 1992, **26**, 1955.
13.   B. Xing and J. J. Pignatello, *Environ. Sci. Technol.*, 1998, **32**, 614.
14.   W. Huang and W. J. Weber, Jr., *Environ. Sci. Technol.*, 1997, **31**, 2562.
15.   B. Xing, J. J. Pignatello and B. Gigliotti, *Environ. Sci. Technol.*, 1996, **30**, 2432.
16.   P. M. McGinley, L. E. Katz and W. J. Weber, Jr., *Environ. Sci. Technol.*, 1993, **27**, 1524.

17. Z. Chen and S. Pawluk, *Geoderma*, 1995, **65**, 173.
18. W. R. Vieth, 'Diffusion In and Through Polymers', Oxford University Press, New York, 1991.
19. K. M. Carroll, M. R. Harkness, A. A. Bracco and R. R. Balcarcel, *Environ. Sci. Technol.*, 1994, **28**, 253.
20. M. H. B. Hayes, in 'Humic Substances in Soils, Peat and Waters: Health and Environmental Aspects', M. H. B. Hayes and W. S. Wilson (eds.), The Royal Society of Chemistry, Cambridge, UK, 1997, p. 3.
21. B. Xing and J. J. Pignatello, *Environ. Sci. Technol.*, 1997, **31**, 792.
22. C. T. Chiou and D. E. Kile, *Environ. Sci. Technol.*, 1998, **32**, 338.
23. E. R. Graber and M. D. Borisover, *Environ. Sci. Technol.*, 1998, **32**, 258.
24. W. J. Weber, Jr. and W. Huang, *Environ. Sci. Technol.*, 1996, **30**, 880.
25. W. Huang, T. M Young, M. A. Schlautman, H. Yu and W. J. Weber, Jr., *Environ. Sci. Technol.*, 1997, **31**, 1703.
26. W. Huang and W. J. Weber, Jr., *Environ. Sci. Technol.*, 1997, **31**, 2562.
27. J. Crank, 'The Mathematics of Diffusion', 2nd edn. Clarendon Press, Oxford, 1995.
28. H. de Jonge and M. C. Mittelmeijer-Hazeleger, *Environ. Sci. Technol.*, 1996, **30**, 408.
29. E. J. LeBoeuf and W. J. Weber, Jr., *Environ. Sci. Technol.*, 1997, **31**, 1697.
30. M. Schnitzer, H. Kodama and J. A. Ripmeester. *Soil Sci. Soc. Am. J.*, 1991, **55**, 745.
31. B. Xing and Z. Chen, *Soil Sci.* (accepted).
32. H.-R. Schulten, *Intern. J. Environ. Anal. Chem.*, 1996, **64**, 147.
33. H.-R. Schulten, *Fresenius' J. Environ. Anal. Chem.*, 1995, **351**, 62.

ADSORPTION OF A PLANT- AND A SOIL-DERIVED HUMIC ACID ON THE COMMON CLAY KAOLINITE

E. A. Ghabbour, G. Davies, K. O'Donaughy, T. L. Smith and M. E. Goodwillie

The Barnett Institute and Chemistry Department, Northeastern University, Boston, MA 02115, USA

1 INTRODUCTION

Composts[1,2] and soils[3] are dynamic mixtures of fauna, flora, decomposing animal and plant remains, clays, minerals, metal species and natural and anthropogenic solutes. Recent work[2] supports long-held conclusions[1] that, given sufficient time, water for mobility and required mechanisms, the transient organic materials in soils are converted to relatively long lived ("refractory", "recalcitrant") brown biopolymers called humic substances (HSs).

Humic acids (HAs) are the carbon-rich class of HSs isolated from plants[4-6] and soils[3,7,8] that dissolve in aqueous base and form waterlogged gels on lowering the pH. Although HAs have been studied for centuries, their structures still are unknown.[8-10] HAs strongly retain water, adhere to clays and minerals[3,7,9,10] and selectively bind metals[11-13] and other solutes.[14-17] As oxidized biopolymers, HAs have a strong tendency to aggregate[18] and they retain and somehow protect polysaccharides and proteins.[10,19,20]

Progress in understanding HAs depends on having purified samples. Finding HAs in live plants[4-6] suggests that HA synthesis is biochemically controlled. We are contributing to HA research by working with highly purified HAs isolated from composts,[14-16] plants,[5,6] and peats and soils[13] from different countries. Analytical and other high purity HA properties indicate that they are not too different.[13,16] For example, while purified HA total acidities vary, their R-COOH contents often are about 3 mmol/g HA.[13]

HAs effectively sorb hydrophilic[14-16] and hydrophobic[17] solutes. Hydrophilic nucleic acid constituents (NACs) adsorb in sequential steps on highly purified solid compost-,[14,15] German peat-[16] and mercury(II)-loaded, German peat-derived[16] HAs. The isotherm data fit the Langmuir adsorption model, which generates a stoichiometric site capacity v_i and an equilibrium constant K_i for adsorption of a given solute in a given step i = A, B or C. The site capacity data v_i indicate that some solutes can free up more HA sites as a result of primary adsorption on the HA surface. Plotting the enthalpy change ΔH_i vs the entropy change ΔS_i for each solute and adsorption step gives a straight line, indicating highly correlated enthalpy and entropy changes. For example, endothermic adsorption on a solid HA results a correlated entropy increase. HA particles open like flowers to adsorb hydrophilic solutes. Adsorbed water and HA disaggregation feature prominently in the proposed adsorption mechanism.[21]

The versatile sorptive properties of solid HAs reflect their macrostructures, which can be altered with the gel drying method[22,23] and can change as a result of metal binding[13] or solute adsorption.[21] As expected, bound metals have dramatic effects on HA adsorptive properties. For example, solid HA-bound Hg(II) centers selectively bind uracil.[16]

HAs' attachment to clays and minerals prevents them from being washed away and thus helps to preserve soils.[8–10] This paper asks how dissolved, highly purified HAs isolated from the free living alga *Pilayella littoralis* (PHA)[5,6] and a New Hampshire bog soil (NHA)[13] interact with the fixed surface area of the common clay kaolinite, $Al_2Si_2O_5(OH)_4$ (Kao). Adsorption isotherms at seven temperatures from 5.0–35.0 °C give averaged data that fit the Langmuir model with two sequential steps. HA monolayer formation step A is followed by formation of a second layer by association of HA with adsorbed HA in step B. Most of the steps are endothermic and result in an entropy increase. Isotherm parameter comparisons suggest different conformations of adsorbed HAs on the kaolinite surface with hydration–dehydration of the components as important factors in adsorption.

2 MATERIALS AND METHODS

2.1 Materials

2.1.1 Humic Acids. HA sample NHA was isolated with the protocol in Figure 1 from a geologically young bog soil collected in the White Mountain National Forest in Rumney, New Hampshire. The humic acid PHA was isolated by the same method with added steps[24] that remove alginic acids, the structural polysaccharides of alga, from free-living alga *Pilayella littoralis* collected in ocean water off Nahant Beach, Massachusetts as previously described.[6] We have compared our HA samples with each other,[6,13] with compost-derived HAs[14,15] and with International Humic Substances Society (IHSS) HA standards.

Humic acid solutions were made by magnetically stirring 100.0 mg of freeze dried solids NHA or PHA in 1.00 L of 0.050 M NaOH at room temperature until each solid was completely dissolved. Each solution was adjusted to either pH 7.0 or pH 3.5 with concentrated HCl. The stock solutions were diluted at fixed pH to give solutions containing 0.010, 0.020, 0.030, 0.040, 0.050, 0.060, 0.070, 0.080 and 0.090 g HA/L. These solutions were used to plot optical absorbance (at 280 nm) *vs.* concentration for calibration of the absorbance measurements at 280 nm used to construct adsorption isotherms at seven fixed temperatures in the range 5.0–35.0 °C.

2.1.2 Kaolinite. A large sample of Kao was supplied by the Geology Department reference mineral collection at Northeastern University. The sample was finely ground in an agate mortor and pestle and then fractionated with graduated geological sieves to give a fraction with average particle size 20 ± 2 μm, which was dried overnight at 110 °C. The surface area of the Kao sample was measured by ASTM standard method D-3663-92 in a custom built instrument with adsorbent N_2 at liquid nitrogen temperature.

2.1.3 Other Reagents. All other chemicals were analytical reagent grade and doubly deionized water was employed throughout.

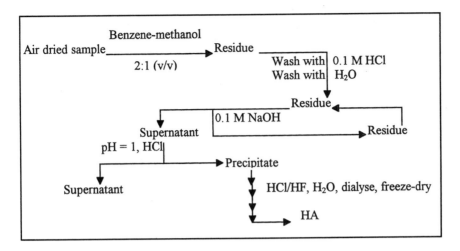

Figure 1 *Humic acid isolation protocol*

2.2 Adsorption Measurements

2.2.1 Isotherm Construction. Adsorption of NHA and PHA on Kao was measured as follows. Separate vessels containing 20.0 mg of solid Kao were treated with 20.0 mL of each of the ten HA solutions listed above. The mixtures were shaken for 48 h at fixed temperature and the absorbance of the supernatant of each mixture was measured after centrifugation for 1 h at 2000 rpm in a Damon bench centrifuge. The calibration curve for each HA was used to calculate the concentration c (g/L) of the HA in equilibrium with solid Kao and the amount of HA adsorbed from each mixture (A, g HA/g Kao). The results were plotted as the isotherm A *vs.* c for each HA at each temperature. Each isotherm measurement was repeated twice to improve data precision.

In practice, unbuffered Kao-solution pH 7.0 was harder to maintain than pH 3.5 used for most of the work. The supernatant absorbances of mixtures of NHA and PHA with Kao were checked over the whole experimental concentration range at times from 24–240 h after mixing. Equilibration of each HA with Kao occurred within 24 h of mixing in all cases.

2.2.2 Isotherm Data Analysis. The Langmuir model (eq 1)[15,25] used to analyse the adsorption isotherms predicts that data for a single step adsorption process on solid Kao will obey eq 2. Here, v is the stoichiometric site capacity (g HA/g Kao) of the Kao surface for the HA, and K is the equilibrium constant (L/g HA) for the adsorption process.

$$Kao(s) + HA(aq) \Leftrightarrow Kao{\bullet}HA(s) \, K \tag{1}$$

$$1/A = 1/v + 1/Kvc \tag{2}$$

At fixed temperature, eq 2 predicts a linear plot of $1/A$ *vs.* $1/c$ with positive intercept $1/v$ and slope $1/Kv$, from which separate values of v and K can be calculated. Linear segments in plots of eq 2 indicate sequential adsorption steps with increasing c.[14,25] The site capacity of an adsorbent with a fixed surface area such as kaolinite is independent of solute molar mass and temperature.[25] Eq 3 predicts the dependence of equilibrium constant K on absolute experimental temperature T (R is the ideal gas constant) for reversible adsorption of an HA on a specific Kao adsorbent site. The slope is $-\Delta H/2.303R$ and the intercept is $\Delta S/2.303R$.

$$\log K = -\Delta H/2.303RT + \Delta S/2.303R \tag{3}$$

2.2.3 Kaolinite Solubility and HA Desorption Measurements. Kaolinite solubility and desorption of NHA and PHA from Kao in 0.10 M NaOH were checked by absorbance measurements at 280 nm at the highest experimental temperature (35.0 °C) with continuous sample shaking for seven days. The supernatants' pHs were reduced to 3.5 with concen-trated HCl and examined for cloudiness or precipitation after standing at room temperature overnight.

3 RESULTS AND DISCUSSION

HA characteristics (see Introduction) frustrate HA molecular structure determination.[8] However, studies with metal cations help to establish common metal binding sites.[11-13] Solid HA adsorptive characteristics reflect HA macrostructures.[14-17] Linear correlation of enthalpy and entropy changes on adsorption has been attributed to the need to dehydrate hydrophilic HA surface functional groups and NAC solutes for adsorption to occur.[21] Adsorption of hydrophilic HAs on a hydrophilic clay surface such as kaolinite[26] should have the same requirements.

3.1.1 Properties of the Kaolinite Adsorbent. Kaolinite (Kao) is a common clay that results from rock weathering. Kao occurs as clusters of rhombic or hexagonal plates with little compositional variation. It has hydrophilic surface OH groups.[26] Scanning electron micrographs (not shown) reveal that adsorption of NHA and PHA on Kao results in Kao agglomeration due to interactions beween HA on separate Kao particles. X-Ray powder patterns (not shown) indicate that HA-treated Kao has the same pattern as Kao itself. HAs and fulvic acids are intercalated by layered clays like montmorillonite[27] but there is no indication of Kao structure intercalation by NHA or PHA.

3.1.2 Humic Acid Characteristics. Refs. 6 and 13 give comprehensive analytical and spectral information (including amino acid and carbohydrate contents after strong acid hydrolysis) for NHA and PHA. Table 1 summarizes basic analytical data.

Humic acid NHA has C, H, N contents that are typical of purified HAs.[3,7-10] It has a much lower ash content than IHSS standard HAs (typically 2% w/w). As noted previously,[5,6] PHA isolated from *Pilayella littoralis* has lower C and O and higher N and ash contents than IHA and NHA. However, NHA and PHA have similar R-COOH contents (Table 1). PHA is more acidic than NHA because of its higher apparent[28] phenolic-OH content (Table 1).

The [13]CPMAS spectra of NHA and PHA show that they are quite similar materials. Percentages of different carbon types in NHA and PHA measured with a conventional[8,11,29] pulse sequence indicate that, on average, 9% of the detected carbon is carboxylic, consistent with the analytical data (Table 1). However, PHA is less aromatic than NHA (Table 2).

Table 1 *Analytical data for freeze-dried solids NHA and PHA*

	Elemental analyses[a]				% Ash[b]	Total acidity	R-COOH[c]	Phenolic-OH[c,d]
	%C	%H	%N	%O				
NHA	52.9	5.40	2.0	37.0	0.25	8.4	2.7	5.7
PHA	42.6	5.20	5.2	38.5	2.4	10.5	3.1	7.4

[a] On a dry, ash-free basis; [b] determined by combustion of 100.0 mg HA samples in air at 850 °C for 2 h; [c] units are mmol/g HA; [d] taken as the difference between columns 7 and 8 (refs. 9 and 28).

Table 2 *Assignments of ^{13}CPMAS spectra[a]*

	alkyl	O-alkyl	aromatic	carboxyl	carbonyl
NHA	29.6	33.3	25.8	9.2	2.2
PHA	40.0	36.3	11.3	10.8	1.7

[a] expressed as percentages of detected total carbon; Assignments: 0-50 ppm alkyl; 50-107 ppm O-alkyl; 107-165 ppm aromatic; 165-190 ppm carboxyl; 190-220 ppm carbonyl; data courtesy of Dr. Baoshan Xing and co-workers, University of Massachusetts.

The major effects of adsorbing NHA, PHA and other purified HAs on minerals is loss of the 2500 and 1715–730 cm^{-1} features characteristic of highly purified HAs,[13] a shift to lower frequency and broadening of the doublet that includes the latter feature and changes in the region below 1500 cm^{-1}. These effects are consistent with adsorption on minerals through R-COO- functional groups of HAs.[30, 31]

3.1.3 Stoichiometry and Thermodynamics of HA Adsorption on Kaolinite. Plots of optical absorbance at 280 nm *vs.* HA concentration were linear for solutions of NHA and PHA to at least 100 mg HA/L. The plots had nearly the same slope for a given HA at pH 3.5 and 7.0. HA aggregation did not complicate isotherm measurement and analysis in the above HA concentration range. The slope of the calibration plot for PHA was nearly the same at pH 3.5 and 7.0 but only about half the magnitude of the slope for NHA.

Isotherm measurements require reproducible absorbances of supernatants in equilibrium with solid Kao. This necessitated long centrifugation of Kao–HA mixtures to eliminate suspended fine particles. Isotherms were measured three times. The results were averaged to give precise HA adsorption data for analysis.

Figures 2– show adsorption isotherms obtained with these precautions. Kao adsorbs more PHA per g than NHA (Figure 2). Adsorption of NHA (Figure 3) and PHA (Figure 4) on Kao is endothermic. The isotherms all showed evidence for sequential steps A and B for HA adsorption with increasing HA equilibrium concentration *c*, as verified by fitting the data to Langmuir eq 2 (Figures 5 and 6). Site capacities v_i and equilibrium constants K_i were calculated from the slopes and intercepts of the segmented plots. The respective thermodynamic parameters ΔH_i and ΔS_i from plots of eq 3 (Figure 7) are collected with averaged site capacity and K_i data for adsorption of dissolved NHA and PHA on Kao in Table 3.

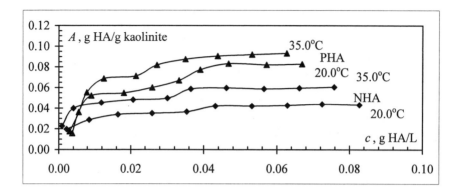

Figure 2 *Isotherms for adsorption of NHA and PHA on kaolinite*

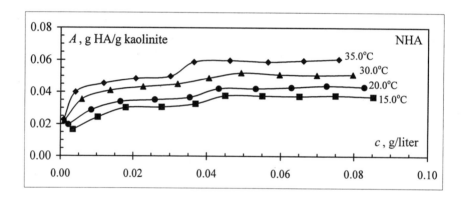

Figure 3 *Isotherms for adsorption of NHA on kaolinite*

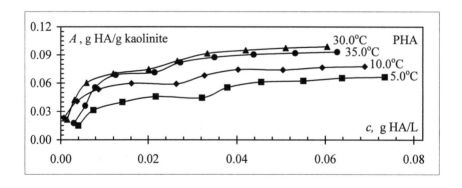

Figure 4 *Isotherms for adsorption of PHA on kaolinite*

3.1.4 Site Capacities. The surface area of the Kao fraction used was 52 ± 3 m^2/g (three determinations). This fixed Kao surface area was presented for HA adsorption at temperatures in the range 5.0–35.0 °C. We assume that the first detected step A in each system is formation of a monolayer of NHA or PHA on the Kao surface. The site capacity data v_A thus measure the stoichiometry of monolayer formation by a given HA.

Except for two systems (PHA at 5.0 and 10.0 °C), site capacities v_A for NHA and PHA adsorption were independent of temperature within experimental precision and averaged well. This is consistent with a fixed Kao surface for HA monolayer formation. However, the average $<v_A>$ for NHA is much less than $<v_A>$ for PHA adsorption (Table 3). The larger capacity of Kao for PHA indicates that PHA molecules cover less of the Kao surface. The simplest explanation for this result is that adsorbed PHA molecules are longer than they are wide. The limiting possible surface HA conformations are disk or spherical and pencil or sausage-shaped. The latter conformation is Wershaw's membrane-micelle model for HS interactions with minerals through HS carboxylate groups.[30] Wershaw's model has -COOH groups at the sausage end that adsorbs on the clay or mineral surface, with relatively fewer functional groups on the hydrophobic sides.[30,31]

Detailed studies of NHA and PHA reactions with other clays and minerals (*e.g.* bauxite, Al$_2$O$_3$•xH$_2$O)[31] will reveal whether NHA molecules always occupy more of the adsorbent surface than PHA. In the meantime we note that $<v_A>$ increases in the order NHA < PHA and with increasing R-COOH and phenolic-OH content (Table 1).

Site capacity v_B for a given HA may increase slightly with increasing temperature. However, within measurement precision the v_B data for NHA or PHA average well over the experimental temperature range to give average $<v_B>$ data in Table 3. The site capacity trends ascribed to HA monolayer formation from $<v_A>$ comparisons also are observed in adsorption step B, with $<v_B>$ increasing with NHA < PHA (Table 3).

Different adsorbed HA molecular conformations are consistent with the trends in $<v_A>$, $<v_B>$ and the total site capacities $\Sigma<v_i> = <v_A> + <v_B>$. The latter increase with NHA < PHA (Table 3). Based on present information, PHA seems to form structured bilayers on Kao.

3.1.5 Equilibrium Constants. Equilibrium constants K_A for monolayer adsorption on Kao vary from 175 to 590 L/g HA at 25.0 °C and increase in the order PHA < NHA (Table 3). The K_B at 25.0 °C for adsorption of HA on monolayer HA ranges from 58 to 140 L/g HA but in the order NHA < PHA (Table 3). The different orders suggest different conformational and/or functional group interactions between different HAs and the Kao surface and between HA molecules themselves. HAs from different sources seem to have different shapes and different functional group distributions but comparisons of adsorption equilibrium constants at fixed temperature give limited mechanistic information.

3.1.6 Thermodynamic Parameters. Stepwise adsorption of NHA and PHA on Kao and on adsorbed HA is mostly endothermic and results in an entropy increase (Table 3). An entropy *decrease* is expected for adsorption of a dissolved solute on a solid surface. Sorption of solutes by HAs[17,21] and of dissolved HAs by Kao has humic acids as a common component. With indications that hydration and dehydration of interacting solid HA sites and hydrophilic solutes are factors in adsorption,[21] ongoing work seeks to determine whether linear correlations are observed for adsorption of HAs from different sources on different clay and mineral surfaces.

3.1.7 Kaolinite Solubility and HA Desorption. Kaolinite solubility and the desorption of NHA and PHA from Kao in 0.10 M NaOH were checked at 35.0 °C with continuous sample shaking for seven days. No cloudiness or precipitation was observed on reducing

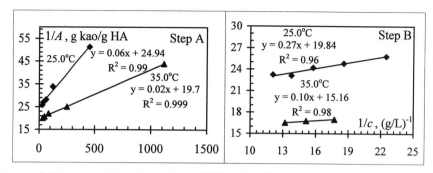

Figure 5 *Plots of eq 2 for adsorption of NHA on kaolinite*

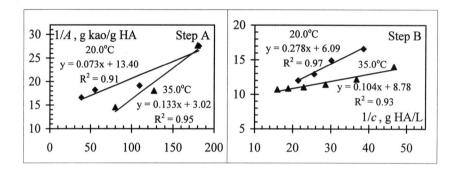

Figure 6 *Plots of eq 2 for adsorption of PHA on kaolinite*

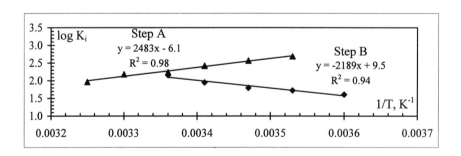

Figure 7 *Plots of eq 3 for adsorption of PHA on kaolinite*

Table 3 *Averaged data for adsorption of NHA and PHA on kaolinite*

HA	Step A				Step B				
	$<v_A>^a$	K_A^b	ΔH_A^c	ΔS_A^d	$<v_B>^a$	K_B^b	ΔH_B^c	ΔS_B^d	$\Sigma<v_i>^a$
NHA	0.045	585	9.5	45	0.067	58	11.1	46	0.11
PHA	0.18	175	-11.4	-28	0.13	140	10.0	43	0.31

[a] Units are g HA/g kaolinite; [b] units are L/g HA at 25.0 °C; [c] units are kcalmol^{-1};
[d] units are caldeg^{-1}mol^{-1}.

the supernatant pHs to 3.5 with concentrated HCl and room temperature standing overnight. Kao–HA composites dissolve very slowly in 0.1 M NaOH and the HAs desorb very slowly from Kao in 0.1 M NaOH at 35.0 °C. Adsorption of NHA and PHA on Kao thus has a dramatic effect on HA solubilities and the adsorption data indicate strong interactions in monolayer and bilayer formation.

ACKNOWLEDGEMENTS

We thank undergraduate research students Nadeem Ghali, Nichole Smith and Marcy Vozzella, who are supported by grants from and administered by Northeastern University, for many contributions. Reverend Jurgen Liias led us to the source of NHA. Professor R. J. Willey measured the surface area of the kaolinite sample and Noralie Barnett provided technical assistance. We thank Amjad Fataftah, Malcolm Hill, Susan Jansen and Daman Walia for useful discussions. This is contribution number 731 from the Barnett Institute at Northeastern University.

References

1. M. de Bertoldi, P. Sequi, B. Lemmes and T. Papi (eds.), 'The Science of Composting', Parts 1 and 2, Blackie Academic and Professional, Glasgow, 1996.
2. B. Chefetz, P. G. Hatcher, Y. Hadar and Y. Chen, *Soil Soc. Sci. Amer. J.*, 1998, **62**, 326.
3. D. L. Sparks, 'Environmental Soil Chemistry', Academic Press, San Diego, 1995.
4. Z. Filip and J. J. Alberts, *Sci. Total Environ.*, 1989, **83**, 273.
5. E. A. Ghabbour, A. H. Khairy, D. P. Cheney, V. Gross, G. Davies, T. R. Gilbert and X. Zhang, *J. Appl. Phycol.* 1994, **6**, 459.
6. A. Radwan, G. Davies, A. Fataftah, E. A. Ghabbour, S. A. Jansen and R. J. Willey, *J. Appl. Phycol.*, 1997, **8**, 553.
7. P. MacCarthy, C. E. Clapp, R. L. Malcolm and R. R. Bloom (eds.), 'Humic Substances in Soil and Crop Sciences: Selected Readings', American Society of Agronomy, Madison, Wisconsin, 1990.
8. M. H. B. Hayes, P. MacCarthy, R. L. Malcolm and R. S. Swift (eds.), 'Humic Substances II: In Search of Structure', Wiley-Interscience, New York, 1989.
9. F. J. Stevenson, 'Humus Chemistry: Genesis, Composition, Reactions', 2nd Edn. Wiley, New York, 1994.
10. H.-R. Schulten and M. Schnitzer, *Naturwiss.*, 1995, **82**, 487; *Soil Sci.*, 1997, **162**, 115 and references therein.
11. F. J. Stevenson and Y. Chen, *Soil Sci. Soc. Am. J.* 1991, **55**, 1586.
12. N. Senesi, *Trans.15th World Congress Soil Sci.*, 1994, **3a**, 384.
13. G. Davies, A. Fataftah, A. Cherkasskiy, E. A. Ghabbour, A. Radwan, S. A. Jansen, S. Kollar, M. D. Paciolla, L. T. Sein, Jr., W. Buermann, M. Balasubramanian, J. Budnick and B. Xing, *J. Chem. Soc. Dalton Trans.*, 1997, 4047.
14. A. H. Khairy, G. Davies, H. Z. Ibrahim and E. A. Ghabbour, *J. Phys. Chem.*, 1996, **100**, 2410 and references therein.
15. A. H. Khairy, G. Davies, H. Z. Ibrahim and E. A. Ghabbour, *J. Phys. Chem.*, 1996, **100**, 2417.

16.   E. A. Ghabbour, G. Davies, A. Fataftah, N. K. Ghali, M. E. Goodwillie, S. A,
       Jansen and N. A. Smith, *J. Phys. Chem.*, 1997, **101B**, 8468.
17.   B. Xing, J. J. Pignatello and B. Gigliotti, *Environ. Sci. Technol.*, 1996, **30**, 2432
       and references therein.
18.   A. Piccolo, S. Nardi and G. Concheri, *Chemosphere*, 1996, **33**, 595; *Eur. J. Soil
       Sci.*, 1996, **47**, 319.
19.   B. E. Watt, T. M. Hayes, M. H. B. Hayes and R. T. Price, in 'Humic Substances
       and Organic Matter in Soil and Water Environments', C. E. Clapp, M. H. B. Hayes,
       N. Senesi and S. M. Griffith (eds.), International Humic Substances Society, St.
       Paul, Minnesota, 1996, pp. 81-91.
20.   S. M. Shevchenko and G. W. Bailey, *J. Mol. Struct. (Theochem)*, 1996, **364**, 197.
21.   G. Davies, E. A. Ghabbour, A. H. Khairy and H. Z. Ibrahim, *J. Phys. Chem.*, 1997,
       101B, 3228.
22.   A. Radwan, R. J. Willey and G. Davies, *J. Appl. Phycol.,* 1997, **9**, 481.
23.   R. J. Willey, A. Radwan, M. E. Vozzella, A. Fataftah, G. Davies and E. A.
       Ghabbour, *J. Non-Cryst. Solids*, 1998, **3806**, in press.
24.   J. N. C. Whyte, in 'Experimental Phycology: A Laboratory Manual', C. S. Lobban,
       D. J. Chapman and B. P. Kremer, (eds.), Cambridge University Press, New York,
       1988, pp. 168-173.
25.   A. W. Adamson, 'Physical Chemistry of Surfaces', 4th edn, Wiley, New York,
       1982, p. 371.
26.   B. K. G. Theng, 'Formation and Properties of Clay–Polymer Complexes', Elsevier,
       Amsterdam, 1979, p. 8.
27.   M. Schnitzer and S. U. Khan, 'Humic Substances in the Environment', Dekker,
       New York, 1972, Ch. 7.
28.   M. Schnitzer, *Proc. Intern. Mtg. on Humic Subst. Wageningen*, 1972, pp. 293-310.
29.   J. Mao, W. Hu, K. Schmidt-Rohr, G. Davies, E. A. Ghabbour and B. Xing, this
       volume, p.79.
30.   R. L. Wershaw, *J. Contamin. Hydrol.*, 1986, **1**, 29; R. L. Wershaw, *Environ. Sci.
       Technol.*, 1993, **27**, 814; R. L. Wershaw, 'Membrane-micelle Model for Humus in
       Soils and Sediments and its Relation to Humification', *US Geological Survey
       Water-Supply Paper 2410,* 1994
31.   R. L. Wershaw, E. C. Llaguno and J. A. Leenheer, *Colloids Surf.*, 1996, **108**, 213
       and references therein.

EFFECT OF DISSOLVED ORGANIC MATTER ON THE MOVEMENT OF PESTICIDES IN SOIL COLUMNS

K. M. Spark[1] and R.S. Swift[2]

Department of Soil Science, University of Reading, Whiteknights, Reading Berkshire, RG6 2AA, UK
[1] Present address: Australian Water Quality Centre, Private Mail Bag 3, Salisbury, S.A. 5108, Australia
[2] Present address: CSIRO Land and Water, Private Mail Bag No. 2, Glen Osmond, S.A. 5064, Australia

1 INTRODUCTION

A considerable amount of work has been done on the sorption characteristics of pesticides by whole soils and the major soil components such as clays and soil organic matter.[1-3] For many pesticides there is a strong correlation between the extent of adsorption and the amount of organic matter present in the soil at other than low levels of organic matter[1,4,5] when most of the mineral surfaces are coated with organic matter.

Most of the studies so far have concentrated on the relationship between the total organic carbon (TOC) fraction and the pesticide sorption properties.[6-8] The TOC in the soil includes an insoluble fraction (particulate organic carbon, POC) and a soluble fraction (dissolved organic matter, DOM). It has been reported that the DOM fraction may interact with pesticides and enhance their solubility.[9-13] This interaction may enhance the pesticide mobilisation and transport with subsequent contamination of ground waters.

A pesticide may exist in the soluble form in the soil solution and in an insoluble form associated with the colloidal fraction of the soil solution, or sorbed to other less mobile soil components. The relative amount of the pesticide in these two states depends on its chemical and physical properties. However, similar to the soil organic matter, the pesticide is unlikely to be in a static condition but rather is involved in a variety of processes with other states of the compound or with other soil components.

The extents to which the soluble organic matter influences the movement of pesticides may be studied by determining the effect of increasing the concentration of dissolved organic matter on the proportion of the pesticide transported through the soil compared with that remaining in the soil matrix.

Observations of the ways in which soluble soil organic matter influences the migration of organic chemicals were made with lysimeters. As many pesticides are known to be strongly sorbed by the solid state organic matter in the soil, it was believed that choosing a soil low in total organic matter and increasing the amount of soluble organic carbon in the leachate would minimise the ratio of $C_{soil} : C_{leachate}$ and hence maximise the interaction between the pesticide and the soluble carbon in the soil solution. From the results of these experiments it was possible to determine the importance of dissolved organic carbon relative to total organic carbon on the fate and hence transport of pesticides in the soil. The data presented here are the preliminary results of research investigating the importance of soluble organic matter on the fate of pesticides in a soil environment.

2 MATERIALS

2.1 Lysimeters

The soil used for the intact cores was a free-draining, brown, coarse, sandy-loam (Sonning series) soil from the University of Reading farm, Sonning, UK. X-Ray diffrac-tion analysis of the clays in the soil indicated moderate amounts of both mica and kaolinite and a small amount of smectite. The particle size distribution determined using the literature method[14] was 59 % in the size range 63–600 μm (associated with coarse sand), 26% < 2–63 μm and 13 % < 2 μm (the size range associated with fine clays). The total organic carbon of the soil was 1.25%. The supernatant of a 1:2 (w/v) soil to solution suspension shaken for 16 hours had a pH of 7.6, conductivity 0.19 mS/cm, dissolved organic carbon (DOC) 23 ppm, Ca^{2+} 2.6 μmol/g and absorbance (400 nm) of 0.05.

The field from which the intact soil cores were taken had been under regular cultivation and maize cropping for several years. It had been laying fallow for the previous three months and had been ploughed to approximately 20 cm depth 2–4 months prior to sampling. The final dimensions of the soil core in the lysimeter were 17.5 cm wide by approximately 11 cm deep. The total pore volume of the lysimeter was estimated to be 650 mL.

2.2 Leachate Solutions

The background electrolyte used for the leaching solution was 0.0005 M Ca^{2+} at pH 7. The DOM leachate solution contained 50 ppm Norfolk soil fulvic acid in 0.0005 M Ca^{2+} at pH 7. This fulvic acid (FA) was isolated from a field drain in the Methwold Fens, Norfolk.[15] The FA solutions were made up daily at pH = 7 and stored in air tight volumetric flasks. Fulvic acid was chosen as the source of DOM as it tends to be more soluble than humic acid. Tritium was used to determine the break-through curves of the lysimeters.

2.3 Pesticides

The pesticides used in the lysimeter studies were: atrazine (6-chloro-N^2-ethyl-N^4-isopropyl-1,3,5-triazine-2,4-diamine), technical grade (99.0%) kindly supplied by Ciba-Geigy; 2,4-D [(2,4-dichlorophenoxy)acetic acid] technical grade (99.0%) grade supplied by Sigma; isoproturon [3-(4-isopropylphenyl)-1,1-dimethylurea], technical grade (99.0%) kindly supplied by Rhône-Poulenc Agriculture Ltd.; paraquat (1,1'-dimethyl-4,4'-bipyridinium dichloride), technical grade (98%) supplied by Aldrich and DDT {1,1'-(2,2,2-trichloroethylidine)bis[4-chlorobenzene]}, technical grade (98%) supplied by Sigma.

The radio-labelled versions of the above pesticides were atrazine (atrazine-ring-UL-^{14}C), kindly supplied by Ciba-Geigy, 2,4-D [(2,4-dichlorophenoxy)acetic acid-carboxyl-^{14}C, > 98% purity] supplied by Sigma, isoproturon (isoproturon-ring-UL-^{14}C), kindly supplied by Ciba-Geigy, paraquat (paraquat-methyl-^{14}C dichloride, > 98% purity), supplied by Sigma and DDT (4 4'-DDT-ring-UL-^{14}C) technical grade (98%) supplied by Sigma.

3 METHODS

3.1 Leaching of Lysimeters

The soil cores were conditioned by leaching with approximately 5 pore volumes of solution prior to the addition of the pesticide. Leaching solution (200 mL) was applied daily to the centre of the surface sand over a period of about 10 minutes. The sand on the surface of the lysimeter was removed, the suction was turned off, the tritium solution was applied as a tracer compound and then the pesticide was applied at the rate of 1kg/ha (total of 1.77 mg or 2 μCi per lysimeter) to the soil at greater than 1 cm from the edge of the core. Then the sand was replaced and the suction turned on. The cores were subsequently leached with approximately 10 pore volumes of electrolyte solution, with or without FA.

Four methods of treatment (in triplicate) were used in these leaching studies.
1.   (EE)   Conditioned and leached using the background electrolyte.
2.   (FF)   Conditioned and leached using a mixture of background electrolyte and fulvic acid.

3.2 Analysis of Leachate

Immediately prior to the daily addition of the leaching solution, the drainage leachate from the lysimeter was removed. This leachate was weighed and centrifuged at 16,000 rpm for 10 minutes. The supernatant was analysed to determine the pH, absorbance at 400 nm and, after addition of the pesticide to the lysimeter, the presence of tritium and pesticide (^{14}C) in solution.

3.3 Analysis of Lysimeter Cores

At the end of the leaching process the lysimeter cores were dissected into approx-imately 1 cm layers, and dried at 40 °C for a week. The dry soil was sieved to 1.0 mm and part of the < 1.0 mm fraction was lightly ground with a mortar and pestle and then resieved using a 250 μm sieve.

3.3.1 Cores used for atrazine, isoproturon and 2,4-D leaching studies: 10 g of the < 250 μm fraction was suspended in 20 g methanol, sealed and shaken for 16 h on an end-over-end shaker and then centrifuged at 16,000 rpm for 10 min. 10 mL of the supernatant was placed in a scintillation vial in a fume cupboard overnight to allow the methanol to evaporate to dryness. Five mL of scintillation cocktail was added to the vial, which was then shaken to disperse the soil residue (precipitated salts, *etc.*) before analysis for ^{14}C and ^{1}H.

3.3.2 Cores used for paraquat and DDT leaching studies: 0.025g of < 250 μm soil fraction was suspended in 0.75 mL water and combined with 2.5 mL scintillation cocktail. The suspension was allowed to stand overnight to remove the bubbles and then analysed for ^{14}C and ^{1}H.

To standardise the method for each pesticide, soil suspended in known amounts of pesticide solution was shaken overnight and centrifuged at 16,000 rpm for 10 min. The solid was washed, and the dried soil was then treated using one of the above procedures, depending on the pesticide. The supernatant and washing solution were combined and analysed to determine what fraction of pesticide was still in the soil. This allowed a correlation between the values obtained from scintillation analysis and the concentration of pesticide in the soil.

4 RESULTS

4.1 Leachate Analysis

For the lysimeter experiments carried out using paraquat and DDT, there was no measurable ^{14}C detected in leachate after eluting the lysimeter core with 10 pore volumes. For the other three pesticides atrazine, 2,4-D and isoproturon, ^{14}C was detected in the leachate. The total amount of the pesticide collected in the leachate is given in Table 1. The order of elution of the pesticides as a function of pore volumes was 2,4-D < isoproturon < atrazine. The total amount of pesticide eluted is in the order 2,4-D > atrazine > isoproturon >>> DDT ~ paraquat.

Table 1 *Pesticide eluted (% of applied) versus treatment[a]*

	Pesticide	EE	FF
Pesticide eluted (mg)	Atrazine	64±10	62±12
	2,4-D	95±33	80±25
	Isoproturon	48±6	50±16
	DDT	nd	nd
	Paraquat	nd	nd
Pesticide in core (mg)	Atrazine	41±8	45±9
	2,4-D	8±2	6±1
	Isoproturon	25±5	21±4
	DDT	92±10	102±10
	Paraquat	82±10	86±10
Total pesticide	Atrazine	105±9	107±10
Accounted (mg)	2,4-D	103±18	86±13
(eluted + core analysis)	Isoproturon	71±6	71±10
	DDT	92±10	102±10
	Paraquat	82±10	86±10

[a] 1.77 mg/lysimeter of pesticide was applied to the soil.

These results indicate that the presence of FA in the leachate solution has no significant effect on the total amount of pesticide eluted. Fulvic acid leachate recovered from the lysimeter core contained up to 75% of initial colour, increasing with total amount of leachate recovered up to approximately 1 pore water volume. Leachate from cores conditioned with electrolyte/fulvic acid had up to 2–3 times more colour than those not leached with fulvic acid, the difference increasing with total amount of leachate recovered up to approximately 1 pore water volume.

4.2 Soil Core Analysis

Results of the soil core analysis showing the concentration of the pesticide remaining in the soil core as a function of the core depth are given in Figure 1. Using the EE treatment (core conditioned and leached using electrolyte solution) after eluting the lysimeter core with 10 pore volumes, for paraquat, DDT and 2,4-D only the top (1 cm) layer had measurable ^{14}C present. The total concentration of pesticide in this top layer was 85% (5.7–6.9 µg/gm), 99% (5.2–5.5 µg/gm) and 7.5% (0.12–0.14 µg/gm) respectively, of the amount initially applied. For the other two pesticides, atrazine and isoproturon, the pesticide was found at all depths in the lysimeter with the total amount present being 41% and 25% of the amount initially applied. In general, the concentration of atrazine increases with depth, whilst the concentration of isoproturon decreases with depth. A mass balance of the fate of the pesticides is shown in Table 1.

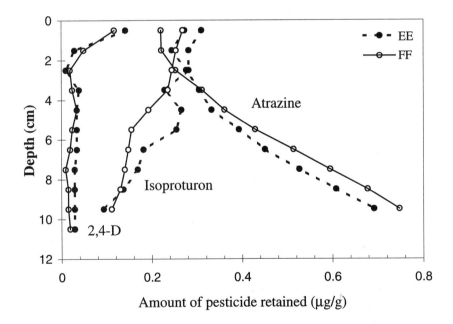

Figure 1 *Concentration of pesticide in the soil core as a function of depth after leaching with 10 pore volumes of solution for the pesticides atrazine, 2,4-D and isoproturon*

5 DISCUSSION

The order of elution from the soil column (2,4-D > atrazine > isoproturon >>> paraquat ~ DDT) correlates inversely with the amount of pesticide left in the lysimeter. This order is not related to the solubility of the pesticides because paraquat is very soluble in water but DDT is practically insoluble. This observation is expected for a soil core conditioned and leached with electrolyte solution only and for a soil low in organic matter. However, as can be seen from the results in Figure 1, the presence of high concentration of soluble fulvic acid had no significant effect on the transport characteristics of any of these pesticides. Even though it has been found that humic substances increase the solubility of some pesticides, and that there appears to be an interaction between the pesticide and some humic substances, this interaction may not necessarily be significant in the soil environment. From the results of this work it appears that the interaction with the solid state components of the soil environment for this soil may be more important than the interaction with the soluble organic matter in the soil solution.

The mass balance of the pesticides is able to account for the fate of 2,4-D, atrazine and DDT, but the fate of part of the paraquat and isoproturon is unknown. The unaccounted for isoproturon may have been lost via the suspended colloidal matter leached from the lysimeter and which was not collected and analysed.[16] But Laird *et al.*[17] have found that isoproturon was more associated with the larger particulate fraction and had a greater affinity than it did for the clay fraction and a greater difficulty to desorb. The loss is not associated with the type of core treatment so it does not appear to be related to the organic matter content, either soluble or insoluble. However, loss by degradation cannot be disregarded as the time elapsed over which the cores were leached was 3.5 weeks.

6 CONCLUSION

The preliminary results from this research work suggest that even though dissolved organic matter interacts with pesticides and in some cases increases the solubility of these compounds, the DOM may not have a significant effect on the fate of the pesticides in a soil environment.

References

1. F. J. Stevenson, *J. Environ. Qual.*, 1972, **1**, 333.
2. S. U. Khan, in 'Soil Organic Matter', M. Schnitzer and S. U. Khan (eds.), Elsevier, Amsterdam, 1978, Ch. 4.
3. I. J. Graham-Bryce, in 'The Chemistry of Soil Processes', D. J. Greenland and M. H. B. Hayes (eds.), Wiley, London, 1981, Ch. 12.
4. O. Francioso, E. Bak, N. Rossi and P. Sequi, *Sci. Total Environ.*, 1992, **123/4**, 503.
5. S. Raman and P. L. Narayana, *Agric. Sci. Digest*, 1989, **9**, 1.
6. E. M. Murphy, J. M. Zachara, S. C. Smith and J. L. Phillips, *Sci. Total Environ.*, 1992, **117/118**, 413.
7. C. A. Seybold, K. McSweeney and B. Lowrey, *J. Environ. Quality*, 1994, **23**, 1291.
8. Z. Gerstl, *J. Contam. Hydrol.*, 1990, **6**, 357.
9. R. Saint-Fort and S. A. Visser, *J. Environ. Sci. Health*, 1988, **A23**, 624.
10. C. T. Chiou, R. L. Malcolm, T. I. Brinton and D. E. File, *Environ. Sci.Technol.*, 1986, **20**, 502.
11. D. Y. Lee and W. J. Farmer, *J. Environ. Qual.*, 1989, **18**, 468.

12. C. Maquedo, E. Morillo, F. Martin and T. Undabeytia, *J. Environ. Sci. Health. B.,* 1993, **28**, 655.
13. K. M. Spark and R. S. Swift, *Sci. Total Environ.,* 1994, **152**, 9.
14. B. W. Avery and C. L. Bascomb (eds.), 'Soil Survey Laboratory Methods', Harpenden Press, London, 1982.
15. B. E. Watt, R. L. Malcolm, M. H. B. Hayes, N. W. E. Clark and J. K. Chipman, *Water Res.,* 1996, **30**, 1502.
16. F. Worrall, A. Parker, J. E. Rae, A. C. Johnson, A. Walker, R. Allen, S. W. Bailey, A. M. Blair, C. D. Brown, P. Gunther, C. R. Leake and P. H. Nicholls (eds.), 'Pesticide Movement to Water', Proceedings of a symposium held at the University of Warwick, Coventry, UK, 3–5 April 1995, 129; BCPC Monograph No. 62, 1995.
17. D. A. Laird, P. Y. Yen, W. C. Koskinen, T. R. Steinheimer and R. H. Dowdy, *Environ. Sci. Technol.,* 1994, **28**, 1054.

GENERATION OF FREE RADICALS BY HUMIC ACID: IMPLICATIONS FOR BIOLOGICAL ACTIVITY

Mark D. Paciolla,[1] Santha Kolla,[1] Lawrence T. Sein, Jr.,[1] James M. Varnum,[2] Damien L. Malfara,[1] Geoffrey Davies,[3] Elham A. Ghabbour[3] and Susan A. Jansen[1*]

[1] Chemistry Department, Temple University, Philadelphia, PA 19122
[2] Kimmel Cancer Center, Thomas Jefferson University, Philadelphia, PA 19107
[3] Chemistry Department and the Barnett Institute, Northeastern University, Boston MA, 02115

1 INTRODUCTION

Humic acids (HAs) are remarkably diverse biopolymers with multiple chemical function-alities. Two principal models exist for the biosynthesis of HA: (1) biodegradation of lignins,[1] and (2) *in vivo* plant biosynthesis.[2] The first model is supported by the high degree of correlation between the HA building block and lignan structures; the second model seems more credible now that HAs have been isolated from numerous genera of live plants.[3] Regardless of "synthetic" mechanism, HAs are versatile materials. The diversity of HAs is illustrated by the many studies of their metal binding capacities, metal exchange rates and affinities.[4–6] These processes are crucial to biomineralization, environmental remediation and agricultural applications of HA.[7] Additionally, HAs can be ozonized and chlorinated to produce potentially carcinogenic compounds.[8–11]

HAs are photo-active. Irradiation studies show that HAs produce a large number of characteristic radical species, most of which are quinone- or catechol-based, after short exposure to UV. They are bleached by irradiation from 300 to 350 nm.[12] Previous studies have been limited by a lack of understanding of HA structure and difficulty in obtaining homogeneous HA materials. Careful isolation of HAs from soil has provided materials with consistent and highly correlated chemical properties; therefore, these materials serve as good substrates for our analyses. The role of free radical species is critical in under-standing many biological and environmental functions of HAs. The radical chemistries of HA are evaluated by a series of experiments in this work.

2 EXPERIMENTAL

2.1 Materials

HA samples from soil sources (German, Irish and New Hampshire) and from kelp were isolated, extracted, and purified by sequential Soxhlet extraction with ether, acetone, ethanol, and dioxane followed by extraction with water and aqueous base-mineral acid dialysis.[3] The solubility of HA samples was extremely limited in $CDCl_3$, C_6D_6, deuterated DMSO and pyridine; however, a mixture of deuterium oxide and NaOD of pD ~ 9.0 was found to be suitable for NMR studies. All NMR spectra were obtained with 20 mg of HA in 1 mL of D_2O–NaOD mixtures. Deuterium oxide (99.96 atom %D) and deuterated sodium hydroxide (99.96 atom %D) were from Aldrich. 5,5-Dimethyl-1-pyrroline-N-oxide (DMPO) from Aldrich was purified further by treatment with activated carbon. Hydrogen peroxide (30% w/w), dimethyl sulfoxide (DMSO), and sodium phosphate were also from Aldrich. The soil samples are denoted as Irish, New Hampshire and German, indicating the source.[13]

Metal loading was performed for Fe and Cu with common soluble salts of each. HA samples were suspended in a solution of the metal salt at pH ~ 4.5. The solutions were stirred overnight at room temperature. The HA was then removed by filtration and washed several times with distilled water to remove unbound metal.[14]

2.2 Photo-irradiation Studies

Solid HA samples were irradiated with UV using an Oriel Model 66087 instrument with a Hg/Xe lamp adjusted for 200 W output. Irradiation times were varied incrementally from fifteen minutes to three hours.

2.3 Electron Paramagnetic Resonance (EPR) Measurements

2.3.1 Photo-irradiation Samples. All EPR spectra of solid, photo-irradiated HA materials were recorded with a Bruker-EMX spectrometer operating at 9.8 GHz at room temperature.

2.3.2 Spin Trapping Analyses. The solutions prepared for the spin trapping experiments contained 10 mg of humic acid, 100 mM of DMPO and 30% DMSO (when added) in 25 mM phosphate buffer at pH 7.4. The phosphate buffer was continuously treated with Chelex® resin for several hours prior to the experiments to remove trace metals commonly found in phosphate salts. All solutions except the phosphate buffer were prepared immediately prior to the start of an experiment.

A Hewlett-Packard 8453 UV–visible spectrometer with a diode array detector was used to determine the concentrations of the stock solutions of DMPO and hydrogen peroxide. The concentration of DMPO was determined from its absorbance at λ_{max} 234 nm ($\varepsilon = 7700$ M^{-1} cm^{-1}). The hydrogen peroxide solution was prepared by diluting the 30% (w/w) stock solution, the concentration of which was measured using $\varepsilon = 43.6$ M^{-1} cm^{-1} at 240 nm. Hewlett-Packard spectrometer cells were UV-grade quartz with a 10 mm path length and a volume of three milliliters. For spin trapping experiments, the EPR spectra were recorded at 295 K with the Bruker EMX spectrometer operating at 9.7 GHz using a flat cell with a total volume of 150 μL in a TM_{110} cavity. A modulation frequency of 100 kHz was used with a scan range of 100 G and scan time of ~2.5 minutes. Microwave power of 20 mW and a modulation amplitude of 1.0 G were typically employed for the experiments. All experiments were run in triplicate.

3 RESULTS AND DISCUSSION

3.1 HA Composition

HAs found in soils are of plant origin and extraction techniques are specific for species stable to both acid and base hydrolysis. Thus, a high correlation of chemical composition between plant and soil derived HAs is not surprising. Other "homologies" exist and have been described elsewhere.[3,15] Table 1 shows the chemical formulas of a series of soil HA samples. The first twelve were selected at random from the literature and the latter group was collected at various locations, isolated and analysed for the present work. Taking the simple compositions, computing the chemical formula and scaling the formula from that composition to a consistent structural unit leads to a series of remarkably similar "empirical formulas." Accounting for additional oxygen and hydrogen as "surface water", "waters of hydration" or "losses on condensation" provides a remarkable homologous set as shown in Table 1.

Table 1 *Composition and empirical formula for soil derived HAs*

Source	BB formula	Formula	BB M_W	C/H	Hydrate formula
A_1M_1	$C_{72}H_{80}O_{40}N_5$	$C_{72}H_{80}O_{41}N_4$	1654	0.90	$C_{72}H_{58}O_{30}N_4.11H_2O$
A_1M_1	$C_{72}H_{80}O_{40}N_7$	$C_{72}H_{80}O_{43}N_4$	1961	0.90	$C_{72}H_{59}O_{30}N_4.13H_2O$
A_1M_1	$C_{72}H_{88}O_{40}N_7$	$C_{72}H_{88}O_{43}N_4$	1694	0.82	$C_{72}H_{62}O_{30}N_4.13H_2O$
A_1M_1	$C_{72}H_{88}O_{43}N_5$	$C_{72}H_{88}O_{44}N_4$	1710	0.82	$C_{72}H_{60}O_{30}N_4.14H_2O$
A_1	$C_{72}H_{61}O_{30}N_4$	----	1461	1.18	$C_{72}H_{61}O_{30}N_4.0H_2O$
A_1	$C_{72}H_{57}O_{31}N_4$	----	1473	1.26	$C_{72}H_{55}O_{30}N_4. H_2O$
B	$C_{72}H_{65}O_{37}N_4$	----	1581	1.08	$C_{72}H_{51}O_{30}N_4.7H_2O$
B/B	$C_{72}H_{74}O_{41}N_4$	----	1654	0.97	$C_{72}H_{52}O_{30}N_4.11H_2O$
B	$C_{72}H_{79}O_{39}N_4$	----	1627	0.91	$C_{72}H_{61}O_{30}N_4.9H_2O$
SCP	$C_{72}H_{120}O_{68}N_3$	$C_{72}H_{120}O_{67}N_4$	2102	0.60	$C_{72}H_{46}O_{30}N_4.37H_2O$
M_1	$C_{72}H_{119}O_{55}N_6$	$C_{72}H_{119}O_{57}N_4$	1941	0.60	$C_{72}H_{65}O_{30}N_4.27H_2O$
M_2	$C_{72}H_{119}O_{68}N_4$	----	2117	0.60	$C_{72}H_{43}O_{30}N_4.38H_2O$
B	$C_{72}H_{113}O_{39}N_4$	----	1647	0.64	$C_{72}H_{95}O_{30}N_4.9H_2O$
SNF	$C_{72}H_{88}O_{40}N_2$	$C_{72}H_{88}O_{38}N_4$	1624	0.82	$C_{72}H_{72}O_{30}N_4.9H_2O$
SGF	$C_{72}H_{86}O_{44}N_2$	$C_{72}H_{86}O_{42}N_4$	1568	0.84	$C_{72}H_{62}O_{30}N_4.12H_2O$
F	$C_{72}H_{100}O_{48}N_4$	----	1778	0.72	$C_{72}H_{64}O_{30}N_4.18H_2O$
SCFT	$C_{72}H_{124}O_{67}N_4$	----	2116	0.58	$C_{72}H_{50}O_{30}N_4.37H_2O$
SCFM	$C_{72}H_{121}O_{70}N_4$	----	2161	0.59	$C_{72}H_{49}O_{30}N_4.36H_2O$
SCFB	$C_{72}H_{121}O_{57}N_4$	----	1953	0.59	$C_{72}H_{67}O_{30}N_4.27H_2O$
HA-freeze	$C_{72}H_{117}O_{57}N_8$	$C_{72}H_{117}O_{61}N_4$	2005	0.61	$C_{72}H_{55}O_{30}N_4.31H_2O$

The data in Table 1 have not been corrected for amino acid or sugar content and the collection and isolation procedure varies with research group. Therefore, there is no reason to expect that this homology is simply an artifact of any particular isolation/purification process. This is provided as evidence of the homogeneity of HA materials throughout various soil types and geographic locations, supporting our contention that HAs have characteristic intrinsic chemistries.

Further corrections for amino acid content and sugar components lead to a general reduced empirical formula of $C_{36}H_{39}O_{15}N_2$. Figure 1 shows a suggested average building block consistent with all available chemical analyses. Spectral evidence suggests similar structure types for soil derived HAs. ^{1}H NMR and IR data are shown in Figure 2 for Irish, German and New Hampshire soil derived HAs. Even the fine details in these spectra are reproduced between samples. Other spectral data and analytical data demonstrate a high level of structural correlation between plant and soil derived HAs.

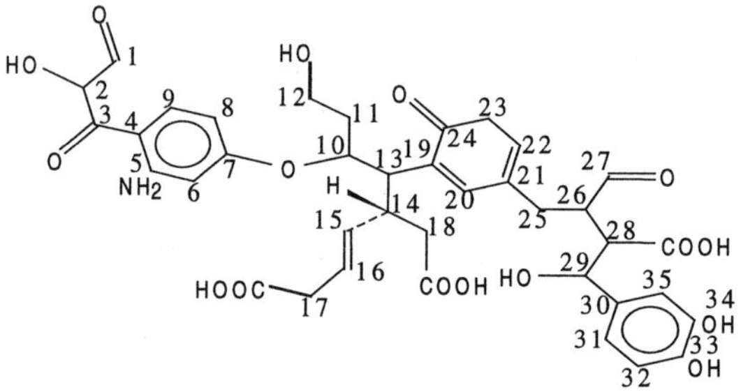

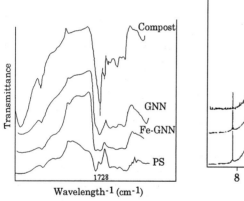

Figure 1 *TNB humic acid structure*

Figure 2a *IR spectra of different HA*   **Figure 2b** *^{1}HNMR spectra of different HAs*
samples and alginic acid (PS)

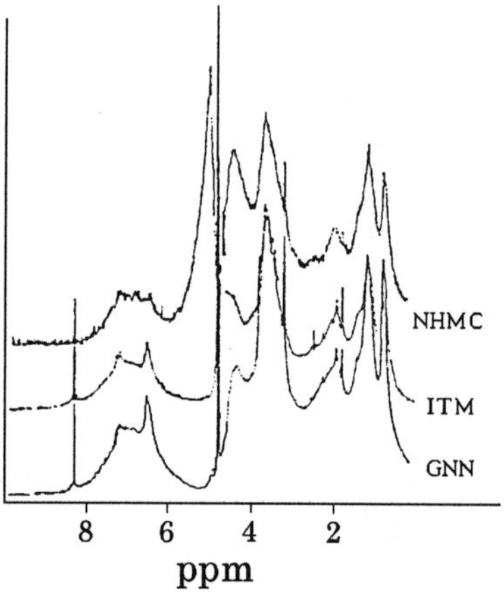

From the empirical formula derived from analytical data, analysis of existing spectral data, chemical degradation pathways and biosynthetic trajectories, the structure in Figure 1 proposed as a model for the humic acid apomonomer is similar to Steelink's 1985 structure and not too distant from lignin models. It is more consistent with available experimental data and mechanistic theories.[2] From this model, multiple metal binding sites can be identified, one involving carboxylates and another involving catecholic functions. Both quinoid and catecholic groups form stable radicals that can be thermally or photo-chemically induced within the HA polymeric structure. Stereochemistry and van der Waals interactions define the secondary structure. Amines and carboxylates are poised for condensation as a means of HA polymerization. HA polymers based on such an apomonomer give rise to a rectangular helix in which two metal binding functions are exposed on two sides of the helix. The more hydrophobic groups remain on the interior as shown in Figure 3. This suggests that both the photo-active quinoids and metal binding functions are on the HA surface and therefore both can be active in radical chemistries.

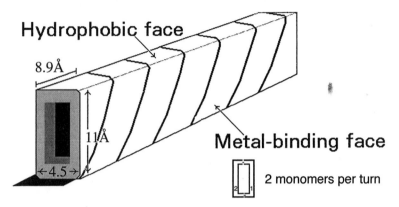

Figure 3 *Representation of secondary HA structure*

3.2 Solid State Photo-irradiation

All photochemical studies were performed in the solid state and therefore involve processes occurring at the humic surface, where catecholic functions and metallic species are well represented. The presence of quinoid and catecholic species is well documented in HAs, even by investigators supporting widely-variant structural models.[16] The photochemistry of quinones and catechols has been well studied.[17] Generally, photo-irradiation is believed to cause the photoreduction of quinones to semiquinones and eventual production of the dianionic species. Analogous intermediates are identified for catecholic moieties. All species can exist in equilibrium; specific preferences are defined by the local environment and irradiation conditions.

Related photochemical studies have been reported for humic materials in both acidic and basic media. It appears from these studies that the photochemistry of HA is not dissimilar to the known photochemistry of free quinones and catechols. The electron source in the solution studies is not clear, but may be the solvent matrix. Once the dianion is created, further electron transfer can occur from the dianion to an acceptor, with concomitant production of the semiquinone or quinone, and is local site dependent. Therein lies the interest in HA photochemistry. In the solid state, intramolecular processes intrinsic to HA can be studied directly.

The binding of metallic species, in particular iron and copper, in HA has been analysed by many groups;[4–6,14] catechols and quinones are strongly implicated. Our modelling experiments suggest that the catecholic sites are strong iron binding sites that are capable of binding both ferric and ferrous species.

Both *ex situ* and *in situ* EPR experiments showed an initial increase in semiquinone or catecholic radical species with UV irradiation or moderate thermal treatment. Continued UV irradiation leads, paradoxically, to a complete loss of the radical species which can be restored with time. EPR spectra demonstrating the photo-induction of semiquinone species are shown in Figure 4. However, determination of the rate of induction and equilibrium between radical species is precluded since the number of surface quinoid/catecholic groups per HA sample is unknown and likely to vary somewhat between samples. The induction, saturation, and formation of the dianionic species are illustrated in Figure 4. It is clear from these experiments that there are well defined photochemical processes occurring at the humic acid surface.

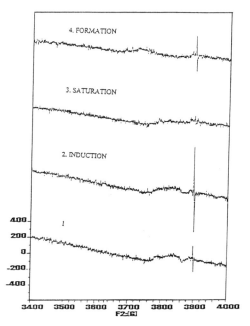

Figure 4 *Photo-induction EPR spectra of HA: 1. EPR spectrum of HA before subjection to UV; 2. EPR spectrum of HA subjected to UV for 30 min.; 3. EPR spectrum of HA subjected to UV for 2 hours; 4. EPR spectrum of (3) after 18 hours*

Other mechanisms of photoreduction are known for quinoid species. In solution, reduction of quinones and catechols can be achieved under either acidic or basic conditions. As solids, HAs exist as hydrates with part of their weight due to lattice water. Surface water can contribute up to 70% of the total HA weight. The acidic nature of the HA surface is well documented. However, both acidic and basic functional groups are likely to be at the HA polymer surface and metals are likely bound at or near these optical chromophores. Therefore, one can imagine numerous mechanisms for photoreduction of HAs, several of which may be active at any one time. However, all HAs studied, regardless of origin, demonstrate the same qualitative photochemical effects. Photo-induced electron transfer generates a greater population of quinoid and catecholate centers. In the presence of oxygen, catechol can reduce Fe(III) to Fe(II) since the reduction potentials for catechol and Fe(III) are of similar magnitude. Electron transfer from catechol to Fe(III), yielding Fe(II) in HA, is suspected and well known to occur in other biological systems,[18] especially since both ferric and cupric species show appreciable binding affinity for catechols and quinones. Equilibrium processes between more reduced dianionic species are likely mediated by the metal center and may lead to the restoration of the semi-dione radical species, as shown in Figure 4.

The catecholic and quinoid species described above are ubiquitous in HA, coal, peat, and almost all known humic materials. Many important functions are linked to the chemistry they contribute to these biopolymers. However, since HAs are known binders of Fe(II), Fe(III) and Cu(II), and also are potential "carriers" of small mineral particles, questions arise concerning HAs' roles in the production of more "unfriendly" radical species. Fe(II) is known to promote the formation of OH• through the well known Fenton reaction and Haber–Weiss chemistry.[19] Other reseachers have shown that hydroxyl radicals generated from small percentages of iron present at mineral surfaces are sufficient to cleave DNA.[20] Chelating agents in these experiments alter the activity of the iron and consequently the concentration of radical species. Both chelation and Fenton effects exist for HAs.

Spin trapping experiments utilizing HAs as substrates show that OH• is produced and trapped by DMPO for all HAs studied in this work. A typical 1:2:2:1 hyperfine splitting pattern is observed from DMPO–OH produced from the Fenton chemistry associated with iron and copper at the HA surface. The observed pattern arises from hyperfine coupling of the nitroxyl electron to both the nitrogen and α-H of the DMPO–OH species. The accidental "equivalence" of the hyperfine coupling constant gives the observed splitting pattern (Figure 5).[21] To verify that the DMPO–OH complex forms by the addition of the radical species and not from the iron catalysed, nucleophilic substitution of H_2O on DMPO, DMSO is added to the solution. Both processes, radical addition and nucleophilic substitution, ultimately lead to the formation of the identical radical species (Figure 5).[22] However, the hydroxyl radical can add to DMSO, displacing the methyl radical. The displaced methyl radical will add to DMPO, creating a unique radical species giving a six line hyperfine pattern (Figure 5) as evidence that DMPO–radical species are generated by a radical mechanism.[23] The increase in OH• species generated from each HA, relative to the background signal, is reported in Table 2.

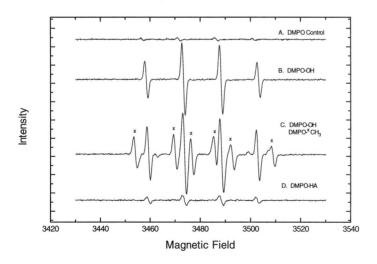

Figure 5 *Spin trapping EPR spectra*

The EPR spin Hamiltonian parameters obtained for DMPO–OH and controls agree well with others reported for similar experiments (Table 2). These parameters confirm the formation of the DMPO–OH species.

Table 2 *EPR spin Hamiltonian parameters for trapped radical species*

Sample	$<g>$	A_N	A_H
Control – DMPO + H_2O	2.0055787	14.9 G	14.8 G
Control – DMPO + H_2O + DMSO	2.0034760	16.7 G	23.1 G
Compost HA	2.0055740	14.7 G	14.8 G
Cu–SIF	2.0055760	14.8 G	14.6 G
Fe–SGF, SNF, SIF	2.0055745	14.6 G	14.7 G
Lignite HA	2.0055763	14.5 G	14.6 G

Each HA sample shows significant generation of radical species with the hydroxyl radical being the predominant form. The data were recorded after the 15–20 minute induction period during the steady state. Typically, the half-life of radical trapping for these HAs is in excess of twenty hours and it is not clear whether radical production or DMPO stability is the determining factor in ending this rather long equilibrium. In any event, radical species are produced continuously for at least twenty hours. A plot of DMPO–OH signal intensity versus time is shown in Figure 6. The compost-derived HA actually provides two radical species: the expected OH• and another undetermined species, possibly superoxide. The secondary radical species is created by direct interaction of HAs

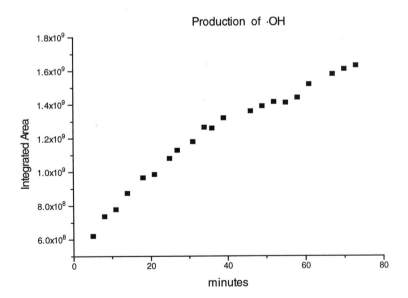

Figure 6 *Plot of DMPO–OH signal intensity versus time*

with DMSO and suggests direct oxidation of DMSO by HA functions. These include bound metals, mineral bound iron or possibly ferryl ion, $(Fe=O)^{+2}$. The presence of the ferryl ion has not been conclusively confirmed in any biological systems but is reported to be an active iron species for many enzymes.

Questions remain concerning the implications of the HA radical chemistry. In this study, soil and compost derived HAs contain mineral impurities. These species are small particles with dimensions ranging from 1–20 μm. Samples demonstrating the largest radical production have the greatest mineral content, which suggests that these HAs may simply be "carriers" of high surface area mineralites.[24] Similar chemistry has been well characterized for iron-containing aluminosilicates.[25] Metals bound in the organic phase of the HA polymer do not appear quite as active toward radical production at high levels. The actual concentration of radical species produced is related to other intrinsic HA functions.

3.3 Photochemical Mediation of Free Radical Production

The solid state photochemical studies define another important property of HAs. The induction of catecholic and quinoid radicals from moderate UV exposure suggests that HAs may possess significant radical scavenging properties. Such properties of quinoid species have been known and exploited for over thirty years.[26] Ubiquinones are known to transfer electrons in photosynthetic processes by mimicking quinone chemistries.[27] Since HAs are produced in plants as part of their life cycle, one can assume that they are produced to serve a function. Plants contain many lignin-like and phenolic complexes of (as yet) undetermined structure that are known to provide protection from oxidative damage from reactive oxygen species, ROS.[28,29] HAs possess similar structural characteristics and demonstrate photo-inducible phenolic, quinoid and/or catecholic radicals with potential to scavenge free radicals.

Initial measurements on the HA samples showed an inverse relationship between the intensity of the HA/quinoid free radical signal at $g \approx 2.00$ and the number of free radical species trapped. The data in Table 3 show that the lowest trapping yields for the free radical species are observed for Aldrich and lignite derived HAs. These samples have much higher iron contents than the soil derived HAs but also demonstrate very high populations of quinoid radical species.

Table 3 *Data summary for spin trapping experiments*

Sample	OH–DMPO signal intensity	Rel. intensity
Control (DMPO, H_2O_2, buffer)	3.97×10^{-6}	1.00
Humic Acid (Aldrich)	6.62×10^{-5}	16.9
Lignite derived HA	6.91×10^{-5}	17.4
Cu–SIF (copper loaded Irish peat HA)	8.70×10^{-4}	219
Fe–SIF, SGF, SNF (iron loaded Irish, German and New Hampshire HA)	5.14×10^{-5}	376
ITM (Irish peat HA)	5.22×10^{-4}	131
Compost (Egyptian compost HA)	4.82×10^{-4}	121

Therefore, the possibility of HA quenching or trapping radical species through this quinoid function is proposed. To assess the role of quinoid species, a series of experiments was performed to determine whether the population of quinoid radical was a determinate in free radical quenching. Samples of HAs were photo-treated for one hour, which induced the maximum increase in the quinoid population. These samples were then used (in the dark) in standard spin trapping experiments. Hydroxyl radicals were generated *in situ* from the normal HA Fenton chemistry, as shown in Table 4. Note that all signal intensities are scaled to the controls, which were the standard spin trapping experiments described previously.

Table 4 *Results of free radical quenching studies by HA*

HA sample and treatment conditions	Rel. int. (after irr./before irr.)	Increase in [OH•]^a
ITM, Compost, Aldrich (No irrad.)	1.00	1.00
HA–Compost	1.50	1.05
HA–ITM	1.40	1.06
Aldrich HA	1.20	1.01

^{a} Increase in radical species trapped, relative to control when HA substrate is photo-irradiated

As further evidence of the radical scavenging ability of HAs, direct radical trapping studies have also been performed. In these experiments, high concentrations of hydroxyl radicals were produced photolytically from H_2O_2 from a 100 mM solution. Ten mg of HA was then added into this solution. The intensity of the quinoid radical was recorded by EPR spectrometry. Coupling the hydroxyl radical with quinoid radical species reduces the resonance of the quinoid. Measurements on soil derived HAs as solid suspensions have shown that direct coupling of OH• with quinone quenches the EPR signal and that the process is a first order or pseudo first order process with a rate constant of $> 10^{-2}$ M sec^{-1}.

From these data and general observations, one concludes that HAs are an effective scavenger of OH• radicals in the solid state. Solution measurements are made difficult by solubility but are proceeding. The high quantities of phenolic, quinoid or catecholic species quench hydroxyl radicals produced at or near the HA surface. Production of OH• from HA surface chemistry appears attributable only to certain iron species within the HA matrix, likely those of mineral origin. Furthermore, electron transport between metal centers appears to be mediated by the quinoid species, and therefore it is affected by photo-irradiation. The photo-induction of quinoid species suggests a potential HA function in biochemical and environmental applications.

4 CONCLUSIONS

In this work we have observed important radical reactions occurring at the HA surface and demonstrate a mode of interaction likely responsible for some of the effects described for HAs. All involve intrinsic and extrinsic properties of HAs. Photochemically formed free quinoid radicals may offer protection from free radical damage. Interestingly, these damaging ROSs are created in plants during photosynthesis. Concurrently, HA-like species may be activated simultaneously for the plant's protection. Clearly, both effects have significant implications for plant biochemistry. It should be emphasized that HAs have been found in a wide variety of plants where biosynthesis is toward "design and function." Fenton chemistry, *i.e.* hydroxyl radical formation, may result from mineralite "impurities" characteristic of the source, method of sampling or isolation. That is, it may be an "extrinsic" HA property. That being the case, greater care in removing or minimizing mineral species should be undertaken. Clearly, the functional properties of additional HA standards need to be investigated at a molecular level. The existence of HAs in living materials almost ensures highly correlated structures and functions across plant species and therefore a source of homogeneous materials. This needs to be investigated further.

References

1. N. G. Lewis and L. B. Davin, in 'Isopentanoids and Other Natural Products: Evolution and Function', W.D. Nes (ed.), American Chemical Society, Washington, D.C., 1994, p. 203.
2. L. T. Sein, Jr. and S. A. Jansen, submitted to *Soil Science*.
3. E. A. Ghabbour, A. H. Khairy, D. P. Cheney, V. Gross, G. Davies, T. R. Gilbert and X. Zhang, *J. Appl. Phycology,* 1994, **6**, 459.
4. S. Golbs, M. Kuchnert and V. Fuchs, *Z. Ges. Hyg.*, 1984, **30**, 720.
5. Z. Tao and J. Du, *Radiochim. Acta*, 1994, **64**, 225.
6. G. Hall, A. MacLaurin and J. Vaive, *J. Geochem. Explor.*, 1995, **54**, 27.
7. J. S. Gaffney, N. A. Marley and S. B. Clark, (eds.) 'Humic and Fulvic Acids: Isolation, Structure and Environmental Role', American Chemical Society, Washington, D.C., 1996.
8. T. Sato, Y. Ose and H. Nagase, *Mutation Res.*, 1986, **162**, 173; T. Sato, Y. Ose, H. Nagase and K. Hayase, *Sci. Total Envir.*, 1987, **62**, 305.
9. H. Gulyas, R. von Bismark and L. Hemmerling, *Water Sci. Tech.*, 1995, **32**, 127; see also R. Beckett (ed.), 'Surface and Colloid Chemistry in Natural Water and Water Treatment', Plenum, New York, 1990, p. 3.
10. J.-K. Lin and S.-F. Lee, *Mutation Research*, 1992, **229**, 217.
11. A. J. Ghio and D. R. Quigley, *Am. Physiol. Soc.*, 1994, **267**, L382.
12. G. G. Choudry, *Toxicol. Environ. Chem.,* 1981, **4**, 209 and 261.
13. E. A. Ghabbour, G. Davies, A. Fatafah, N. K. Ghali, M. E. Goodwillie, S. A. Jansen and N. A. Smith, *J. Phys. Chem. B*, 1997, **101**, 8468.
14. G. Davies, A. Fatafah, A. Cherkasskiy, E. A. Ghabbour, A. Radwan, S. A. Jansen, S. Kolla, M. D. Paciolla, L. T. Sein, Jr., W. Buermann, M. Balasubramanian, J. Budnick and B. Xing, *J. Chem. Soc., Dalton Trans.*, 1997, 4047.
15. A. Radwan, Doctoral Dissertation, Northeastern University, 1996.
16. R. L. Wershaw, 'Membrane-Micelle Model for Humus in Soils and Sediments and its Relation to Humification', U.S. Geological Survey Water-Supply Paper 2410, 1994; H.-R. Schulten and M. Schnitzer, *Naturwissenschaften,* 1995, **82**, 487; L. T. Sein, Jr., S. Kolla, J. M. Varnum, G. Davies, E. A. Ghabbour and S. A. Jansen, in 'The Role of Humic Substances in the Ecosystems and in Environmental Protection', J. Drozd, S. S. Gonet, N. Senesi and J. Weber (eds.), Polish Society of Humic Substances, Wroclaw, Poland, 1997, p. 73.
17. E. Hideg and I. Vass, *Plant Science*, 1996, **115**, 251.

18.   T. J. Kappock and J. P. Caradonna, *Chem. Rev.*, 1996, **96**, 2659; D. N. R. Rao and
       A. I. Cederbaum, *Free Radical Biol. Med.*, 1997, **22**, 439; S. J. Gould and C. R.
       Melville, *Tet. Lett.*, 1997, **38**, 1473; W. S. Murphy, P. Neville and G. Ferguson,
       Tet. Lett, 1996, **37**, 7615; H. M. Mukai, S. Kikuchi and S. Nagaoka, *Biochem.
       Biophys. Acta,* 1993, **1157**, 313.
19.   W. D. Flitter and R. P. Mason, *J. Biochem.*, 1989, **261**, 831.
20.   K. Hiramoto, N. Ojima, K.-I. Sako and K. Kikugawa, *Biol. Pharm. Bull.*, 1996, **19**,
       558.
21.   K. Hagi, A. Ide, H. A. Murakami and M. Nishi, *Can. J. Chem.*, 1992, **70**, 2818.
22.   Y. Mura, J. -I. Ueda and T. Ozawa, *Inorg. Chim. Acta*, 1995, **234**, 169.
23.   P. M. Hanna and R. P. Mason, *Archiv. Biochem. Biophys.*, 1992, **295**, 205.
24.   G. T. Babcock, R. Floris, T. Nilsson, M. Pressler, C. Varotsis and E. Vollenbroek,
       Inorg. Chim. Acta, 1996, **243**, 345.
25.   J. Moser, S. Punchihewa, P. P. Infelta and M. Grätzel, *Langmuir,* 1991, **7**, 3012.
26.   S. Pekonen, R. Siefert, S. Webb and M. R. Hoffman, *Environ. Sci. Technol.*, 1993,
       26, 2056.
27.   See for examples: S. Patai and Z. Rappaport (eds.), 'The Chemistry of Quiniod
       Compounds', Vol. I, 1974; II, 1988, Wiley, New York, and references therein.
28.   L.-E. Andersson and T. Vaangard, *Ann. Rev. Plant Physiol. Plant Mol. Biol.* 1988,
       39, 379; J. Barber and B. Anderson, *Nature,* 1994, **370**, 31; J. Barber (ed.), 'The
       Photosystems: Structure, Function and Molecular Biology', Elsevier, 1992.
29.   C. H. Foyer, p. 427; F. Navari-Izzo, M. F. Quartacci and C. M. L. Sgherri, p. 447;
       O. Leprince, G. A. F. Hendry, N. M. Atherton and C. Walters-Vertucci, p. 451; M.
       Schraundner, C. Langebartels and H. Sandermann, Jr., p. 456; A. R. Wellburn and
       F. A. M. Wellburn, p. 461; R. H. Burdon, D. O'Kane, N. Fadzillah, V. Gill, P. A.
       Boyd and R. R. Finch, p. 469, G. A. F. Hendry, M. M. Khan, V. Greggains and O.
       Leprince, p. 484, in 'Free Radical Processes in Plants', Biochemical Society
       Transactions, 1996.

HUMIC ACID AS A SUBSTRATE FOR ALKYLATION

Santha Kolla,[1] Mark D. Paciolla,[1] Lawrence T. Sein, Jr.,[1] John Moyer,[1] Daman Walia,[2] Harley Heaton[2] and Susan A. Jansen[1*]

[1] Department of Chemistry, Temple University, Philadelphia PA 19122
[2] Arctech, Inc, 14100 Park Meadow Dr., Chantilly, VA 22116

1 INTRODUCTION

Humic acids (HAs) are well known environmental materials. New studies of purely organic (>99.8% C,H,O,N) HAs suggest that HA may be a single but complex biopolymer or a homogeneous complex of plant-derived biomaterial.[1] HAs have been extracted from plant sources and organic soil, and they can be obtained from mild oxidation of coal. All such materials are chemically similar.[2] These HAs possess extensive functional chemistry.[3] Table 1 shows results of functional group analysis of HAs.[4,5] Spectral data (Figure 1) show virtually identical NMR and IR spectral profiles and characteristic sugar and amino acid distributions.[2–6] These data suggest remarkable homology between soil-derived HA samples regardless of their origin or the extraction technique. HAs demonstrate a number of remarkable physiological affects owing to their ion exchange properties and polyanionic surface.[7] Some sites in HAs bind metal species tightly and other sites can be exploited for ion exchange.[8]

Table 1 *Functional group data (meq/g) for humic acids*

Source	Acidity	COOH	Ar-OH	C=O	Ar=O
Marine	2.8-5.8	1.8-3.9	0.9-1.5	1.2-3.0	0.7-2.1
Marine	3.0-7.0	2.0-5.0	0.5-2.5	3.0-6.0	
Soil	6.0-6.7	2.7-3.3	3.5-3.9	3.8-5.2	1.4
Soil	6.0-9.7	3.9-5.7	2.1-4.4		
Soil	7.3	3.8	3.5		
Soil	6.7	3.6	3.9		
Average	6.3	3.6	2.7	3.7	1.4

These characteristics of HAs have been well established for many years even though there presently exists no unambiguous structure for HA. Proposed HA structures include oxidized coal structures, structures reminiscent of common biosynthetic transformation, and lignin-like frameworks.[9] The latter two structure types are quite similar, as they derive from some common amino acid building blocks and similar enzymatic processes.

However, HA-like materials are found in plants and alga that lack the necessary enzymes for lignin biosynthesis, and indeed, from plants containing no lignins at all. As a result, these HA and/or HA-like materials retain some distinct chemical characteristics relative to lignin, even though proposed HA building blocks bear striking similarities to simple lignans.

Sample structures of each type of HA are shown in Figure 2. Other differences in the structure of these HAs likely depend on the state of humification of the organic material or redox state of the polymer. Materials obtained directly from plant extracts are similar to the latter structure types.

Composted plants and soil materials show striking similarities to plant materials, with a unique distribution of metallic components as harbingers of enzymatic species of soil microorganisms. Materials obtained from coal are further along the progression of organic to inorganic, from the living (plant) to the dead (coal). They are characterized by fused rings that have undergone a fairly significant number of mild oxidation processes. All these materials possess centers suspected to be active toward various chemistries including substitution, addition, alkylation, reduction and oxidation.

Previous studies of HA alkylation focused primarily on the use of substitution reactions as a tool to aid in the characterization of humic acids by quantitating hydroxyl sites. Carboxylic acid and phenolic sites were identified by their chemical shift using ^{13}C labelled reagents.[10]

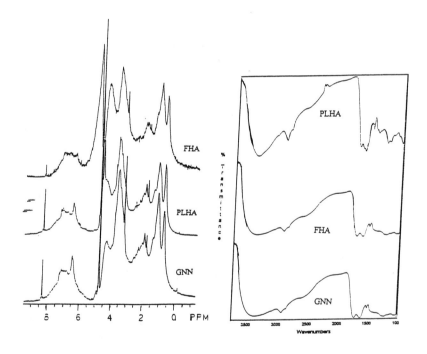

Figure 1 a *^{1}H NMR spectra of HA samples*   **b**. *IR spectra of HA samples*
(FHA) HA derived from Forsythia leaves; (PLHA) plant-derived HA; (GNN) HA from German peat

Methylation of HA bound polysaccharides was also observed in the soil derived materials.[10] This earlier work focused on HAs extracted from soil samples in which metal concentrations were characteristic of the source and varied somewhat.[11] In the present work we describe effective experimental protocols to remove or reduce alkylating species from the environment using humic acids as the reaction substrates. Studies are performed on metal-free material as well as the commercial material Humasorb™ produced by Arctech, Inc. The effect of metal binding and metal content on product distribution is an important consideration for both structural chemistry and applications of HA chemistry.

All HA substrates used are insoluble, natural polymers that can immobilize a variety of environmental contaminants, including energetic materials and alkylating agents. Functional groups available for reaction with alkylating species include carboxylic acid sites, phenolic oxygen centers, phenolic carbon centers (though less probable) and free amines. Once reacted, the solubility, mobility and ion exchange characteristics may be affected, calling into question the ability of the reacted humic material to serve in other environmental applications.

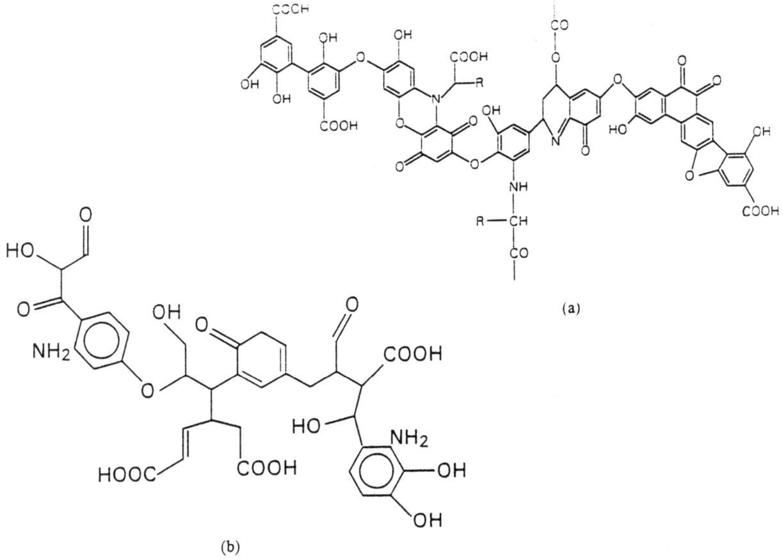

Figure 2 *Model coal-derived (a) and plant-derived (b) humic acid structures*

In this work we describe the reactivity of a plant-derived HA with methyl iodide as a model for simple alkylating agents. Two methods were employed. The first method is simply a base catalysed alkylation utilizing the carboxylic acid functional groups of HA. The anionic nature of the humic acid surface enables the formation of methyl esters. Formation of methylphenylethers can occur at the phenolic sites as well. All processes are thermally activated. Photochemical activation may induce ether formation by promoting C-I bond cleavage and activating quinoid radicals for O-methylation. A diagram for these processes is shown in Scheme 1. Verification of the methylation of HA is easily achieved by using [13]CH$_3$I. Multiple methylation sites are observed for each HA treatment. Coal and plant-derived HAs were used for comparison. Experimental conditions were optimized to reduce reaction time, maximize yield and limit the use of strong base toward a more "green" reaction process.

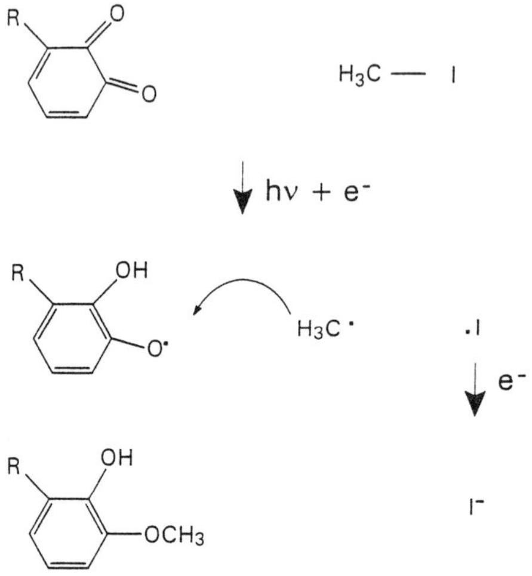

Scheme 1 *Reaction pathway of CH₃I reaction with a catechol moiety of HA*

2 EXPERIMENTAL

2.1 Chemicals

Deuterium oxide (99.96 %D), deuterated sodium hydroxide (99.96 %D), deuterated and regular dimethyl sulfoxide, deuterated chloroform, methyl iodide, ^{13}C labelled methyl iodide, ferric nitrate trihydrate and sodium hydroxide were obtained from Aldrich and used without further purification. The organic compounds were used in NMR and reaction studies. Ferric nitrate trihydrate and sodium hydroxide were used in HA metal loading.

Our humic acids were extracted with a very simple procedure. The first step is grinding and sieving the peat material. This is followed by Soxhlet extraction with a 2:1 benzene/methanol mixture, leaching with 0.1 M hydrochloric acid and stirring overnight in 0.1 M NaOH and precipitation at pH 1. The final product is extracted with dimethyl-formamide (DMF) to remove bound proteinaceous material. Because of the organic nature of the starting material, HF is not used for metal removal. The DMF step is unique to this procedure and is applied to yield a more protein-free extract. The empirical formula for the product is reproducibly $C_{36}H_{38-40}O_{15}N_2$. This material is extracted in 15–22% yield with less than 1% of the total weight due to water and less than 0.3% due to chloride from hydrochloride complexes formed during acid precipitation. Materials thus extracted have no detected metal. Organic constituents account for 99.7–99.8% of total mass (Galbraith Laboratories). Iron, copper and aluminum were detected at less than 0.03% of total weight. Commercial humic acids were obtained from Aldrich and are known to be derived from low rank coals.

2.2 Metal Loading and Atomic Absorption Spectroscopy

100 mg of HA was suspended in 50 mL of 0.1M $FeNO_3.3H_2O$ solution and stirred for 1hr. It was then filtered, washed several times with water to remove unbound iron and dried. Washing was deemed sufficient when the supernatant contained no iron species. A Buck 210 Atomic Absorption Spectrometer was used to estimate Fe using an air/acetylene oxidizing flame. Iron standards with concentrations in the range of 1–5 ppm were used for calibration. Absorbance readings were recorded at 249.7 nm against water as a blank. The results also were compared with chemical analysis performed by Galbraith Laboratories, which are in agreement within 1 ppm. A known amount of iron loaded HA was dissolved in 0.1M NaOH solution and absorbances were recorded in triplicate. The amount of Fe uptake by plant and soil derived HAs was found to be in the range 0.4–4.65% w/w.

2.3 NMR Spectroscopy

NMR spectroscopic studies were performed with GE-QE 300 MHz NMR and GE-OMEGA 500 MHz NMR instruments, the former being used for proton experiments and the latter for the ^{13}C experiments. The ^{1}HNMR spectra were obtained with presaturation at 4.8 ppm to suppress the HOD resonance. The presaturation pulse length was 2 seconds with a line broadening of 2 Hz. 256 scans were collected for each sample. ^{13}C NMR off resonance and proton decoupled spectra were obtained for all HA samples after methylation. All accumulations were overnight runs with 4192 scans.

2.4 Photo-irradiation Studies

HA samples treated with methyl iodide were irradiated with UV using an Oriel Model 66087 instrument with a Hg/Xe lamp adjusted for 200 W output. The actual photo-power striking the sample was 145 ± 5 Watts. Irradiation times were varied incrementally from 15 to 60 minutes.

2.5 Alkylation

Direct alkylation of HA with methyl iodide was achieved under various experimental conditions. $^{13}CH_3I$ was used to identify the methylated sites of HA by ^{13}C NMR. A series of reactions was also performed to estimate the amount of methyl iodide consumed as a function of reaction conditions. Both methods are described below.

2.5.1 Method (a): Base catalysed alkylation. 50 mg of metal-free HA and iron loaded HA were dissolved separately in 4 mL of 0.1 M NaOH and stirred into 3mL of methyl iodide. The reaction period was varied and the product distribution obtained is shown in Table 2. This reaction was also performed using deuterated solvents and ^{13}C labelled methyl iodide. The resulting reaction product was filtered through glass wool/cotton into an NMR tube. Proton-coupled ^{13}C NMR demonstrated successful alkylation. NMR spectra of methylated HAs are shown in Figures 3a and 3b. Multiple methylation can be identified for HA.

2.5.2 Method (b): Radical coupling. Duplicate reactions were also conducted under UV irradiation. Homolytic cleavage of the C-I bond and photoreduction of the quinoid/catecholate humic species readily gives the alkylated products.

Table 2 *Kinetic studies of methylation*

| Time (min) | Metal-free HA | Methyl iodide calculated (% w) in | | |
		Metal-free HA/UV	Fe loaded HA	Fe loaded HA/UV
5	2.4	6.2	0.5	1.6
10	4.8	12.1	1.2	3.4
15	6.4	18.4	1.7	4.7
25	8.6	30.1	4.6	14.2
35	9.8	32.2	5.2	16.8
60	11.8	32.3	5.8	18.2

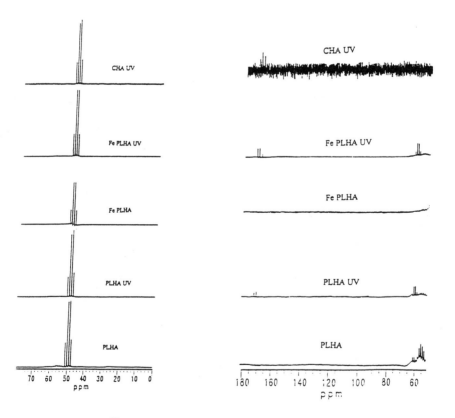

Figures 3a and 3b ^{13}C *NMR spectra of HA samples*

3 RESULTS

For the metal-free HAs in a sodium hydroxide medium under normal conditions at room temperature, methylation was achieved at only three sites (Figures 3a, 3b and Table 3). The carboxylic acid and phenolics were the most predominant sites. Figures 3a and 3b show a quartet (due to $^{13}CH_3$) between 50 and 60 ppm. The primary site was observed at 50.16 ppm. This peak is assigned to methylation at the carboxylate site. The other two methylated sites were found at 55.4 and 60.8 ppm, the former due to phenolic OH. The latter is associated with O-methylation of sugars.[12] Methylation of iron loaded HA under normal conditions at room temperature shows only one active site, the carboxylic acid one. However, the intensity of this carboxylic acid methyl is considerably lower when compared with data for methylated, metal-free HA.

Table 3 ^{13}C *NMR chemical shift data for each methyl group*

Sample	Chemical shift (ppm)	Methylated sites on HA	Intensity of methyl peak at 50.16 ppm
PLHA	51(major),	carboxylic OH	2.54
	55.4	phenolic OH	
	60.8	ether/acetal of PS	
PLHAUV	51(major),	carboxylic OH	6.74
	55.4	phenolic OH	
	60.8	ether/acetal of PS	
	171.2	see text	
Fe PLHA	51	carboxylic OH	1
Fe PLHAUV	51(major)	carboxylic OH	3.9
	55.4	phenolic OH	
	60.8	ether/acetal of PS	
	171.2	see text	
CHA	51	carboxylic OH	undetectable
CHAUV	51(only)	carboxylic OH	1.85

Methylation of metal-free HA and iron loaded HA under UV radiation at room temperature increases the overall yield and rate (Table 4). The yield of methylation under UV increases more than 3.5 times. A new methylated site is seen at 171.2 ppm (Table 3). This peak is likely due to secondary structure effects in which a methyl species folds into the ring current of one of the aromatic rings and, therefore, it is difficult to determine unambiguously the nature of the species.

3.1 Quantitation of Reaction Products: Estimation of Yield by ^1H NMR Integration

The amount of methyl iodide reacted with HA was estimated by integration of the ^1HNMR signal at 2.106 ppm, which is due to the methyl group of CH_3I in $CDCl_3$. Standards of known concentrations of CH_3I in the range 2–10% v/v in $CDCl_3$ were prepared. Integration readings were recorded keeping the same gain for all samples and a calibration plot was prepared. A known volume of unreacted methyl iodide from the methylation process from the various HA samples was dissolved in an appropriate volume of $CDCl_3$, and integration values were recorded. The amount of reacted methyl iodide was calculated by difference

Table 4 *Amount of methyl iodide reacted with HA samples*[a]

Sample	Amount of HA(mg)	Amount of CH₃ found %w/w			Average %w/w
		Trial I	*Trial II*	*Trial III*	
PLHA	50	12.3	11.9	12.2	12.3
PLHAUV	50	32.8	32.5	33.2	32.8
FePLHA	50	5.8	5.9	5.7	5.8
FePLHAUV	50	18.2	18.0	18.1	18.1
CHA	50	0.5	0.6	0.6	< 1
CHAUV	50	6.2	6.3	6.1	6.2

[a] Plant-derived humic acid (PLHA); HA irradiated with UV (PLHAUV); Fe loaded HA (FePLHA); Fe loaded HA irradiated with UV; Commercial HA (CHA); Commercial HA irradiated with UV (CHAUV)

using the calibration plot described above (Table 3). ^{1}HNMR and ^{13}C NMR were used to characterize the methylation products of HA. An estimate of reaction yield and reaction time is included in Tables 2 and 3. The data in these Tables suggest that the ester resulting from methylation of the carboxylic acid is the primary product regardless of reaction protocol. Other species are also methylated, but the assignments cannot be made unambiguously. These include a signal that arises from the methylation of phenolic groups, producing ethers. ^{13}C NMR chemical shift assignments for each methylation site are given in Table 4. This chemistry suggests that multiple acid and phenoxide species are present in the polymeric surface of HA and can be exploited using in its anionic form utilizing mildly basic conditions or free radical processes.

This model reaction was also tested on iron loaded HA, as metal binding occurs at quinoid sites with high affinity. Metal loading involves a few processes. Previous work shows that metal species can be "exchanged" under mild basic conditions from carboxylic acid sites and that quinoid sites show high binding potential. The latter are influenced by the redox state of the quinoid that can be augmented thermally or photochemically. Therefore, the loading conditions for the metal are critical for reproducibility. At ambient temperature, the iron uptake for HA suspensions reaches a maximum at more than 6 hours.[13] This rather long "time constant" is indicative of an equilibrium between multiple processes that must overcome solid state diffusion, ion exchange and quasi-reversible redox conditions.

Under any circumstances, the maximum Fe loading for the plant and soil derived HAs is in the range of 3–5 w/w%. The iron loadings for a plant-derived HA material are reported in Table 5. Significant differences are noted between the metal loaded materials and pristine HA. It is likely that the phenoxide anion is stabilized by ferric species. Binding constants for Fe(III) at quinoid or catecholic sites are quite high, and therefore compromise effective methylation.[14] Photochemical reduction at the quinoid sites promotes Fe(III) reduction to Fe(II) and therefore weakens M–HA binding, promoting reactivity at that site. The data for the time evolution of products supports this analysis, in that iron loaded materials react somewhat more slowly than HA. Similarly, for soil derived HAs iron uptake is a slow process under normal conditions.

For completeness, iron loading was attempted exhaustively on methylated "pure" organic HAs. For these materials, the iron uptake was drastically reduced. A typical metal uptake profile gives a 1–2 w/w% of metal in HA. The total iron uptake for the methylated HA is ~ < 0.5 w/w% Results for a series of HA samples are given in Table 6.

Other effects of metal complexation have been noted. Aggregate size, solubility and molecular weight all increase with metal-loading, especially for trivalent metals such as Al(III) and Fe(III).[15] This is due to the increase in interchain interactions that are mediated by metal ions. This means that the average size of the HA polymer decreases as the relative number of surface or solvent accessible functional groups decreases. This latter effect can be described as a structural effect, but may significantly alter the HA chemistry by affecting the concentration of reactive sites.

A commercial lignite derived HA also has been analysed as a methylation substrate. This material possesses a high iron concentration. It is coal-derived, and therefore has a predominance of phenolic-OH groups accessible for binding. The results for this HA substrate are given in Tables 3 and 4. A single carboxylic acid site appears to be the only methylation site under the mild conditions discussed here. This is likely due to the bound ferric species and morphology introduced by metal cross linking that leaves only the carboxylic acid site available.

The properties of the methylated HA are important especially if this methylated species is to remain in our environment. The methylated HAs are insoluble in alcohol and aqueous media from pH 1–10. In addition, the materials still show appreciable metal binding capacity, which suggests that alkylating HA may not compromise its utility in other applications. This work suggests that the most useful HAs may be metal-free and that metal uptake by some HAs may be compromised by alkylation.

Table 5 *Iron uptake by plant HA*

Time	Amount of Fe, %w/w
10 min	0.78
20 min	1.63
30 min	3.26
1hr	3.46
2hrs	3.89
4hrs	4.22
6hrs	4.75
24hrs	5.20

Table 6 *Fe loading of methylated humic acid (treatment period 20 minutes)*

Sample ID	Experimental conditions	Uptake, %w/w
Plant HA	Control (w/o methylation)	0.38
	UV (w/o methylation)	1.52
	dark	0.19
	UV irradiated	0.43
Lignite HA	dark	0.10
	UV irradiated	0.20

The extension of this chemistry to other alkylating agents will likely follow similar mechanistic pathways. Nitrogen and sulfur mustards frequently react as stable carbonium ions at anionic sites. Such sites are available and accessible in plant and coal derived HAs.

Free radical chemistry on these mustards will likely cleave the C-Cl bonds, liberating Cl radical and leaving the free radical species for coupling. Since these alkylating agents are bifunctional, cross linking reactions in the HA are likely, thus "knotting" the alkylating agent in the HA network and eliminating the hazard of the chemical agent.

4 CONCLUSIONS

Our results clearly suggest that HA functional chemistry is characteristic and can be applied to important environmental problems. HAs analysed include samples from plant materials and coal processing. Both show significant reactivity, with the major methylation product being the methyl ester. Other products form, consistent with some functional group variation, metal binding and structure morphology. In all cases ^{13}C NMR studies demonstrate that the methyl species is actually bound in the HA network, with the methylated HA remaining an insoluble, immobile material. The presence of HA-bound metal in plant-derived HAs appears to affect the methylation run time and yield, suggesting that metal-free HA is a more active substrate. The methylated HAs have reduced affinity for metal ions; therefore, methylation may compromise the HA for subsequent use as an ion exchange material. In contrast, the coal derived HAs are less reactive toward alkylation; therefore, metal uptake and ion exchange properties are not significantly altered. The extension of this chemistry to other, more significant, alkylating agents is imminent.

References

1.   L. T. Sein, Jr., M. D. Paciolla, S. Kolla, E. A. Ghabbour, A. Fatafah, G. Davies, M. H. B. Hayes, J. M. Varnum and S. A. Jansen, submitted to *Chemistry of Materials*.
2.   W. Ziechmann, 'Humic Substances', BI Wissenschaftsverlag, Mannheim, 1993; P. MacCarthy, C. E. Clapp, R. L. Malcolm and R. R. Bloom, 'Humic Substances in Soil and Crop Science: Selected Readings', Madison, Wisconsin, 1990.
3.   G. Davies, A. Fatafah, A. Cherkasskiy, E. A. Ghabbour, A. Radwan, S. A. Jansen, S. Kolla, M. D. Paciolla, L. T. Sein, Jr., W. Buermann, M. Balasubramanian, J. Budnick and B. Xing, *J. Chem. Soc., Dalton Trans.*, 1997, 4047; G. Davies, E. A. Ghabbour, S. A. Jansen, J. Varnum, in 'Advanced New Materials and Emerging Technologies', P. N. Prasad, J. E. Mark and T. J. Fai (eds.), Plenum, NY, 1995, p. 677.
4.   S. Kolla, L. T. Sein, Jr., M. D. Paciolla and S. A. Jansen, 'Recent Developments in Physical Chemistry', **2**, 1998; S. A. Jansen, M. Malaty, E. Johnson, E. A. Ghabbour, G. Davies and J. M. Varnum, *Mat. Sci. Eng.*, 1996, **C3**, 173.
5.   N. Senesi and T. M. Miano (eds.), 'Humic Substances in the Global Environment and Implications for Human Health', Elsevier, Amsterdam, 1994.
6.   S. A. Jansen, S. Kolla, L. T. Sein, Jr., M. D. Paciolla, E. A. Ghabbour, A. Radwan, G. Davies and J. M. Varnum, in 'The Role of Humic Substances in the Ecosystems and in Environmental Protection', J. Drozd, S. S. Gonet, N. Senesi and J. Weber (eds.), Polish Society of Humic Substances, Wroclaw, Poland, 1997, p. 157.

7.   A. H. Khairy, *Acta Medica Emperica*, 1981, **11**, 898; A. H. Khairy, *De Natura Rerum*, 1989, **3**, 229; R. Klockling, U. Eichhorn and T. Blumohr, *Fres. Z. Anal. Chem.*, 1978, **292**, 408.

8.   A. Fatafah, Doctoral Dissertation, Northeastern University, 1997; M. Schnitzer, Proceedings of the International Humic Substances Society, Wagenegen,1972, 293; J. Ephraim and J. A. Marinsky, *Environ. Sci. Technol.,* 1986, **20**, 367.

9.   N. G. Lewis and L. B. Davin, in 'Isopentanoids and Other Natural Products: Evolution and Function', W. D. Nes (ed.), American Chemical Society, Washington, D.C., 1994, p. 203.

10. C. Steelink, M. Mikita and K. Thorn, *Aquat. Terr. Humic Mater.*, 1981, 83; J. Del Rio, F. J. Gonzalez-Vila, F. Martin and T. Verdejo, *Org. Geochem.*, 1994, **22**, 885.

11. H.-R. Schulten and M. Schnitzer, *Naturwissenschaften*, 1995, **82**, 487; *Soil Sci*, 1997, **162**, 115; N. Senesi and G. Calderoni, *Org. Geochem.*, 1988, **13**, 1145; N. Senesi, G. Sposito, G. R. Bradford and K. M. Holtzlaw, *Water, Air, Soil Pollut.*, 1991, **53**, 409.

12. M. Schnitzer and S. Skinner, *Soil Sci.*, 1969, **108**, 383.

13. J. M. Varnum, S. Kolla, M. D. Paciolla, L. T. Sein, Jr., S. Nwabara, P. Kim, G. Davies, A. Smith, E. A. Ghabbour, A. Fataftah and S. A. Jansen, in 'The Role of Humic Substances in the Ecosystems and in Environmental Protection', J. Drozd, S. S. Gonet, N. Senesi and J. Weber (eds.), Polish Society of Humic Substances, Wroclaw, Poland, 1997, p. 741; Y. Cao, M. Conklin and E. Betterton, *Environ. Health Perspect.*, 1995, **29**, 103.

14. A. M. Martell and R. M. Smith, 'Critical Stability Constants', Plenum Press, NY, 1977.

15. A. Piccolo, S. Nardi and G. Concheri, *Chemosphere*, 1996, **33**, 595; *Eur. J. Soil Sci.,* 1996, **47**, 319; S. A. Jansen, M. D. Paciolla, E. A. Ghabbour, G. Davies and J. M. Varnum, *Mater. Sci. Eng.* 1994, **C4**, 181

HUMIC SUBSTANCES FOR ENHANCING TURFGRASS GROWTH

C.E. Clapp,[1] R. Liu,[2] V.W. Cline,[3] Y. Chen[4] and M.H.B. Hayes[5]

[1] USDA-ARS and [2] Department of Soil, Water, and Climate and [3] Department of Horticulture, University of Minnesota, St. Paul, Minnesota 55108, USA; [4] Department of Soil and Water, The Hebrew University of Jerusalem, Rehovot, Israel; and [5] School of Chemistry, The University of Birmingham, Edgbaston, Birmingham B15 2TT, UK

1 INTRODUCTION

Numerous publications since the early years of this century refer to humic substances (HSs) as plant growth promoting factors. The mechanisms involved in growth enhancement and the conditions under which these effects of HSs are expressed either in nutrient solutions or in the field have been extensively debated.

Relations of HSs with plant growth have been critically reviewed by Chen and Aviad[1] and Chen et al.[2] In their review, Nardi et al.[3] described growth promoting effects in cereals such as wheat (*Triticum aestivum* L.), barley (*Hordeum vulgare* L.), and corn (*Zea mays* L.). Stimulation of root growth and enhancement of root initiation have commonly been found, often to a greater extent than that of shoots.[1] Research by Lee and Bartlett[5] indicated that the application of humic acids (HAs) to a sandy soil low in organic matter or to nutrient solutions improved plant growth compared with the control. Hormone-like activity has been suggested by Cacco and Dell'Agnola[6] using a 'cress test' and a 'senescence test', although HA biological activity was 100 times lower than that of indoleacetic acid (IAA) at a 10 mg L^{-1} level. The concept of hormone-like activity has been discussed in detail by Chen et al.[2] and by Nardi et al.[3] An emphasis on micronutrient involvement in growth promotion effects rather than of hormones was stressed by Chen[4] and Chen et al.[2]

Modern agriculture replaced organic manures with chemical fertilizers. This trend has been reversed in recent years and the desire to utilize organics and recycle organic matter using field application has increased. Of special interest is the addition of organic soil amendments to turfgrass, which has become a popular technology in turf management. These amendments can improve soil physical structure by increasing water holding capacity or reducing soil bulk density. They can also improve soil chemical properties by increasing cation exchange capacity or buffering soil pH to provide more suitable soil environments for turfgrass under a wide range of soil conditions.

A variety of organic amendments has been used as fertilizer supplements in turfgrass management. They not only enhance fertilizer efficiency for plant growth, but also reduce the potential for ground water contamination. Major examples are HSs, seaweed products, waste materials, manures, biosolids, composts, and peat moss.[7,8] Among them, HSs are the most commonly used materials in turfgrass management. Humic substances consist of humic acids (HAs), fulvic acids (FAs) and humin. Owing to the insoluble nature of humin, it is rarely used as a soil amendment. In spite of the extensive utilization of HS amendments in turfgrass (*e.g.* golf courses), we were unable to find journal articles on the effects of HAs on turfgrass growth and management.

Therefore, the objectives of this research were to study these effects using newly developed laboratory screening techniques for examining growth enhancement of turfgrass by HSs. A 'pouch' method and a 'microsystem' method adapted from Nelson and Craft[9] were designed for this purpose. We compared commercially-produced and laboratory-prepared HS products with fertilizer controls for plant growth parameters. Our selected results from these experiments can provide essential information for field experiments or for practical users of HSs on golf courses and sports turf.

2 MATERIALS AND METHODS

2.1 Materials

2.1.1 Humic Substances. The following commercially-produced HSs samples were used in the experiments: Spodic soil HA, peat HA, leonardite HA and a leonardite HA–FA mix. Laboratory grade materials included HAs from fibric, hemic and sapric peats in addition to HSs extracted from other soils and plant material as follows: Norfolk peat HA and FA, Kerry soil HA and FA, and Moss HA and FA. Humic material concentrations in our plant growth experiments ranged from 0 to 50 mg L^{-1}.

2.1.2 Fertilizers. The following fertilizers were used: NH_4NO_3, $Ca(H_2PO_4)_2$ and KNO_3 at a ratio of 4:1:4 of N:P:K. The N concentration in the nutrient solutions was 3.6 mg L^{-1} in all experiments investigating the effects of the source and level of HS supplements (Tables 1, 2, 4 and 5). N concentrations varied from 0.9 to 57.6 mg L^{-1} in experiments investigating effects of N rates on plant growth (Table 3).

2.1.3 Grasses. Species used in the experiments were creeping bentgrass (*Agrostis palustris Huds., cv.* Providence) and ryegrass (*Lolium perenne* L., *cv.* Omega).

2.2 Methods

Growth pouches (Mega International) were 16 x 18 cm in size. Fifteen ryegrass seeds were placed in the filter paper pocket in each pouch, which contained solutions of HSs and fertilizers. After a 7-day germination period in an incubator at 20 °C, the seedlings were thinned to 10 plants for each pouch. Growth chambers were set at 20 °C, 16 hours in light and 8 hours in dark. Solutions were added to the growth pouches as required during the growth period. Plants were allowed to grow for three additional weeks before harvesting. At harvest, all plants were removed from each pouch, and shoots and roots were separated and placed into vials. Plant tissues were dried at 65 °C to constant weight (about 2 days), and weighed to measure the tissue mass.

Microplates (Fisher Scientific) with 12 wells, about 5 mL each, were used. Four holes (3 mm) were drilled at the bottom of each well. One of the holes held a filter paper wick extending about 5 mm above and below the bottom of the hole to supply the nutrient solution. Earlier experiments showed that four holes were optimal for solution exchange and root growth. Fine sand (< 1 mm) and coarse sand (1 to 2 mm) were layered into the well to provide a desired level of water-holding capacity. Six creeping bentgrass seeds were placed in the sand and kept moist with a bottom microplate containing water. After 1 week of germination and growth in a 20 °C constant temperature chamber, the upper plate was placed in a plastic tray arranged to hold three microplates, and containing fertilizer only or fertilizer plus HS solutions. Each treatment tray contained a total of 36 wells. Seedlings were thinned to three plants per well. A 'constant-head' bottle was used to provide additional nutrient solution to the plastic tray. Plants remained in the plates for three additional weeks before harvesting.

For harvest, three plants from each well were taken and washed with water to remove the sand. After separating shoots from roots, shoots from five wells were combined into one sample; six replicates were collected from each of the treatments. For root dry weight measurements, roots from two wells were combined into one sample; nine replicates were collected per treatment. Root and shoot samples were placed into vials, dried at 65 °C to constant weight (about 2 days), then weighed for tissue mass. For root length measurement, fresh roots were combined from two wells, washed, and placed into 10% methanol solution. The roots were stained overnight with basic fuschin (0.025%) before image processing.[10] Triplicate samples were measured for each treatment.

3 RESULTS

3.1 Experiments Performed in Growth Pouches

Plant growth data for pouch grown plants are summarized in Tables 1 and 2. Root dry weight obtained for plants grown in control NPK solutions was about the same as for the commercially-produced HS treatments (Table 1). The root dry mass varied between 91 and 104 % compared with the control.

Table 1 *Dry weights of roots and shoots, percentage change in root growth, and root/shoot ratios for some commercial humic acids using the Pouch Method.*[a]

Treatment[b]	Rate (mg L^{-1})	Root (mg)	Change (%)	Shoot (mg)	Change (%)	Root/shoot
Control	3.6	49.7	100	47.7	100	1.04
Peat HA	10	51.9	104	42.7	90	1.22
Peat HA	25	45.5	91	36.9	77	1.23
Peat HA	50	45.3	91	35.7	75	1.27
Leonardite HA	10	49.4	99	42.3	89	1.17
Leonardite HA	25	49.7	100	40.1	84	1.24
Leonardite HA	50	51.8	104	38.2	80	1.36

[a] Values are means of duplicate determinations; [b] N rate of 3.6 mg L^{-1} for control and all HA treatments.

Shoot dry weight, however, decreased with the HS treatments from 75 to 90%. The reduced shoot weight and constant root weight is reflected in a significant change of root/shoot ratio, which increased between 13 and 31%. A slight increase in the root/shoot ratio can be observed with increased HS concentration from 10 to 25 mg L^{-1}. Data in Table 2 show root and shoot dry weight increases compared with the controls. The HSs extracted from Minnesota peats seemed to exhibit a significant enhancement effect on growth of both shoots and roots.

Table 2 *Dry weights of roots and shoots, percentage change in root growth, and root/shoot ratios for some Minnesota peat humic acids using the Pouch Method.*[a]

Treatment[b]	Rate (mg L^{-1})	Root (mg)	Change (%)	Shoot (mg)	Change (%)	Root/shoot
Control 1	3.6	39.0	100	31.0	100	1.26
Fibric peat HA	10	45.0	115	43.3	140	1.04
Hemic peat HA	10	47.6	122	46.5	150	1.02
Sapric peat HA	10	41.2	106	38.6	125	1.07
Control 2	3.6	54.1	100	38.6	100	1.40
Fibric peat HA	25	61.0	113	46.8	121	1.30
Hemic peat HA	25	59.0	109	49.9	129	1.18
Sapric peat HA	25	66.0	122	46.0	119	1.43

[a] Values are means of duplicate determinations; [b] Control 1 for 10 mg L^{-1} treatments, 28-days experiment; control 2 for 25 mg L^{-1} treatments, 40-day experiment; N rate of 3.6 mg L^{-1} for controls and all HA treatments.

3.2 Experiments Performed in Microplates

Bentgrass plant response to N rates showed a steady increase for both root and shoot dry weights with increasing N rates up to 7.2 mg L^{-1} (Table 3). Differences in tissue weight of plants grown at N rates between 7.2 and 28.8 mg L^{-1} were small. When the N rate was as high as 57.6 mg L^{-1}, both root and shoot dry weight decreased. This is consistent with the root length measurement (278 cm root length per well at 57.6 mg L^{-1} N, compared with 305 cm per well for a N rate of 28.8 mg L^{-1}). A notable change was the decreasing root/shoot ratio with increasing N rates between 0.9 and 7.2 mg L^{-1}. Almost constant root/shoot ratios (between 0.7 to 0.8) were observed at the higher N rates between 10.8 and 57.6 mg L^{-1}.

Plant responses to spodic HA rates are shown in Table 4, where the HA concentrations ranged from 1.8 to 50.0 mg L^{-1}. The N rate was the same as that of the control for all treatments. Slight increases in dry root weight were observed while slight decreases in shoot dry weight were recorded for HS rates from 1.8 to 25 mg L^{-1}. Accordingly, the root/shoot ratios increased slightly. Root length increased with increasing concentrations of the HAs. Treatment rates ranging from 10 to 25 mg L^{-1} appeared to induce enhanced root dry weight and root length.

Table 3 *Plant response to nitrogen rates by dry weight of roots and shoots, root/shoot ratios, nitrogen concentration of roots and shoots, and root lengths using the Microsystem Method.*[a]

N Rate mg/L^{-1}	Dry weight			N concentration		Root length (cm well^{-1})
	Root	Shoot	Root/shoot	Root	Shoot	
	(mg well^{-1})			(%)		
0.9	4.5	2.9	1.6	0.96	1.30	180
1.8	7.3	5.7	1.3	0.98	1.59	252
3.6	7.6	8.0	1.0	1.07	2.07	202
7.2	10.2	12.8	0.8	1.45	3.07	234
10.8	12.6	15.4	0.8	1.64	3.11	295
14.4	9.5	12.4	0.8	2.10	3.73	249
28.8	10.8	15.5	0.7	2.91	4.63	305
57.6	8.6	10.8	0.8	3.79	6.16	278

[a] Values are means of triplicate determinations.

Table 4 *Plant response to spodic soil humic acid rates by dry weights, root/shoot ratios, and root lengths using the Microsystem Method.*[a]

HA rate (mg L^{-1})	Dry weight		Root/shoot	Root length (cm well^{-1})
	Root	Shoot		
	(mg well^{-1})			
0.0[b]	8.5	7.6	1.1	180
1.8	8.2	7.6	1.1	220
3.8	7.5	7.0	1.1	273
7.5	6.9	6.2	1.1	234
10.0	9.2	7.7	1.2	279
15.0	9.1	6.9	1.3	271
25.0	8.6	7.2	1.2	274
50.0	9.9	8.2	1.2	252

[a] Values are means of triplicate determinations; [b] Control, N rate of 3.6 mg N L^{-1} for all treatments.

Plant responses to various HSs including HAs and FAs from commercially-produced and laboratory-grade sources are summarized in Table 5. Both root and shoot dry weights increased for most treatments. The root dry weight increased up to 46% (fibric peat HA), and the corresponding shoot dry weight increased up to 65%. For root dry weight, the fibric peat HA exhibited the highest increase; seven other treatments (hemic peat HA, sapric peat HA, Kerry soil HA and FA, Leonardite HA, Norfolk peat FA, and leonardite HA–FA) exhibited enhanced root weight compared with the control. The Norfolk peat HA induced a decrease in root weight. As found for the roots, the highest shoot dry weight was exhibited by the fibric peat HA.

Table 5 *Plant response to several experimental and commercial humic and fulvic acids by dry weights, root/shoot ratios, and root lengths using the Microsystem Method.*[a]

Treatment[b]	Dry weight		Root/shoot	Root length
	Root	Shoot		
	(mg well^{-1})			(cm well^{-1})
Control	11.4	7.3	1.6	135
Hemic peat HA	15.0	8.3	1.8	180
Sapric peat HA	13.2	7.1	1.9	222
Fibric peat HA	16.6	12.1	1.4	207
Norfolk peat HA	9.7	6.1	1.6	172
Kerry soil HA	14.6	9.3	1.6	193
Leonardite HA	12.9	7.6	1.7	172
Moss HA	11.0	7.2	1.5	156
Norfolk peat FA	12.2	8.7	1.4	234
Kerry soil FA	15.9	9.7	1.6	160
Moss FA	11.3	8.1	1.4	174
Leonardite HA–FA	13.4	8.2	1.6	195

[a] Values are means of triplicate determinations; [b] N rate of 3.6 mg L^{-1} for control and all treatments. HA and FA rates 3.6 mg L^{-1}.

Increases were also exhibited by the hemic peat HA and the Kerry soil HA and FA treatments. Plants grown in nutrient solutions containing hemic peat HA, sapric peat HA and the Leonardite HA exhibited enhanced root/shoot ratios compared with the control.

All treatments (excluding the Norfolk peat HA, the Moss HA and the Moss FA) resulted in enhanced growth of both shoots and roots of creeping bentgrass plants grown in nutrient solutions containing fertilizers and supplemented with HSs. This provides encouraging information that supports application of HSs on turfgrasses in addition to chemical fertilization.

4 DISCUSSION

Data for plants grown in pouches (Tables 1 and 2) clearly demonstrated that HSs changed the plant root/shoot ratios in ryegrass. This increase was obvious for the commercial products but not for the purified HSs. However, both root and shoot dry weight increased when the plants were treated with purified HSs.

Higher N rates are known to favor shoot growth. This also was demonstrated by the N application rate experiment (Table 3). Although both root and shoot weight in these experiments increased at higher N levels, the root/shoot ratios decreased. Plant shoots consumed more easily-available nutrients than did the roots. Our results indicate that HSs can affect nutrient flow through a preferential effect on root growth. Four of the six tested humic treatments resulted in enhanced root/shoot ratios in bentgrass. The HSs obviously preferentially promote root growth rather than shoot growth (Tables 3 and 5). This is in accordance with many earlier publications.[1]

Lee and Bartlett[5] indicated that a positive growth response can be observed only when the application of HAs is directed to a soil low in organic matter or to nutrient solution. This was further stressed by Chen and Aviad.[1] Our results (Table 3) agree with this statement. When the N levels were set at 3.6 mg L^{-1}, which is equivalent to 0.025 lb N per month per 1000 ft^2, positive responses were obtained for the HSs. The HAs in this experiment exhibited positive effects on root growth even at very low concentration. This can also be seen in the root length measurements where all treatments gave longer roots than those of the control (Table 3). The physiological mechanisms involved in the plant growth enhancement observed in our study have not been specifically investigated. We believe that maintaining micronutrients in soluble forms of humic complexes, thereby making them available to the plants, as proposed by Chen and coworkers,[1,2] is the major mechanism involved with turfgrass growth enhancement.

Strong evidence[3] indicated that HSs stimulate root cell elongation and prolifer-ation. This was demonstrated by scanning electron micrographs that showed abundance of root hairs in wheat in HS treated roots, and also by transmission electron microscopy, which showed that HSs induce a higher differentiation rate. Our present results show that HS treatments enhanced root growth in preference to shoot growth in bentgrass. This information may be of particular significance to golf course managers. Most commonly, bentgrass is grown in 80 to 100% sand. Application of the appropriate amount of HSs may increase the plant's tolerance to environmental stress due to enhanced root growth. Our results obtained in the 'pouch' and 'microplate' systems may serve as screening methods for larger tests. Obviously, greenhouse and field experiments are needed for further confirmation of positive effects.

References

1. Y. Chen and T. Aviad, in 'Humic Substances in Soil and Crop Sciences: Selected Readings', P. MacCarthy, C. E. Clapp, R. L. Malcolm and P. R. Bloom (eds.), American Society of Agronomy, Madison WI, 1990, p. 162.

2. Y. Chen, H. Magen and J. Riov, in 'Humic Substances in the Global Environment and Implications on Human Health', N. Senesi and T. M. Miano (eds.) Elsevier, Amsterdam, 1994, p. 427.

3. N. Nardi, G. Concheri and G. Dell'Agnola, in 'Humic Substances in Terrestrial Ecosystems', A. Piccolo (ed.), Elsevier, New York, 1996, p. 361.

4. Y. Chen, in 'Humic Substances in Terrestrial Ecosystems', A. Piccolo (ed.), Elsevier, New York, 1996, p. 507.

5. Y. S. Lee and R. J. Bartlett, *Soil Sci. Soc. Am. J.*, 1976, **40**, 876.

6. G. Cacco and G. Dell'Agnola, *Can. J. Soil Sci.*, 1984, **64**, 225.

7. J. Roche, *Landscape Management*, Nov., 1994, p. 28.

8. J. F. Wilkinson, *Golf Course Management*, March, 1994, p. 80.

9. E. B. Nelson and C.M. Craft. *Phytopathology*, 1992, **82**, 206.

10. R. H. Dowdy, A. J. M. Smucker, M. S. Dolan and J. C. Ferguson, *Plant Soil*, 1998, **200**, 91.

GREENHOUSE GAS DILEMMA AND HUMIC ACID SOLUTION

D. S. Walia, A. K. Fataftah and K. C. Srivastava

ARCTECH, Inc., Chantilly, VA 20151

1 INTRODUCTION

Five years ago, U.S. Vice President Al Gore wrote a book entitled "Earth in Balance" and highlighted adverse global changes since the Industrial Revolution and their impact on our very existence. He pointed out the impacts of emissions of greenhouse gases, ever increasing loss of top soil and pollution of our waters. He made a passionate plea for worldwide actions to create a balance of our human needs with our planet Earth's ecology. Many experts and the general public are recognizing this need but are concerned with costs to worldwide economics, especially today, when we address the basic needs of large and growing populations.

The Industrial Revolution that started about two centuries ago has been energized with ever increasing use of our fossil fuels, especially coal. This has resulted in unprecedented economic growth worldwide and probably for ever has changed our relationship with our planet Earth and its resources. A major impact now being recognized is global warming resulting from the build up of greenhouse gases, especially carbon dioxide (CO_2), due to the burning of fossil fuels. The greenhouse effect, by which a small amount of solar heat is retained near the surface of our planet, is critical to maintain a fragile ecology. However, unacceptable levels of accumulation of these greenhouse gases in our atmosphere are now believed to be gradually heating up the planet. This effect has the potential to cause drastic adverse ecological imbalances for inhabitants throughout the world.

In December 1997, at the global climate convention held under the auspices of the United Nations at Kyoto, Japan, 160 nations gathered for landmark negotiations to establish a treaty to reduce greenhouse gas emissions below the 1990 level between the years 2008–2012. For example, this will require the U.S. to reduce 7% of CO_2 below the 4.8 billion tons emitted in 1990, and thereafter maintain a yearly level of emission at 4.5 billion tons. Considering the continuing increase in emissions, the net reduction will be 30%. However, the U.S. Senate must ratify the provisions of this treaty before they become mandatory requirements for U.S. industry. Today, the U.S. contains only 5% of the world population but emits 25% of the worldwide greenhouse gases. Thus, U.S.

industry and government must respond to this challenge and contribute in (1) creating solutions which will continue the industrial revolution of the past two centuries well into the future, and (2) sustain market based economic growth needed in today's rapidly changing world economics. This is reflected in the recent announcement by President Clinton to establish a $6 billion program to achieve this vision. President Clinton has put forth a challenge that economic growth must be achieved while implementing solutions to curb the greenhouse effect.

2 THE MicGAS™ TECHNOLOGY SOLUTION

Strategies for CO_2 reduction being considered include: (1) improved energy efficiency, thus requiring less fuel; (2) forestation and reforestation of lands to increase CO_2 absorption; (3) captured CO_2 reuse and disposal in land and ocean reservoirs; and (4) switching to less CO_2 producing fuels. Coal, which is the least efficient fuel in terms of Btus of heat energy/ton and the highest CO_2 producing fuel compared to natural gas and petroleum fuels, faces a serious challenge and may become obsolete.

Coal is the most abundant and the least costly fossil fuel. With increasing economic growth and world population, increased use of all fossil fuels is critical well into the next century and thereafter to meet the growing energy demand. Sustainability of abundant coal resources as viable fuels in the strategy is needed to produce low cost energy and also to sustain the enormous economic infrastructure that millions of people depend on to support their jobs.

The power industry depends upon coal for more than one-third of its fuel needs and constitutes the largest market for coal. The conventional method of generating electricity with coal involves spraying finely pulverized coal with hot air into a furnace chamber lined with water-filled coils. Coal burning inside the chamber converts the water in the tubes to steam, which is then used to rotate a turbine power generator. This process, devised more than a century ago, is termed the "Rankine Cycle." Control equipment has recently been mandated to remove pollutants such as sulfur dioxide and fly ash from being discharged into the atmosphere. Only limited success has been achieved in developing synfuels and coal gasification due to the high costs involved in these conversion technologies.

The novel MicGAS™ technology offers a new way of utilizing our vast coal resources to produce energy with lower CO_2 emissions, replenish soils with humic matter for increased food production and cost-effectively clean our water and soils.[1-4] A broad based implementation of this integrated technology solution will propel significant economic growth and in fact can be considered as an antidote to the problems highlighted by Vice President Gore in his "Earth in Balance" treatise.

The common element to each of the conventional processes that utilize coal for the generation of electricity is that the coal carbon is combusted. By doing so, power is generated but CO_2 is also emitted in large quantities, thereby contributing to its accumulation in the Earth's atmosphere.

The MicGAS™ technology is radically different from the current use of coal in the power industry. It is based on biological conversion of coal to produce clean fuel gas along with valuable humic acid byproducts, while reducing the CO_2 emitted into the atmosphere. In one innovative stroke this technology will address the important issues related to the development of a strategy for climate control. It will:

- achieve significant reduction of CO_2 greenhouse gas emissions from coal use by creating carbon-rich humic acid products for use on the planet Earth and producing hydrogen-rich fuel gas for energy generation;
- enhance agricultural growth and environmental cleanup through the use of unique humic acid products;
- revitalize the coal industry and propel economic growth; and
- constitute a least regrets strategy for CO_2 control.

The innovative MicGAS™ technology is based upon applying natural micro-organisms adapted to convert coal into clean fuels under anaerobic conditions. Unlike the conventional coal gasifier, the solid residue from the MicGAS™ anaerobic treatment is not a waste but is very rich in humic acid.

The residual coal from this treatment is further subjected to an aerobic/anoxic bio-chemical process for extraction of humic acid, a valuable byproduct that has applications as a fertilizer material and in environmental remediation.

ARCTECH has developed this technology by adapting microorganisms (derived from the wood eating and humus eating termites) to coal conversion in the presence of appropriate nutrient components. The process conditions have been optimized so that the technology can be applied in typical sewer treatment systems. The technology can also be adapted for applications *in situ* or in above ground bioreactors.

The bioconversion is accomplished in three major steps. In the first step, (the *hydrolytic and fermentation* process), microbes convert the coal into volatile organic liquids (primarily acetic acid) and CO_2. In the second step, the liquid from the first step and the gases produced are contacted with methanogenic ("methane producing") microbes that hydrogenate the acetic acid and CO_2 to methane. The methane produced is separated and then in the third step the residual product undergoes an aerobic/anoxic biochemical conversion. In this step the coal residue is converted into humic acid products (Figure 1).

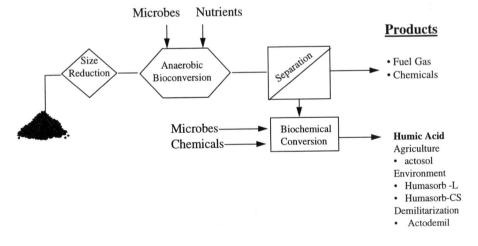

Figure 1 *MicGAS™ coal bioconversion technology schematic*

In its simpler version, the MicGAS™ technology converts carbon in coal into two primary components. One is hydrogen-rich, clean fuel gas and the second is carbon rich, humic acid products. This efficient utilization of coal carbon into useful products results in lower CO_2 emission for power generation and increased value derived from products than the conventional power generation Rankine Cycle approach (Figure 2). This coal utilization concept is similar to that used in an oil refinery. As a result of cracking and refining, a barrel of crude oil provides a multitude of products that are valued at several times higher than the price of the crude oil itself. The higher value obtained from the sale of these byproducts (petrochemicals) enables the oil industry to sell #6 oil at almost the same or below the price of crude coming into the refinery. MicGAS™ technology, while creating a solution to global warming, will result in converting the U.S. coal industry's income from about $25 billion to over one trillion dollars. Even though the strategy of MicGAS™ follows the oil refinery multiple product scenarios, it benefits because of products based on unique humic acids developed from coal and will serve large agricultural and environmental market sectors. Thus, it will not compete for petrochemical markets.

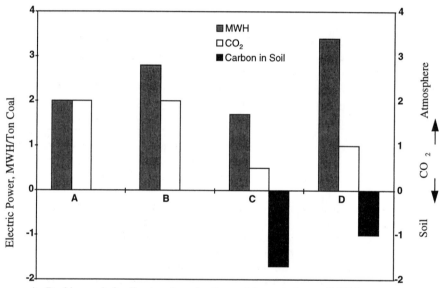

A. Rankine cycle for direct coal combustion and flue gas cleanup-DOE data
B. Thermal IGCC: Integrated gasification combined cycle-DOE data
C. Bio-IGCC/Fuel Cell: Based on laboratory data -at ARCTECH, 25% carbon to fuel gas
D. Bio-IGCC/Fuel Cell: Theoretical possible -at ARCTECH, 50% carbon to fuel gas

Figure 2 *Electric power and CO_2 gas produced from different coal utilization strategies*

3 HUMIC ACID PRODUCTS APPLICATIONS

Widespread application of humic acid products is therefore needed to mitigate CO_2 emissions and at the same time create high value from our vast coal resources. In the past ten years, ARCTECH has developed several innovative applications of humic acids for both agriculture and environmental markets. As a water retainer, metal binder and sorbent, humic acid is essential to maintain fertile soils. Humic acids' water retention property gives the Earth a thermal buffer capacity that prevents catastrophic climates. The versatile charactertistic properties of humic acids include a high cation exchange capacity, the ability to chelate metals, the ability to adsorb organics, a high water holding capacity, an ease of precipitation at low pH or in the presence of coagulants, and an ease of combustion due to its organic nature. These versatile properties are useful for agricultural and environmental purposes. The agricultural applications include a slow release source of the micronutrients for plant and microbial growth, a high water-holding capacity, and a buffering capacity that results in plant growth stimulation.

The environmental applications include metal removal by chelation, removal of organics by adsorption, neutralization of acidic water streams, removal of anions, reduction of metal species [*e.g.* Cr(VI) to Cr(III)] and explosives and chemical agent destruction.

It is these remarkable properties of humic acids that make them extremely useful materials for multiple applications. In turn, large scale applications of HAs will enable us to sequester carbon within our planet and avert the consequences of a greenhouse effect on our global ecology instead of carbon being emitted as CO_2 into the atmosphere.

Recognizing the above mentioned potential of the MicGAS™ technology, ARCTECH scaled up a pilot-production facility to produce humic acid from inherently-humic rich lignite. ARCTECH produces humic acid and formulates it into commercial Actosol® fertilizer products. Actosol® fertilizer is being successfully marketed in the U.S. for golf courses, landscaping, erosion control and agricultural crops such as corn, wheat and soybean. In the Middle Eastern Gulf countries, Actosol® is being used for alfalfa, palm trees and multitudes of crops grown in the greenhouses of the harsh desert climate. In Mauritius (a prosperous island in the Indian Ocean), Actosol® is being used for increasing yields of sugar cane. In South Korea, Actosol® has been introduced in the marketplace for golf courses and huge greenhouse markets. Presently, Actosol® is being tested in the Chinese market.

The primary benefits demonstrated in these applications are that yields of crops and plants increase by 20–100% (Figure 3),[5] which results in the net value gain of about $100/acre. This estimate is based on input costs of about $50/acre with use of 5 gallons/acre per season and at an average price of $10/gallon. Thus, the full-scale implementation of a MicGAS™ technology-based plant, the production of large volumes of humic acid and the subsequent potential for humic acid at the projected low prices are expected to generate applications of humic acid in the large agricultural market sectors, both in the U.S. and world-wide.

ARCTECH has successfuly developed a novel adsorbent named HUMASORB-CS™ (a water insoluble polymer).[6,7] This novel adsorbent has been shown to remove both inorganic and organic contaminants in a single step from water (Table 1). HUMASORB-CS™ is being evaluated for emplacement as a subsurface barrier at large groundwater contaminated sites, thus presenting a permanent low-cost solution.

ARCTECH is also successfully demonstrating applications of its humic acid-based ACTODEMIL™ technology for recycling of nitrogen-containing explosives from conventional munitions into usable fertilizers.[8] This fertilizer product has met all regulatory requirements and was approved for use by the Nevada Department of Environmental Protection for application by an alfalfa farmer in Nevada. Recently, the

U.S. Army selected ACTODEMIL™ technology for further evaluation for safe disposal of chemical munitions that contain both explosives and chemical agents.

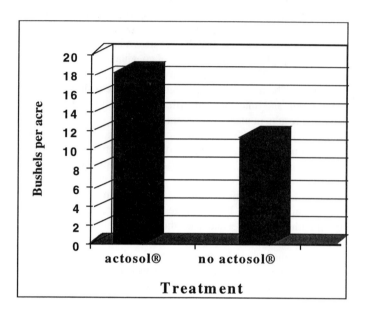

Figure 3 *Increased yield of soybean with use of Actosol®*

Table 1 *HUMASORB-CS™ is effective for mixed waste remediation*

CONTAMINANT	SIMULATED WASTE Concentration		
	Initial (ppm)	Final (ppm)	Removal (%)
Chromium (III)	88	< 0.5	> 99
Copper	98	< 0.5	> 99
Lead	18	< 0.5	> 97
Trichloroethylene (TCE)	140	1	>99
Perchloroethylene (PCE)	26	N.D.	>99

N.D.= Not detected.

4 CONCLUSIONS

In summary, ARCTECH's vision and strategy of MicGAS™ technology clearly provides a means to create an efficient use of coal carbon on our planet Earth, rather than emitting it into the atmosphere and causing increased global warming. The significant impact of the implementation of the MicGAS™ technology is based on sound scientific principles and the process feasibility is already proven for producing clean fuel gas and humic acid products from coal. MicGAS™ technology will also enable efficient utilization of our vast coal resources for energy production, while sequestering carbon in the form of humic acid products on our planet Earth to meet our large needs of vegetation growth and to remove toxic and harzardous contaminants from the environment. In addition, the MicGAS™ technology has the potential to mitigate greenhouse CO_2 gas emissions by 50–75% in the near term from the existing Rankine Cycle based power generation systems. Economical, clean fuel gas from the MicGAS™ plants will also easily integrate with new higher efficiency cycles and fuel cells, thereby enhancing their economics for long-term implementation in new power plants and permanently reducing CO_2 emissions.

References

1. K. C. Srivastava and D. S. Walia, US Patent 5670345, 1997.
2. D. S. Walia and K. C. Srivastava, 'Proceedings of the Coal-Fired Power Systems: Advances in IGCC and PFBC Review Meeting', U.S. Department of Energy, 1994, Vol. 1, p. 376.
3. K. C. Srivastava and R. J. Manolov, Fourth International Symposium on Biological Processing of Fossil Fuels, Alghero, Italy, 1994.
4. D. R. Quigley, 'Microbial Transformations of Low Rank Coal', CRC Press, Boca Raton, FL, 1992, p. 28.
5. P. MacCarthy, C. E. Clapp, R. L. Malcolm and R. R. Bloom., 'Humic Substances in Soil and Crop Sciences: Selected Readings', American Society of Agronomy, Madison, Wisconsin, 1990.
6. H. G. Sanjay, K. C. Srivastava and D. S. Walia, US Patent, in process.
7. H. G. Sanjay, K. C. Srivastava and D. S. Walia, 'Proceedings of the Environmental Technology Through Industry Partnership Conference', U.S. Department of Energy, 1995, Vol. 1, p. 411.
8. H. Heaton, J. J. Stashick and D. S. Walia, US Patent 5538530, 1996.

Subject Index

Absolute mobilities, 110
Abundance, 2, 16, 22, 23, 44, 88, 236
Acetonitrile, 3, 48, 49, 62
Acid–base properties, 61, 113
Acid–base titration, 71
Acidification, 72, 123
Acids, 32, 163
Actinide, 113
Activated sludge, 69, 70
Adhesion, 138
Adsorption, 24, 48, 66, 71, 92, 147, 173,
    174, 178, 179, 185–193, 195, 239
  endothermic, 185, 186, 189, 191
  isotherms, 174–181, 185–187, 189, 190
  mechanism, 185
  process, 187
  sites, 180, 185, 191
  step, 185, 191
Aerobic/anoxic biochemical conversion,
    237
Affinity, 133, 141, 143, 173, 200, 208,
    222, 224
Aggregates, 3, 6, 8, 9, 23, 62, 134, 140–
    142
Agriculture, 69, 70, 76, 77, 227, 239
  applications, 203, 239
  growth, 237
  productivity, 3
  sustainable, 173
Alcohols, 22, 32, 59, 71, 72, 152, 155
Aldaric acids, 58
Aldehydes, 22, 23, 32, 41, 66
Aldonic acids, 58, 59
Alicyclic ring, 32
Aliphatic, 23, 32, 34, 35, 39, 53, 58, 63,
    70–72, 74–76, 84, 88, 152, 154, 155
  alcohols, 41, 47, 51
  esters, 44

hydrocarbons 14, 19, 49
hydroxy acids, 47
protons, 11
substituents, 14
tricarboxylic acids, 22
Aliphaticity, 69, 72, 156
Alkaline hydrolysis, 32, 65
Alkylation, 216, 217, 219, 223, 224
Alkylsubstituted phenyl structures, 66
Aluminosilicates, 211
Amide, 11, 14, 16–19, 38, 69, 71, 74,
    152, 154, 155
Amide carbonyl groups, 38
Amide II bonds, 74
Amino acids, 2, 3, 13, 16, 19, 21, 23, 29,
    32, 34, 51, 53, 63, 72, 75, 188, 205,
    206, 215, 216
Amino groups, 51, 53
Amorphism, 10
Amperometric titrations, 167
Ampholite, 7
Animal tissues, 61, 118
Animal wastes, 70
Anion exchange resins, 5
Annealing algorithm, 134
Anode, 109, 110
Anomeric carbon, 19, 21, 35, 36, 38, 41,
    51, 52, 58, 65, 82
Anthropogenic organic chemicals, 3, 24
Anthropogenic substances, 118
Aquatic organic matter, 148
Aromatic, 3, 17–19, 24, 36–38, 43, 56,
    60, 71, 72, 75, 79, 82, 84, 91, 138,
    152, 154, 175, 180, 188, 189, 221
  acids, 58
  aliphatic ethers, 23
  amide protons, 11
  bands, 35, 41, 52

Aromatic, *continued*
  compounds, 16, 53, 66, 76
  esters, 44, 58
  polymers, 142
  region, 47, 63, 66, 67
  rings, 22, 23, 51, 66, 74, 88, 154
  structures, 22, 23, 65, 66, 128, 135
Aromaticity, 19, 21, 23, 71, 72, 76, 152, 156
Ash content, 5, 9, 71, 125
Associations of molecules, 8
Atom excitation, 147
Atomic absorption, 219
Atomic charges, 134, 141
Atrazine, 143, 174, 180, 181, 196–200
Availability, 2
Availability and mobility of metals, 147
Average building block, 206

Background electrolyte, 196, 197
Background subtraction, 101
Basic fuschin, 229
Batch equilibration, 175
Bauxite, 191
Bentgrass, 228, 229, 230, 233
Benzene, 143, 218
  monocarboxylic acid, 66, 67
  carboxylic acids, 21, 22
  dicarboxylic acids, 22
  hexacarboxylic acids, 22
Benzoic acid, 8, 115
Bilayer formation, 193
Binding of metals, 24
Binding potential, 222
Binding profiles, 148, 151, 158
Bioavailability, 133, 144, 182
Biochemical stability, 165
Bioconversion, 237
Biodegradation, 2, 69, 123, 130, 175, 203
Biological activity, 3, 127, 227
Biological importance, 165
Biomineralization, 203
Biomolecules, 2, 10, 29, 104
Biopolymer, 29, 185, 203
Biosolids, 227
Biosynthesis, 203, 206, 213, 216
Blue shift, 128

Borate, 7, 8
Bovine serum albumin (BSA), 150
Branched alkyl chains, 72
Bridges, 4, 14
Buffered systems, 123, 239
Buffering soil pH, 227
Building blocks, 9, 91, 203, 216
Butanoic acid, 8

C/N ratio, 69, 70, 75, 126, 127
Calibration plot, 189, 221, 222
Calibration range, 150
Capillary electrophoresis (CE), 108, 109, 112
Carbohydrates, 1, 3, 24, 31, 35, 36, 41, 43, 44, 47, 51, 52, 58, 59, 62, 63, 67, 69, 71, 72, 75, 76, 84, 134, 135, 137–143, 188
Carbohydrate chains, 141
Carbon percentages, 175
Carbonaceous residue, 179
Carbonate minerals, 114, 123, 173
Carbon–carbon bonds, 21
Carbonium ions, 223
Carbonyl, 18, 32, 38, 44, 51, 52, 57, 58, 59, 60, 63, 65, 67, 71, 79, 82, 84, 139, 175, 189
Carboxylate salts, 38
Carboxylic acid, 5, 9, 21–23, 38, 41, 47, 57, 58, 63, 66, 67, 70–72, 74–76, 84, 91, 103, 127, 139, 142, 152, 155, 165, 175, 188, 189, 196, 217, 221–223
  sites, 216
Carboxyphenol, 10
Carcinogenic compounds, 203
Catchment basins, 124
Catechol, 57, 203, 207, 208, 218
Cation bridges, 142
Cation exchange, 2, 3, 48, 166, 174, 239
Cation exchange capacities (CEC), 9, 227
Cationic surfactant, 92, 93
Cations, 5, 109, 141, 158, 165, 188
  Al^{3+}, 4, 151
  Ca^{2+}, 4, 142, 196
  Fe^{3+}, 4, 208, 209, 222, 223
  Mg^{2+}, 4, 150, 151

Cattle manure, 76
Cavities, 180
Cell membranes, 59
Cellulose,
  crystalline cellulose, 35
  hemicelluloses, 35, 51, 59, 65, 138
  microfibrils, 51, 138
  noncellulosic components, 52
  noncrystalline cellulose, 35
  oligomer, 138
  ribbon, 137
C–H bonds, 16
Chain branching, 63
Chain shortening, 47
Charge density, 4, 9
  differences, 3, 4, 6, 7
Chelation, 209, 239
Chemical
  analyses, 80, 81, 88, 206
  changes, 29, 39
  contaminants, 133, 144
  degradation, 21, 91
  fertilizers, 227
  oxygen demand, 69
  reactions, 133, 135
  reactivity, 134, 135, 143, 144
  shifts, 16, 18, 35, 37, 38, 41, 44, 51,
    52, 53, 57
  treatments, 24
Citrate, 34, 63
Clays, 24, 61, 174, 185, 186, 188, 191,
  194–196, 200
Clean fuels, 237
Climate control, 236
CO_2 emissions, 236–238, 241
Coagulation, 6
Coal, 88, 209, 215, 216, 223, 235–238,
  240, 241
  conversion, 237
  derived humic acids, 224
Coastal waterways, 1
Colloidal-size organic matter, 140
Column,
  elution volume, 7, 8
  interactions, 148
  resin,
    anion exchange, 48
    cation exchange, 48, 166
  hydrophobic, 69
  XAD, 2, 3, 5, 10, 19–21, 48, 62, 114
  XAD-4, 3, 5, 10, 19–21, 48
  XAD-8, 2, 3, 5, 19, 21, 48, 62, 70–
    72, 75, 92
  void volume, 7, 8, 148, 151
Commercial HAs, 80, 84, 88
Competition, 173, 179, 180, 182
Competitive sorption, 89, 173–175,
  179–181
Complexation, 92, 113, 114, 147, 165,
  167, 223
Composition, 3, 4, 19, 24, 35, 38, 61,
  63, 67, 75, 105, 130, 174, 182, 205
Compositional differences, 18
Compost, 29, 149, 152, 155–157, 159,
  185, 210, 211, 227
Computational chemistry, 133, 143
Computer simulation, 51, 133
Conditional binding constants, 113
Conformations, 4, 134, 136, 137, 139,
  140, 168, 186, 191
  shrunken, 4
Connectivities, 14
Contaminants, 76, 79, 113, 133, 143,
  144, 217, 241
Contamination, 149, 195, 227
Continuum, 2, 21, 23, 62
Coordinative bonds, 165
Co-precipitation, 5
Core humic structures, 23
Correlated enthalpy and entropy
  changes, 185
Counter current distribution, 6
Crop growth, 3
Crop irrigation, 69
Cross linking, 135, 179, 223, 224
Cross polarization magic-angle spinning
  (CPMAS), 10, 18–20, 30, 32, 34, 37,
  49, 57, 62, 71–73, 79, 188, 189
Cross-polarization (CP), 30, 49, 62, 79–
  82, 84–86, 88, 89, 175
  contact time, 175
Crystallographic structure, 140
Crystallography, 136
Cutin, 32, 34, 38, 41, 63
Cyclic voltammetric measurements, 166
Cytochrome-c, 150

D_2O, 14, 16, 17, 49, 204
DDT, 196–200
Decanedioic acid, 22
Decarboxylation, 21, 22
Decay time distribution, 117
Deconvolution, 39, 51, 52, 58, 59, 63,
    65, 81, 82, 84, 86, 115
Deep sea environments, 1
Degradation, 2, 21–23, 29, 41, 45, 47,
    53, 56, 59, 61, 95, 152, 200, 206
Degradative enzymes, 2
Dehydrogenation polymerization, 35
Depolymerization, 21, 41, 44, 66, 67,
    138, 144
Desorption, 91–93, 95, 99–106, 142,
    147, 188, 191
  resonant desorption, 101
Desorption ionization methods, 92
Detection limits, 147
Detergents, 69, 118
Detoxification, 173
Deuterium, 14, 204
Diagenesis, 47, 59, 62, 67
Dialysis, 5, 7, 203
Dianionic species, 207, 208
Diastereomeric forms, 136
Dicarboxylic acids, 21, 22, 41
Differentiation rate, 233
Diffuse reflectance Fourier transform-
    infrared (DRIFT-IR), 9, 149, 152,
    156, 162
  spectra of HA, 152, 153, 155
Diffusion coefficient, 179
Diffusion rates, 178
Digestion process, 70
Dimethyl sulfoxide, 2, 204, 218
Diode array detector, 150, 204
Dipolar decoupling, 10
Dipolar dephasing, 34
Dipolar interactions, 8, 31, 34, 173
Discrete component approach (DCA),
    115, 116
Dissociable protons, 23
Dissolution, 147, 162, 173
Dissolved organic carbon (DOC), 5, 47,
    48, 59, 69, 70, 72, 76, 114, 123, 124,
    195, 196
Dissolved organic matter (DOM), 1, 6,

    76, 148, 195, 196, 200
Distribution, 32, 35, 82, 95, 115–119,
    123, 128, 133, 177, 180, 181, 196,
    216, 217, 219
Distribution analysis, 115, 116
Distribution coefficients, 173
Ditrigonal cavities, 140
Donor–acceptor orbital, 120
DPMAS, 79–82
Droplet simulation, 134
Dual mode model, 178–180, 182
Dual reactive domain model, 179

E_4/E_6 ratios, 10, 70, 75, 126, 152
Ecological imbalances, 235
Ecology, 235, 239
Economic growth, 235–237
Electroactive species, 165
Electron Paramagnetic Resonance
    (EPR), 204, 207–210, 212
  spin Hamiltonian parameters, 210
Electron spin resonance (ESR), 9, 10,
    165
Electron transfer, 207, 208
Electro-osmotic flow (EOF), 109, 110
Electrophoresis,
  isoelectric focusing, 7
  isotachophoresis, 7
  moving boundary, 7
  polyacrylamide gel, 7
  zone, 7
Electrophoretic mobility, 109
Electrostatic interactions, 134, 143
Electrostatic repulsion, 142
Elemental analyses, 30, 38, 71, 75, 76,
    79, 83, 84, 86, 91, 125, 130, 147,
    149, 156, 189
Elemental compositions, 7, 80, 81
Elemental numbers, 82, 83
Emission, 113–115, 118–120, 125–127,
    129, 235
Emission wavelength, 118–120, 125
Empirical formulas, 205, 206, 218
Empirical method, 134, 141, 206
Energy donor, 120
Energy efficiency, 236
Energy transfer, 100, 103, 117, 119–121
Enols, 4, 5

Enthalpy change, 185
Entrapment, 2, 24
Entropy change, 185
Environmental,
  fate, 133, 173, 195, 196, 199, 200
  importance, 165
  problem, 70
  quality, 173
  stress, 233
Environments, 1, 2, 8–10, 23, 30, 66,
    76, 88, 91, 107, 117, 126, 133, 139,
    144, 165, 173, 194, 196, 200, 207,
    217, 223, 224, 227, 241
  aquatic, 128
Enzymes, 211, 216
  α-chymotrypsinogen, 150
  dioxygenases, 66
  endogenous, 45
  exogenous, 45
  extracellular, 29
  α-lactalbumin, 150
  lipoxygenase, 41
  mono-oxygenases, 66
  peptide hydrolases, 29
  polyphenol oxidases, 44
Equilibrium constant, 185, 187–189, 191
Essential information, 228, 239
Ester linkages, 34, 38, 44, 52, 58, 60,
    63, 65, 67
Esters, 21, 23, 32, 38, 58, 59, 71, 72, 82
  gallic esters, 35
Estuaries, 1
Ethanedioic acid, 22
Ethanoic acid, 8
Ethanol, 3, 6, 8, 203
Ether, 18, 22, 57, 135, 154, 155, 203,
    217, 221
  bands, 41, 51
  extractable, 69
  linkages, 21, 41
Evaporation, 48, 66
Excitation, 10, 81, 93, 113–115, 120,
    125, 126, 129
Exponential decay, 114–116
Exponential series method (ESM), 114–
    119
Exposure, 133, 182, 203, 211
Extended helix, 137

Extraction, 2, 5, 10–12, 19, 61, 63, 65–
    67, 70, 72, 80, 149, 152, 156, 179,
    205, 215, 216, 218
Extrinsic fluorescence, 113, 114

Fast-atom bombardment (FAB), 92–95,
    99–101, 105
Fate of pesticides, 199, 200
Fats, 2, 32
Fatty acids, 3, 21, 23, 29, 32, 41
  polyunsaturated, 29
Fenton reaction, 209
Fermentation, 237
Ferricyanide, 165–167, 170
Ferrocyanide, 165
Filtration, 5, 7, 124, 204
Fingerprint, 10, 18, 79, 91
Fission, 47, 51, 59
Flavones, 52, 57
Flavonoids, 32
  polyflavonoids, 35, 38
Flocculation, 123
Fluorescence anisotropy, 114
Fluorescence decay, 113–121
Fluorescence quenching, 113
Fluorescing structures, 117
Fluorophores, 10, 113, 115–117, 120,
    128
Food production, 236
Fossil fuels, 235, 236, 241
Fractionation, 3, 6, 7, 9, 11, 47, 49, 56,
    61, 124, 127, 166
Fractions, 2, 3, 5–7, 10, 18, 19, 48, 49,
    51–60, 62–64, 75, 109, 110, 111,
    112, 126–130, 148, 151, 158–162,
    166, 169, 179
  operationally defined, 19
Fragmentation, 92, 95, 100, 103
Free energy, 8, 173, 174
Free radical, 10, 29, 203, 211, 212, 222,
    224
Freeze-dried, 7, 48, 49, 62, 70, 71, 124,
    125, 186
Freundlich isotherm, 175
Frictional ratio data, 9
Fructose, 37
Fuel cells, 241
Fuel gas, 236–238, 240, 241

Fulvic acids, 3, 10–18, 20, 23, 24, 61–
      67, 70, 71, 73–76, 91–94, 100, 101,
      105, 106, 116–119, 126, 133, 142,
      165, 168, 179, 188, 196–200, 213
   acidic fraction, 2
   aquatic, 5
   commercial, 232
   compost, 75
   extraction, 103
   IHSS, 93, 95
   isolation, 103, 114
   Kerry soil, 228, 231, 232
   leaf derived, 66
   model, 75
   moss, 228, 232
   Norfolk peat, 228, 231, 232
   soil, 21, 75
Functional group analysis, 71, 91, 215
Functional groups, 11, 32, 44, 47, 56,
      82, 83, 103
   acidic, 4, 9, 61, 91, 208
   aliphatic hydrocarbon, 19
   dihydroxyphenyl, 66
   methoxyl, 16, 19, 21–23, 34, 41, 51,
      53, 63, 72
   phenolic, 53
Functionalities, 3, 5, 8, 10, 11, 15, 18,
      91, 133
   alcohols, 22
   aldehydes, 22
   amides, 19
   anionic, 4
   aromatic, 14, 19
   carbonyl, 19
   carboxylic acids, 19
   esters, 19, 21
   ether, 19, 23, 34, 44
   peptide, 19
Fused silica capillaries, 109, 110

Gallic acids, 57
Gas chromatography, 175
Gas phase reactions, 91, 103
Gaussian distribution, 7, 81, 178
Gel drying, 186
Gel permeation chromatography, 6
   effectiveness, 6, 7
Global changes, 235

Global minimum conformers, 136
Globular protein, 124
Glucopyranoside, 137
Glycerol, 94, 101
Goethite, 143
Golf course, 233
Graphite furnace atomic absorption
      spectroscopy (GFAAS), 148
Greenhouse effect, 235, 236, 239
Greenhouse gases, 235
Ground water, 119
Growth pouches, 228
Growth promotion, 227
Guaiacylpropanoids, 57, 136

Hazardous waste, 143
H–C couplings, 16
Heavy metals, 6, 76, 79, 165
   loading, 123
Helix, 137, 138, 143, 207
Hematology mixers, 175
Heterogeneity, 1, 56, 101, 105, 114–
      117, 119, 120, 178, 179
Heteronuclear HMQC, 16
Heterotrophy, 130
Hexagonal unit, 141
Hexane extraction, 175
Hexoses, 52
HF/HCl, 174, 175
H–H couplings, 16
High performance liquid
      chromatography (HPLC), 150
High performance size exclusion
      chromatography (HPSEC), 8, 148,
      151, 156, 158, 159, 162
   autosampler, 151
High surface area carbonaceous
      materials (HSACM), 178, 179
Hole (site)-filling mechanisms, 178,
      179, 182
Homogeneity, 4
Homologous series, 111
Hormone-like activity, 227
Humic acids, 3–10, 13, 18, 20–24, 53,
      61–71, 73, 76, 79–89, 91, 133–136,
      139–143, 148–151, 154–159, 168,
      175–176, 180, 186, 187, 192, 203–
      213, 216–220, 227, 236–241

adsorptive properties, 186
aggregation, 189
Aldrich, 60, 68, 80, 84, 88, 196, 204, 211, 212, 218
aquatic, 5, 21, 88
aqueous, 23
ARC, 84
ash content, 149, 152, 188
association, 23
building blocks, 216
characterization, 149
chemistry, 217, 223
commercial, 84, 218, 222, 229, 232
complexes, 233
compositions, 19, 21, 152
compost, 75, 152, 155, 156, 158, 161, 162, 186, 212
disaggregation, 185
dissolving, 2
equilibrium concentration, 189
extraction, 237
Fe loaded, 222
fibric peat, 230–232
fractions, 2, 5, 6, 10, 18, 23, 91, 158, 159, 174, 179
fractionation, 148
functional groups, 152, 155, 188, 191, 217
    acidic, 188
    phenolic, 188, 191
functions, 211, 212
homogeneity, 205
immature, 158
iron loaded, 221
irradiated with UV, 222
Kerry soil, 228, 231, 232
leaf derived, 66
leonardite, 70, 155, 162, 228, 229, 231, 232
matrix, 212
metal binding, 147, 165
metal loading, 218
methylated, 223, 224
models, 75, 88
molecular structure, 188
moss, 228, 232
Norfolk peat, 228, 231, 232
oxidative degradation, 21, 88

peat, 155, 158, 160, 162, 176, 178, 228–230
photochemistry, 207
plant, 80
polymer, 207, 208, 211
polymerization, 207
role, 209
samples, 204, 215, 216
sapric peat, 230–232
sites, 185, 191
sludge, 70
soil, 21, 75, 84, 148, 149, 156, 158, 162, 239
solubilities, 193
spodic soil, 228, 231
standards, 213
structures, 16, 203, 207, 215, 217
surface, 185, 207–209, 212, 217
synthesis, 185
treatments, 217, 230, 232
Humic type materials, 2, 91, 92, 123, 236
Humic molecules, 22
  compositions, 9
  shapes, 7
  sizes, 7
  source, 4
Humic substances, 11, 25–27, 62, 66–68, 77, 79, 92, 116–123, 127, 128, 131, 134–136, 144, 152, 173, 177, 178, 182, 193, 200, 213, 224, 225, 227–233
  acid groups, 23
  allochthonous, 1, 123, 126
  amendments, 228
  anionic functionalities, 4
  aquatic, 2, 6, 23, 93, 143, 114, 117, 121, 127, 133, 156
  ash content, 174
  associations of molecules, 9
  autochthonous, 1
  backbone, 21
  building blocks, 9, 21
  characterization, 70, 113
  chemical properties, 75, 76
  chlorinated, 3, 174, 175, 179, 180, 203
  components, 4, 21
  compositions, 2, 9, 10

conformational changes, 23
definitions, 1
degradation digests, 16
dissolved, 166
electrophoretic behaviour, 109
elemental composition, 9
extraction, 5, 6
fractionation, 4, 6, 7
fractions, 2, 5, 6, 8, 10, 24, 110, 111, 112, 165–170
functional groups, 9, 53, 189, 208, 223
gross mixtures, 4
healing effects, 3
heterogeneous, 3, 10, 91
hydrophilic, 3, 5, 8, 9, 48, 72, 75, 185, 188, 191
hydrophobic fractions, 6, 72
in solution, 8
in the oceans, 2
isolation, 4, 5
lakes, 1
lignin characteristic, 19
matrix, 179
molecules, 4, 9
origin, 1, 2, 21, 23, 61, 80, 88, 112–114, 117–119, 123, 205, 208, 212, 215
oxidative degradation, 21–23, 41
alkaline permanganate, 22
separation, 7
soil, 23, 69, 76, 113
solubility, 6
solutions, 9
structures, 10, 22
Humification, 10, 11, 19, 29, 49, 135, 138, 144, 152, 155, 162, 174, 194, 216
humified components, 61
secondary synthesis reactions, 61
Humins, 2, 3, 21, 61, 91, 174, 176, 178, 179, 228
Humus, 2, 25, 29, 47, 61, 67, 77, 90, 106, 162, 163, 182, 193, 213
Humus eating termites, 237
Hydration shells, 111, 134, 139, 140
Hydrodynamic injection, 110
Hydrodynamic size, 111

Hydrogen bonds, 4, 5, 8, 58, 134, 138, 175
Hydrolysates, 21
Hydrolysis, 21, 23, 41, 44, 63, 67, 188
Hydrophobic,
acid, 70
associations, 5
bonding, 4, 8
fractions, 5
neutrals, 3
Hydrophobic organic compounds (HOCs), 173–175, 177, 178
Hydroxides, 24, 61, 114
Hydroxyquinone, 72
Hydroxyl, 14, 22, 23, 29, 57, 58, 66, 70, 209, 216
radical, 209, 210, 212, 213
Hydroxylation, 47
Hyperfine pattern, 209
Hysteresis, 179, 182

Immobilization, 165
Inclusion complex, 138, 144
Indoleacetic acid (IAA), 227
Inductively coupled plasma atomic emission spectrometry (ICP-AES), 147, 156
nebulizer, 147
Inductively coupled plasma mass spectrometry (ICP-MS), 147, 148, 150, 151, 156–158, 160–162
central processing unit (CPU), 151
data acquisition, 151
detector, 151
nebulizer, 150, 151
In-lake processes, 126–130
Inorganic colloids, 3, 24
Inorganic contaminants, 133, 239
Inorganic sulfates, 127
Inositols, 59
Insoluble components, 24
Interchain interactions, 223
Interfacing, 148, 151, 158
Intermolecular interactions, 117, 179
International Humic Substances Society (IHSS), 2, 4, 5, 9, 13, 16, 80, 84, 86, 88, 93, 122, 149, 186, 188
peat, 174

Interpolymeric interactions, 140
Ion chromatograms, 151, 152
Ion exchange, 5, 215, 217, 222, 224
Ionic bridges, 143
Ionic strength, 114, 158
Ionization energies, 138
Iron–humate, 166
Irrigation, 70, 76
Isolation, 9, 61, 127, 187, 203, 205, 213
Isoproturon, 196–200
Isotope-dilution, 148

Kaolinite, 143, 186–193, 196
Keto functional groups, 23, 154, 155
Ketones, 32, 38
Kinetic control, 169
Kubelka–Munk functions, 149

Lactones, 21, 58, 59
Langmuir adsorption model, 185–187,
    189, 214
Lanthanide, 113
Laser desorption (LD), 92, 95–98, 101–
    106
Laser-desorption Fourier-transform
    mass spectrometry (FTMS), 93, 97
Laser-induced fluorescence, 113
Lattice water, 208
Leachability, 144
Leaf litter, 47
Leonardite, 80, 149, 157
Library spectra, 16
Lignans, 135, 216
Lignification, 138
Lignin, 16, 18,19, 23, 31–38, 43, 44, 47,
    51, 59, 60, 114–117, 119, 134–143,
    145, 152, 203, 211, 215, 216
  depolymerization, 65
  fragments, 53, 56, 65, 67
  hydrolysis, 65
  matrix, 138
  models, 206
  oxidative degradation, 53
  oxidized, 142
  structure, 65
  syringylpropanoid units, 41, 63
Lignin tannin-type residues, 19
Lignin carbohydrate complex (LCC),

    137–140, 142
  precursors, 135
Lignin-derived oligomer, 142
Lignin-like structures, 51, 119
Lime, 34, 124, 130
Liming, 123, 124, 130
Linear free energy relationships, 174
Lipids, 32, 34, 38, 44, 46, 59, 67
  hydroperoxidation, 41
  oxidation, 41
Liquid chromatography, 147, 175
Liquid matrix, 95, 100
Liquid scintillation, 175
Long chain acids, 22
Long chain ethers, 11
Long chains, 32, 63
Lysimeters, 195–200

Macroconformers, 138
Macrofauna, 123
Macrostructures, 186, 188
Magic angle spinning (MAS), 79, 175
  spinning rate, 175
Malate, 63
Malic acid, 44
Manures, 227
Mapping, 10
Mass spectrometry, 91
Matrices,
  diethanolamine, 92
  glycerol, 92
  3-nitrobenzyl alcohol, 92
  nitrophenyloctyl ether, 92
  nujol, 92
  polyethylene glycol, 92
  thioglycerol, 92
  triethanolamine, 92
  triethyl citrate, 92
Matrix surface, 92, 93, 99–101, 105
Maximum Entropy Method (MEM), 114
Mechanisms, 4, 22, 41, 47, 101–106,
    109, 142, 155, 162, 173, 180, 182,
    185, 203, 208, 227, 233
Membrane-micelle model, 191
Metal,
  binder, 239
  binding capacities, 203
  binding sites, 113, 188, 206

Metal, *continued*
  cations, 4, 5, 142
  elements
    ^{75}As, ^{9}Be, ^{210}Bi, ^{114}Cd, ^{58}Cu, ^{115}In,
      ^{139}La, ^{7}Li, ^{24}Mg, ^{208}Pb, ^{45}Sc, ^{51}V,
      ^{64}Zn, 150, 151
    ^{27}Al, ^{57}Fe, 151
  exchange, 203
  ions, 113, 114, 121, 142, 147, 155,
    156, 165, 223, 224
  oxides, 174
  species, 147, 159, 185, 215, 222, 239
Metals, 5, 7, 24, 121, 147, 155
  Al(III), 223
  Ba, 156, 157
  Ce, 156, 157
  chromium(III), 240
  Co, 10, 60, 68, 156, 157, 162, 163
  Cr, 156, 157, 162
  Cr^{3+}, 239
  Cr^{6+}, 239
  Cu, 148, 156-159, 162, 204, 209-211,
    240
  Fe, 80, 154, 157, 159, 162, 166, 167,
    169, 170, 204, 210, 211, 219-223
  Fe(II), 208, 209, 222
  Fe(III), 208, 209, 222, 223
  Fe=O^{+2}, 211
  Hg, 157, 186, 204, 219
  lead, 156, 240
  Mg, 142, 149, 157
  Mn, 92, 157, 158, 162
  Mo, 148, 156, 157, 162
  Na, 100, 103, 126, 157
  Ni, 156, 157
  Pb, 156-159, 162
  Sr, 156, 157, 162
  Ti, 156, 157, 162
  V, 26, 46, 68, 71, 77, 107, 136, 139,
    140, 145, 156, 157, 162, 167
  Zn, 156-159, 162
  Zr, 157
Metastable scanning experiments, 100
Methane, 237
Methanogenic microbes, 237
Methanoic acid, 8
Methine, 11, 31
Methyl esters, 34, 51, 217, 224

Methyl ethers, 34, 43, 44, 51, 53, 56, 57,
  59
Methyl iodide, 217-219, 221, 222
Methyl protons, 11
Methylation, 22, 92, 93, 217, 219-224
Methylene carbons, 31
Methylene groups, 32, 34, 63
Methylphenylethers, 217
Metolachlor, 174
Mica, 141, 142, 196
Micelle, 6, 8, 9
  critical micelle concentration (CMC), 8
  like aggregates, 8, 9
  structures, 9
Microbe, 1, 21, 23, 66, 72
  activity, 175
  degradation resistance, 2, 29
  growth, 239
Micronutrients, 227, 233, 239
Microorganisms, 2, 23, 29, 45, 69, 127,
  130, 237
Microplates, 228
Microscopic methods, 134
Migration, 76, 109, 111, 195
  times, 110
Mineral aggregates, 133
Mineral colloids, 23
Mineral surfaces, 133, 134, 140–144,
  173
Mineralites, 211
Mineralogy, 143
Mineral–organic matter associations, 61
Minerals, 2, 3, 5, 23, 62, 66, 67, 123,
  147, 154, 156, 175, 182, 189, 203,
  209, 212, 213
  aluminosilicates, 211
  bauxite, 191
  clays, 24, 61, 174, 185, 186, 188, 191,
    194–196, 200
  goethite, 143
  hematite, 173
  montmorillonite, 188
  muscovite, 134, 140–143
  oxide, 133
  phyllosilicate, 133, 140, 143
  smectite, 196
  surfaces, 143
MM3 force field, 136

MNDO, 134, 141
Models, 75, 76, 88, 116, 120, 133, 134, 137–141, 186, 187, 203, 206, 217, 222
compounds, 10, 114, 117, 119
humic polymers, 139
Modelling, 180, 207
Molecular charge
differences, 110, 111
Molecular diffusion, 179
Molecular dynamics, 134, 139–141
Molecular mechanics, 137, 139
Molecular modeling, 180
Molecular size, 123, 126, 130, 148, 165, 167
differences, 6, 7, 109, 110, 111, 112
homogeneity, 6
Molecular weight, 1, 4, 7, 8, 9, 23, 44, 61, 66, 69, 72, 75, 76, 91–93, 95, 109, 110, 111, 112, 128, 133, 143, 144, 150, 156, 158, 159, 162, 165–167, 169, 173, 223
standards, 148
Monolayer, 100, 133, 191, 193
adsorption, 191
formation, 186, 191
Monosaccharides, 29
Montmorillonite, 188
Multi-spectral analyses, 130
Muscovite, 134, 140–143
Mutual organization, 138

Nanoporosity, 180
Naphthalene, 174, 175, 176
Natural organic matter (NOM), 8, 123–131, 177
New acid groups, 21
Nitrogen content, 126, 130
Nitroxyl electron, 209
NMR, 22, 38, 41, 74, 90, 215
^{13}C, 9, 11, 15, 16, 18, 29–32, 39, 40, 42–44, 47, 49, 50–64, 66–68, 71–73, 76, 79, 91, 219–222, 224
acquisition parameters, 30, 49, 62
band assignments, 53
band intensities, 31
baseline, 16, 31, 71, 81
chemical shifts, 32, 82

chemical shift anisotropy, 79, 82
contact time, 30, 49, 62, 71, 79, 81
coupling times, 31
CP/T$_1$-TOSS, 80, 81, 82, 84–86, 88, 89
dipolar dephasing, 31, 53
Direct Polarization Magic Angle Spinning (DPMAS), 79–81, 86, 89
ether bands, 35
fourier transformation, 30
free induction decay, 30, 49, 62
free rotation of methyl groups, 31
Hahn echo, 81
inflection points, 34
inverse gated decoupling, 49
iterative curve fitting, 30
line broadening, 30, 31, 62, 71, 219
Lorentzian, 30, 81
overlapping bands, 30
pre-echo delay, 81
signal-to-noise ratios, 30, 34, 93
solid-state, 29, 32, 35, 44, 49, 59, 62, 63, 66, 71, 81, 175
solid-state cross-polarization magic-angle-spinning, 30, 49, 62
spectral envelopes, 30, 31
spin-lattice relaxation time, 80
spinning rates, 30
spinning sidebands, 31, 79
spinning speed, 79, 81
^{1}H, 10–17, 206, 216, 219, 221, 222
COSY, 11, 12, 14, 16–18
high field strength (600 MHz) spectrometer, 11
HMQC, 11, 15
liquid-state spectrum, 59
TOCSY, 11, 13, 14, 16
two-dimensional, 10
Nominal molecular weight (NMW), 124, 127–129
Non-bonded attachment, 138
Non-bonding interactions, 138
Nondestructive technique, 29
Nonlinear isotherms, 174, 177, 178, 182
Nonprotonated carbons, 31
Nonquantitative spectra, 79
Nucleic acids, 1, 21
Nucleophilic substitution, 209

Nutrients, 147, 156, 227–229, 232, 233, 237

Oblate ellipsoid model, 9
Odd-electron species, 100
O-demethylation, 47
Oligomers, 35, 136, 142
Oligosaccharides, 37, 101
O-Methylation, 217, 221
Optical brighteners, 118
Optimal structure, 137
Optimized parameters, 137
Organic amendments, 227
Organic contaminants, 239
Organic manures, 227
Organic matter, 1, 61, 67, 69, 123, 126, 130, 143, 147, 148, 178, 194, 200, 227, 233
 dissolved, 1
 soil, 47, 70
Organic oligomers, 142
Organometallic compounds, 95
Organo–mineral aggregates, 134, 139, 142, 143
Orthoquinones, 57
Oxidation, 23, 41, 44, 57, 59, 66, 121, 138–140, 147, 167, 211, 215, 216
Oxidation–reduction reactions, 147
Oxidative degradation, 47
Oxidative fission, 51
Oxides, 133
Oxidized lignin, 141, 142
Oxidized lignin-carbohydrate, 143
Oxygen species (ROS), 211
Oxygen-containing functional groups, 11, 155

Para-hydroxyphenyl structures, 56, 57, 66
Paramagnetic species, 79
Paraquat, 196–200
Particulate organic carbon (POC), 195
Particulate organic matter, 1
Partitioning, 173–175, 178–180, 182
Peak hop mode, 150, 151
Peat, 80, 93, 109, 110, 112, 166, 174–179, 183, 185, 209, 211, 216, 218, 228, 230
Peat moss, 227

Pectins, 32, 35, 38, 44, 51, 59
Peptide hydrolases, 29
Peptides, 1, 2, 19, 32, 38, 44, 51–53, 72, 75, 95
Perchloroethylene, 240
Peroxyl, 29
Pesticides, 76, 201
 atrazine, 143, 174, 180, 181, 196–200
 2,4-D, 196–200
 DDT, 196–200
 isoproturon, 196–200
pH, 2–14, 16–21, 48, 61, 62, 70, 71, 109, 110, 114–119, 123, 124, 126, 127, 130, 139, 142, 158, 166, 185–187, 189, 196, 197, 204, 218, 223, 239
Phenol, 5, 21, 23, 35, 56–60, 63, 70–73, 75, 76, 84, 88, 154, 174, 189, 211, 212, 217, 221–223
 compounds, 128
 ethers, 56
 hydroxyls, 22
 sites, 216
Phenylpropane, 23, 56, 57, 135
Phloroglucinol, 57, 65
Phosphoinositides, 59
Photochemical degradation, 123, 207, 208, 211
Photoreduction, 207, 208, 219
Photosynthetic processes, 211
p-Hydroxyphenylpropane, 57, 136
Phyllosilicate, 133, 140, 143
Physiological effects, 215
Physiological mechanisms, 233
Physisorption processes, 174
Pilayella littoralis, 80, 84, 87, 88, 186, 188
Plant, 1–2, 69, 79, 88, 118–119, 185, 203, 205, 211–213, 215–219, 222–224, 241
 Acalypha wilkesiana, 30, 39, 40, 44
 Acer campestre L., 34, 35, 39–41, 44, 48–56, 59, 62–64, 67
 aerial organs, 32
 alga, 80, 186, 216
 Bixa orellana L, 30, 39
 cell death, 29
 cell wall, 138, 139

composted leaves, 56
cutins, 21
cuticle, 32
deciduous, 29, 30, 124
degradation, 29
  anabolic, 29
  catabolic, 29
derived organic material, 47, 59
derived HA, 206
dicotyledonous, 38
Dombeya wallichii, 39
growth, 227–229,233, 239
fragments, 66
leaf spectra, 34, 35, 44, 49, 51
leaves, 29, 34–45, 59–67, 216, 223
  cuticle layer, 32
  epidermal cells, 32
  leachates, 41, 47, 48, 49–56, 58–60,
    195–199
  nonsenescent, 30–32, 34, 35, 39–44,
    47
  senescence, 29
  tobacco, 34, 63
Lespedeza cuneata, 43
lignified components, 23
litter decomposition, 29, 47
Mak leaves, 48
Malus 'Manbeck Weeper', 31, 34, 39,
  43, 51, 52
material, 228
morphology, 138
non-wooden, 136
nutrients, 3, 147, 156, 227–229, 232,
  233, 237
positive growth response, 233
residue, 152
roots, 66, 147, 228–233
  dry weight, 229–231
  growth, 227, 229, 230, 232, 233
  initiation, 227
  length, 230–232
  shoot ratio, 229–232
rye grass, 38
shoot dry weight, 230–232
shoots, 227–232
soil system, 165
stems, 66
suberins, 21

tissue, 29, 32, 35, 38, 62, 66, 117, 138,
  228, 229, 230
tropical, 30
Ulmus pumila, 30, 41, 42
undecayed, 61
wheat, 38, 227, 233, 239
wood, 35, 41, 138, 139, 237
Zelkova serrata, 31, 34, 35, 41, 42, 48,
  50, 51
Plasma, 100, 103
Podzols, 23, 165
Polarographic analyzer, 166
Pollutants, 130, 134, 143, 236
Pollution, 77, 144, 235
Polyacrylamide gel electrophoresis
  (PAGE), 7, 109
Polyanionic surface, 215
Polycyclic aromatic hydrocarbons
  (PAH), 113
Polydispersity, 24, 109, 110
Polyelectrolyte, 165
Polyester polymer, 32, 63
Polyethyleneglycol (PEG), 110, 111,
  112
Polyfunctionality, 165, 168
Poly-hydroxybenzenecarboxylic acids,
  22
Polyphenolic structures, 65
Polysaccharides, 3, 7, 21, 29, 51, 72, 74,
  75, 158, 185, 186, 217
Polysulfone membrane filter, 48
Polyvalent cations, 4
Polyvinyl chloride, 178
Power density thresholds, 103
Power industry, 236
Precipitation, 2, 6, 123, 127, 130, 147,
  188, 191, 218, 239
Precursor, 21, 56, 135, 179
Preferential effect, 232
Proanthocyanidins, 35, 60
Prodelphinidin, 35
Prometon, 174, 180, 181
Propanoic acid, 8
Protective coatings, 3
Protective functions, 138
Proteins, 21, 29, 34, 69, 72, 75, 76, 150,
  156, 158, 185

Proton rotating-frame spin-lattice
  relaxation time, 79
Protonated carbon, 31, 35
Purification, 80, 205, 218
Pyrolysis mass spectrum, 88, 91

Quantitative spectra, 80, 84
Quantum chemical method, 141
Quinone, 32, 38, 44, 72, 88, 203, 207,
  208, 211, 212

Radical,
  addition, 209
  coupling, 219
  mechanism, 209
  quenching, 212
Ramp sequence, 79
Random coil
  concept, 9, 24
  conformations, 7, 9
Rank analysis, 113
Rate of induction, 208
Reaction time, 22, 218, 222
Red shift, 128
Redox state, 216, 222
Reduction, 41, 44, 48, 51, 59, 147, 167,
  174, 179, 216, 222, 235–237, 239
  potentials, 208
Reference electrode, 166
Reforestation, 236
Relative
  intensities, 32, 39, 43, 52, 58
  stability constants, 166, 167, 169, 170
Remediation technologies, 173
Residence time, 123
Resolution, 9, 14, 16, 30, 91, 106, 125,
  149
Resonance, 11, 15–19, 21, 71, 91, 212,
  219
Retention, 123, 124, 158, 173
  volume, 156
Reverse osmosis, 5, 124, 130
Reversible adsorption, 188
Ring fission, 47, 66
Root growth, 233
Rotary evaporation, 48
Rubbery polymers, 178
Ryegrass *Lolium perenne* L, 228, 232

Saccharides, 38, 59, 65
Salicylic acid, 115–117
Salting out, 6
Sampling, 30, 70, 124, 196, 213
Sand, 174, 196, 197, 229, 233
Scanning electron micrographs, 188,
  233
Scintillation analysis, 198
Seaweed products, 227
Secondary,
  radical species, 210
  structures, 134
Secondary-ion mass spectrometry
  (SIMS), 95
Sediments, 61, 79, 106, 107, 140, 173
Selectivity, 111, 114
Self association phenomena, 9
Semi-arid regions, 69, 70
Semiempirical method, 134, 141
Semi-quantitative analysis (SQA), 150,
  156, 157
Semiquinone, 29, 207
Senescence, 41, 45
  test, 227
Sensitivity, 114, 165
Separation, 6, 7, 47, 48, 109, 111, 147
  1,2,4-Aminotriazole, 7
Sequential adsorption steps, 185–187,
  189
Sequestering carbon, 241
Sewage sludge, 70
  uncomposted aerobic, 76
Sewer treatment, 237
Shock-wave, 101, 103–105
Signals, 10, 14, 16, 95, 100, 105, 158,
  165
Silt, 174
Simulated annealing, 134, 139–142
Single ring compounds, 23
Site capacity, 185, 187, 189, 191
Size exclusion chromatography (SEC),
  147, 148, 150, 151, 158, 162
Skeletal structures, 23
Sludge, 69, 70, 73–75, 88
Smectite, 196
Sodium amalgam, 22
Sodium benzoate, 143
Sodium hydroxide, 2, 3, 5, 19, 61, 62,

65–67, 70, 103, 166, 186, 188, 191, 193, 204, 218, 219, 221
Sodium pyrophosphate, 4, 19, 178
Soft ionization methods, 92
Soil organic matter (SOM), 1–3, 195,196
Soils, 1–7, 9, 10, 13, 16, 18–24, 29, 44, 45, 47, 48, 62, 67, 69, 70, 76, 79, 84, 88, 92, 93, 106, 107, 109, 123, 133, 140, 142–144, 147, 148, 152, 155, 156, 162, 176, 179, 181–183, 185, 186, 194–198, 203, 205, 212, 215, 219, 222, 224, 233, 236, 241
  acidity, 173
  aggregates, 2, 3, 23, 173
  alluvial farm, 80
  amendment, 227, 228
  bulk density, 227
  chemical properties, 227
  column, 199
  components, 195
  conditions, 227
  conservation, 3
  core, 197, 199, 200
  derived humic acids, 206, 211
  dry, 197
  fraction, 21, 197
  fractionation, 7
  genesis, 144
  materials, 216
  matrix, 195
  moisture, 173
  microorganisms, 216
  mineral colloids, 2, 174
  nonhumified components, 61
  organic matter, 1, 61, 139, 195
  physical structure, 227
  plant system, 165
  podzolic, 66
  residue, 197
  samples, 204, 217
  solution, 195, 200
  top, 235
Solid–solution distribution ratio, 175
Soluble organic matter, 1–3, 6, 61, 76, 100, 165, 188, 191, 195, 200, 203, 212, 217, 223
Solute adsorption, 186

Sorbent, 7, 174, 175, 179, 239
Sorption, 5, 62, 155, 177, 182, 195
  coefficients, 173, 180
  competitive, 89, 173–175, 179–181
  1,3-dichlorobenzene, 175, 178, 179
  domains, 179
  energies, 137, 141, 144
  enthalpies, 173
  mechanisms, 173, 175
Soxhlet extraction, 203
sp^2-C, 83, 84, 88
sp^3-C, 83, 84, 88
Space group, 143
Spatial organization, 139
Speciation, 113, 133, 144, 147, 156, 158, 162, 165
Spectroscopy, 11, 29, 44, 68, 71, 162
  fluorescence, 8–10, 115–121, 123–130
    anisotropy, 113, 114
    3-dimensional spectra, 124, 128
    excitation bands, 10
    extrinsic, 113, 114
    steady-state techniques, 113
  FTIR, 22, 29, 47, 55, 59–61, 69, 71, 73, 74, 76, 123, 125, 127, 165
  FT-Raman, 9, 22
  IR, 9, 47, 49, 52, 53, 56–61, 68, 71, 91, 101, 106, 124, 125, 149, 152 166, 206, 215, 216
    carbonyl stretching frequencies, 58
    C–O stretching modes, 63
    ring stretching vibration, 57
  Laser-desorption Fourier-transform mass spectrometry (FTMS), 93, 95
  UV–visible, 10, 125, 126, 147–149, 152, 156, 162, 166, 167, 204
    chromatograms, 148, 158
Spin trapping, 204, 209, 211, 212
Sputtering, 95, 99–101, 105
Stability constants, 169, 225
Standard compound, 18
Standard sample preparation methods, 10
Starch, 51
Stereochemistry, 206
Structural units, 21, 35
Structures, 1, 4, 6, 8, 9, 11, 14, 18, 19, 21–24, 32, 34, 37, 38, 41, 44, 47, 57,

Structures, *continued*
    65–67, 72, 74, 76, 79, 88, 106, 116,
    118, 119, 121, 127, 133–139, 143,
    178, 185, 188, 203, 206, 211, 213,
    215, 216, 221, 224
    1,3,5-ring positions, 16
    1,4-butanedioic acid groups, 51, 59
    3,4,5-trihydroxyphenols, 56
    backbone, 23
    conjugated, 52, 60, 63, 67, 128
Suberins, 23, 135
Substituent groups, 23
Substituted 1,2-benzenedicarboxylic
    acid, 66
Substitution, 11, 16, 57, 60, 209, 216
Sucrose, 37, 38
Sugars, 2, 3, 19, 21, 23, 59, 65, 72, 205,
    206, 215, 221, 239
    hexose sugars, 58
Sulfur mustards, 223
Superoxide, 29, 210
Supporting electrolyte, 166
Suppression effect, 100
Surface active, 8, 105
Surface adsorption, 173
Surface area, 124, 141, 178, 180, 186,
    187, 191, 211
Sustainable agriculture, 173
Synchronous scan fluorescence, 113,
    128

T_1 correction errors, 87
Tannins, 31–33, 35–38, 53, 56, 65, 66,
    69
    ellagitannins, 35
    gallotannins, 35, 57, 58, 63, 67
    hydrolyzable tannins, 35, 38, 43, 44,
        52, 60
    nonhydrolyzable tannins, 33, 35, 43,
        47, 60
    valoneoyl groups, 43
Temperature, 22, 30, 71, 101, 103, 110,
    125, 134, 137, 149, 150, 156, 157,
    162, 180, 186–188, 191, 193, 204,
    221, 222, 229
Terminal methyl groups, 63
Thermal desorption, 101
Thermal vibrations, 137

Thermodynamic parameters, 189
Time-resolved fluorescence, 113
Toluene, 174, 175, 177, 179
Total acidity, 23, 70, 71, 75
Total ion current (TIC), 93
Total organic carbon (TOC), 125, 195
Toxic elements, 165
Trace metals, 147–149, 151, 156, 158,
    162, 204
Transmission electron microscopy, 234
Transport, 76, 133, 144, 162, 195, 200,
    212
Trichloroethylene, 175, 240
Trihydroxyphenyl, 22, 57, 66
Turfgrass growth, 227, 228, 233

Ubiquinones, 211
Ultracentrifuge, 7
Ultrafiltered fractions, 129
Ultrafiltration, 5, 7, 92, 109, 110, 124,
    130, 166
Unit cell, 143
Uracil, 186
Uronic acids, 59

Vacuum drying, 30
van der Waals forces, 4, 5, 23, 206
Vegetation, 123, 124, 241
Voltammetric methods, 165

Water,
    acidic, 239
    brown, 115–118
    fresh, 69, 70, 126
    ground, 48, 76, 92, 195, 227, 239
    lake, 123, 130
    rain, 62
    retention, 239
    sewage, 69, 70, 73–75
    shed, 2, 123, 124
    shortage, 70
    soil, 48, 62, 66
    soil seepage, 117, 118
    stream, 69, 239
    supplies, 1
    surface, 48, 123, 129, 205
    treated, 69
    waste, 69, 70, 75, 76, 118

ways, 1
Waters, 3–5, 18, 24, 25, 67–68, 79, 106,
    114, 123, 129, 147, 148, 183, 235
Waters of hydration, 205
Waxes, 23, 32, 34

XAD-4 acids, 3, 10, 19, 21, 24
XAD-8, 20
Xenobiotics, 113
X-ray diffraction, 180

Yields of crops, 239